CAMBRIDGE LIBRARY COLLECTION

Books of enduring scholarly value

Botany and Horticulture

Until the nineteenth century, the investigation of natural phenomena, plants and animals was considered either the preserve of elite scholars or a pastime for the leisured upper classes. As increasing academic rigour and systematisation was brought to the study of 'natural history', its subdisciplines were adopted into university curricula, and learned societies (such as the Royal Horticultural Society, founded in 1804) were established to support research in these areas. A related development was strong enthusiasm for exotic garden plants, which resulted in plant collecting expeditions to every corner of the globe, sometimes with tragic consequences. This series includes accounts of some of those expeditions, detailed reference works on the flora of different regions, and practical advice for amateur and professional gardeners.

Beeton's Gardening Book

Samuel Orchart Beeton (1831–77), the publishing entrepreneur who made his wife's *Book of Household Management* one of the bestselling titles of the century, gave his name to many other books of domestic, medical and general information for the middle classes. (The 1871 *Book of Garden Management*, published and probably compiled by him, is also reissued in this series.) This work was published in 1874 by Ward Lock, to whom Beeton was forced to sell his own business after a financial collapse in 1866. The book contains 'such full and practical information as will enable the amateur to manage his own garden'. It covers flower, fruit and vegetable gardening, with a section on garden pests and a monthly calendar of tasks. It also contains advertisements for gardening and medicinal products, as well as for other books from the publishers, offering a fascinating insight into social as well as garden history.

Cambridge University Press has long been a pioneer in the reissuing of out-of-print titles from its own backlist, producing digital reprints of books that are still sought after by scholars and students but could not be reprinted economically using traditional technology. The Cambridge Library Collection extends this activity to a wider range of books which are still of importance to researchers and professionals, either for the source material they contain, or as landmarks in the history of their academic discipline.

Drawing from the world-renowned collections in the Cambridge University Library and other partner libraries, and guided by the advice of experts in each subject area, Cambridge University Press is using state-of-the-art scanning machines in its own Printing House to capture the content of each book selected for inclusion. The files are processed to give a consistently clear, crisp image, and the books finished to the high quality standard for which the Press is recognised around the world. The latest print-on-demand technology ensures that the books will remain available indefinitely, and that orders for single or multiple copies can quickly be supplied.

The Cambridge Library Collection brings back to life books of enduring scholarly value (including out-of-copyright works originally issued by other publishers) across a wide range of disciplines in the humanities and social sciences and in science and technology.

Beeton's Gardening Book

Containing Such Full and Practical Information as Will Enable the Amateur to Manage his Own Garden

SAMUEL ORCHART BEETON

CAMBRIDGE
UNIVERSITY PRESS

CAMBRIDGE
UNIVERSITY PRESS

University Printing House, Cambridge, CB2 8BS, United Kingdom

Cambridge University Press is part of the University of Cambridge.
It furthers the University's mission by disseminating knowledge in the pursuit of education, learning and research at the highest international levels of excellence.

www.cambridge.org
Information on this title: www.cambridge.org/9781108072236

This edition first published 1874
This digitally printed version 2014

ISBN 978-1-108-07223-6 Paperback

The original edition of this book contains a number of colour plates, which have been reproduced in black and white. Colour versions of these images can be found online at www.cambridge.org/9781108072236

Selected books of related interest, also reissued in the CAMBRIDGE LIBRARY COLLECTION

Amherst, Alicia: *A History of Gardening in England* (1895) [ISBN 9781108062084]

Anonymous: *The Book of Garden Management* (1871) [ISBN 9781108049399]

Blaikie, Thomas: *Diary of a Scotch Gardener at the French Court at the End of the Eighteenth Century* (1931) [ISBN 9781108055611]

Candolle, Alphonse de: *The Origin of Cultivated Plants* (1886) [ISBN 9781108038904]

Drewitt, Frederic Dawtrey: *The Romance of the Apothecaries' Garden at Chelsea* (1928) [ISBN 9781108015875]

Evelyn, John: *Sylva, Or, a Discourse of Forest Trees* (2 vols., fourth edition, 1908) [ISBN 9781108055284]

Farrer, Reginald John: *In a Yorkshire Garden* (1909) [ISBN 9781108037228]

Field, Henry: *Memoirs of the Botanic Garden at Chelsea* (1878) [ISBN 9781108037488]

Forsyth, William: *A Treatise on the Culture and Management of Fruit-Trees* (1802) [ISBN 9781108037471]

Haggard, H. Rider: *A Gardener's Year* (1905) [ISBN 9781108044455]

Hibberd, Shirley: *Rustic Adornments for Homes of Taste* (1856) [ISBN 9781108037174]

Hibberd, Shirley: *The Amateur's Flower Garden* (1871) [ISBN 9781108055345]

Hibberd, Shirley: *The Fern Garden* (1869) [ISBN 9781108037181]

Hibberd, Shirley: *The Rose Book* (1864) [ISBN 9781108045384]

Hogg, Robert: *The British Pomology* (1851) [ISBN 9781108039444]

Hogg, Robert: *The Fruit Manual* (1860) [ISBN 9781108039451]

Hooker, Joseph Dalton: *Kew Gardens* (1858) [ISBN 9781108065450]

Jackson, Benjamin Daydon: *Catalogue of Plants Cultivated in the Garden of John Gerard, in the Years 1596–1599* (1876) [ISBN 9781108037150]

Jekyll, Gertrude: *Home and Garden* (1900) [ISBN 9781108037204]

Jekyll, Gertrude: *Wood and Garden* (1899) [ISBN 9781108037198]

Johnson, George William: *A History of English Gardening, Chronological, Biographical, Literary, and Critical* (1829) [ISBN 9781108037136]

Knight, Thomas Andrew: *A Selection from the Physiological and Horticultural Papers Published in the Transactions of the Royal and Horticultural Societies* (1841) [ISBN 9781108037297]

Lindley, John: *The Theory of Horticulture* (1840) [ISBN 9781108037242]

Loudon, Jane: *Instructions in Gardening for Ladies* (1840) [ISBN 9781108055659]

Mollison, John: *The New Practical Window Gardener* (1877) [ISBN 9781108061704]

Paris, John Ayrton: *A Biographical Sketch of the Late William George Maton M.D.* (1838) [ISBN 9781108038157]

Paxton, Joseph, and Lindley, John: *Paxton's Flower Garden* (3 vols., 1850–3) [ISBN 9781108037280]

Repton, Humphry and Loudon, John Claudius: *The Landscape Gardening and Landscape Architecture of the Late Humphry Repton, Esq.* (1840) [ISBN 9781108066174]

Robinson, William: *The English Flower Garden* (1883) [ISBN 9781108037129]

Robinson, William: *The Subtropical Garden* (1871) [ISBN 9781108037112]

Robinson, William: *The Wild Garden* (1870) [ISBN 9781108037105]

Sedding, John D.: *Garden-Craft Old and New* (1891) [ISBN 9781108037143]

Veitch, James Herbert: *Hortus Veitchii* (1906) [ISBN 9781108037365]

Ward, Nathaniel: *On the Growth of Plants in Closely Glazed Cases* (1842) [ISBN 9781108061131]

For a complete list of titles in the Cambridge Library Collection please visit:
www.cambridge.org/features/CambridgeLibraryCollection/books.htm

VIRGIN CORK FERNERY
J.C. GRIFFITHS SC

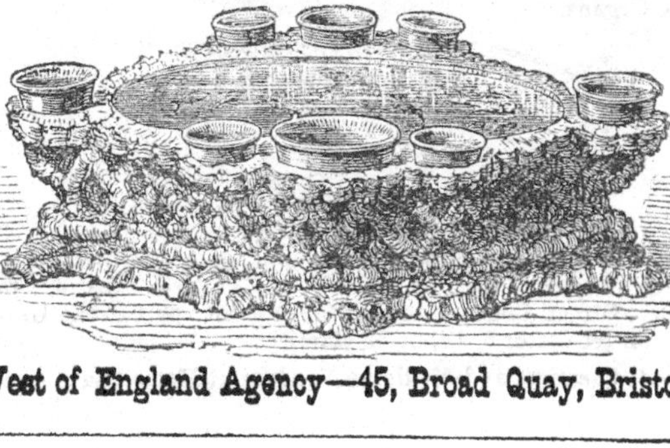

Pl. I.

BEETON'S

GARDENING BOOK.

CONTAINING

SUCH FULL AND PRACTICAL INFORMATION AS WILL ENABLE THE AMATEUR TO MANAGE HIS OWN GARDEN.

AMPLY ILLUSTRATED.

LONDON:
WARD, LOCK, AND TYLER, WARWICK HOUSE,
PATERNOSTER ROW, E.C.

GENERAL CONTENTS.

———:o:———

OUR COLOURED PLATE.—No. 1.

1. *Ageratum.*—This pretty flower is valuable for the length of time it keeps in bloom. Wherever it is grown, fresh blossoms will always be seen until the frost of winter destroys the plant. When cut for bouquets it also lasts well.

2. *Anemone.*—There are many varieties of the Anemone, and they are to be found of almost every shade of colour. The double sorts, however, are generally either scarlet or purple. To produce fine heads of blossom they require a very rich light soil, and, when once planted, the tubers may remain in the soil for several years if care be taken to give it an annual dressing of well-rotted manure.

3. *Arum.*—This curious perennial, so popular as a pot plant, is allied to the Lord and Ladies which grow wild in our hedges and woods. It is propagated by off-sets and suckers, and the secret of its cultivation is to give the plant a long period of rest after it has done flowering; after this, to start it in entirely fresh soil, and to give it an abundance of water while in a growing state. A fine grown specimen makes a handsome window plant.

4. *Azalea.*—Both out of doors and under glass, Azaleas are to be valued. The hardy varieties are of American extraction; the more tender sorts, of which our illustration is one, are natives of India. As show-plants under modern training, nothing can surpass Azaleas. Those who want to see them in perfection should pay a visit in the spring to the Messrs. Veitch's nurseries.

5. *Balsam.*—Balsams, of late years, have ceased to be the fashionable flowers they once were. There is a stiffness and formality about them which displease many persons. Still, a well-grown balsam, in full flower, is very showy, and it may also be turned to a good account, for the

individual blossoms have an excellent effect when used to ornament fruit dishes on a dinner table.

6. *Begonia.*—Some beautiful seedlings, of late years, have been obtained of this favourite plant; all of them, however, require a greenhouse, and many of them stove heat. Such delicate foliage never can be well developed out of doors in our variable climate. Even under glass, a drop of water will soon destroy the beauty of a leaf.

7. *Bourbon Rose.*—This rose, which is a variety of the Rosa Indica, is itself the parent of a numerous progeny. Mr. Paul and other eminent growers have hybridized it with the Chinese, Noisette, and other roses, until the varieties are infinite. All the Bourbon roses are singularly hardy, and noted for their brilliancy and clearness of colour.

8. *Camellia.*—Who is there who does not admire the Camellia? The rich dark green foliage of the plants, and their blossoms so beautiful in shape and colouring, make them universal favourites. Our artist has here illustrated C. Donkelaarii, a variety with very large single flowers, and, generally, a profuse bloomer. This sort is not so useful as a cut flower as many others, but it is extremely showy in the greenhouse.

9. *Canna.*—This beautiful herbaceous plant is far too tender to endure our climate in winter, but when forced on in a hot-house, it may be planted out either singly or in mixed beds. The foliage and blossoms are both highly ornamental. Plants may be obtained from seed, and also from division of roots. The popular name of the Canna is Indian Shot.

BEETON'S

GARDENING BOOK.

GARDEN OPERATIONS.

1.—LAYING OUT A GARDEN.

This subject being far too extensive to be fully discussed in these pages, our observations had better be confined to a moderate-sized garden, such as is generally attached to villa residences. We will therefore treat of a garden of a single acre. Here about two-thirds should be devoted to lawn, flower-garden and shrubberies, and one-third to kitchen-garden, exclusive, we will suppose, of the melon-ground.

The melon-ground ought to be about twenty yards square, walled or fenced round to the height of six feet, with a gateway leading into it large enough to admit a horse and cart. The drainage of the melon-ground should be perfect, the water from the pits and houses falling into a tank placed sufficiently deep in the ground to receive all the drainage from the dung-beds and compost heaps. If this tank is within the kitchen-garden, it will be invaluable in the cultivation of flowers and vegetables. Here also are placed the potting-sheds, and sheds for the preparation of composts, which should always be prepared under cover; and as the yard is by no means ornamental, it should be placed as far as possible from the house.

In the plan, 1 is the house; 2, the conservatory; 3, clump of American plants, consisting of some rhododendrons, ledums, and heaths; 4, roses; 5, flower-beds, with coniferæ in the centre; 6, flower-beds; 7, jardinette, with fountain; 8, borders planted with Alpine plants; 9, vines or ornamental climbers; 10, pears, cherries, &c., trained against the wall; 11, verandah with climbers; 12, carriage-drive; 13, arches over path for climbing roses and other ornamental climbers; 14, fernery; 15, turf lawn; 16, shrubberies 17, summer house; 18, flower-beds, with deodar in the centre, surrounded by turf; 19, shady walks; 20, flower-border fronting conservatory; 21, flower-border fronting shrubberies; 22, melon-ground and compost-yard; 23, back-entrance, wide enough for carts to enter; 24, range of three forcing-pits; 25, vinery and forcing-house; 26, tool-house; 27, frames; 28, manure-bed; 29, garden entrance. The kitchen-garden being thoroughly drained, trenched, and manured, and the walls in order, the following will be its first order of cropping:—*a*, Jerusalem artichokes; *b*, gooseberries; *c*, raspberries; *d*, red, white, and black currants in rows; *e*, strawberries, seakale, rhubarb, and globe

artichokes; *f*, a row of plumtrees, asparagus, horseradish, and more strawberries; *g*, pot-herbs, potatoes, and peas; *h*, a row of pyramid apple-trees, parsnips, carrots, and turnips; *i*, cabbages, celery, broad beans, scarlet runners; *k*, pyramid pear-trees, scarlet runners, broad beans, cauliflower, and early broccoli. On the south border, plums and cherries.

In all theoretical gardening it is forbidden to crop the border on which wall fruit is planted; but this is rare in practice: the crops indicated here generally occupy such borders; but probably a line might be drawn beyond which such crops should not approach the wall. Supposing such border to be 16 feet, 12 feet might be devoted to such crops in the kitchen-garden as require a warm sunny border.

The following is another illustration of a convenient villa garden.

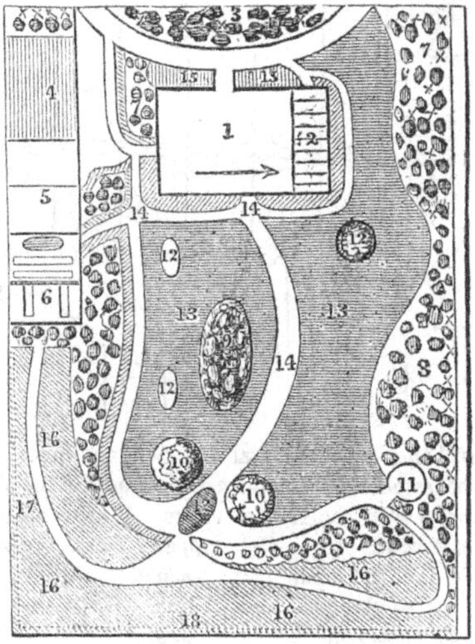

Where it can be so arranged, the garden should be an oblong square; 100 yards from east to west, and 70 yards from north to south, about the proportions laid down in the accompanying plan. This allows the vegetables to range from north to south, which is always to be preferred, otherwise they get drawn to one side by the side-light of the sun.—1, The site of the house; 2, the conservatory, 3, a clump of trees and shrubs fronting the main entrance; 4, coach-house and stables; 5, tool-house; 6, manure and frame-yard; 7, flower-borders and shrubberies; 8, ferns and American plants; 9, rose clumps; 10, circular beds for hollyhocks, dahlias, and other free-blooming plants in summer, and thinly planted with evergreens to take off the nakedness in winter; 11, arbour; 12, flower-beds; 13, lawn; 14, paths; 15, beds for placing out flowers in pots; 16, kitchen-gardens; 17, peach-wall; 18, east wall for plums, cherries, and pears.

It is sometimes advantageous to have buildings and even groups of large trees contiguous to gardens. Where these are situated to the north, they not only break and turn aside the cold winds, but concentrate the heat of the sun; they also preserve the crops during winter. Buildings have this advantage over trees, that they afford the shelter without robbing the soil of the food necessary for its legitimate crop. In the accompanying plan it will be observed that the whole frontage north of the house is laid out as lawn, and to the south, that the breadth of the house and offices is disposed in the same way; a single winding path running through it. East of the house lie the conservatory and offices, sheltered by a belt of shrubbery, which runs round the whole lawn. The kitchen-gardens occupy the north-west side of the ground, and adjoining, at the southern extremity, are vineries, forcing-houses, and orchard-houses. The northern boundary is a dwarf wall with green iron railings.

2.—LEVELLING.

For levelling any tract of country, a theodolite, which is a spirit-level raised on three legs and furnished with a telescope, is the instrument employed. A quadrant is also frequently used for the same purpose, and for determining the level of drains, &c. The following diagram and remarks are taken from

Loudon's "Self-Instruction for Young Gardeners":—"Suppose it were required to run a level through the ground indicated, A B, from the point A. Provide a few staves proportioned in length to the work in hand, and let them have cross-pieces to slide up and down; then, having firmly fixed the staff in the ground, to which the quadrant is attached at the point A, set the instrument in such a position that the plumb-line shall hang exactly parallel to the perpendicular limb of the quadrant: the upper limb will then be horizontal. This done, direct the eye through the sights, and, at the same time, let an assistant adjust the slides on each staff so as exactly to range with the line of vision. Then suppose the height A C to be five feet downwards from the upper side of the slide upon each staff, so shall the dotted line A B represent the level line required. Suppose the operation had been to determine a cut for a drain, to have a fall of 3 inches in every 20 feet, the distance between each staff in the above figure may be supposed to be 20 feet; then 5 feet 3 inches would have to be measured down the first staff, 5 feet 6 inches down the second, 5 feet 9 inches down the third, &c. &c. The dotted line A B would then represent the line parallel to the bottom of the intended drain."

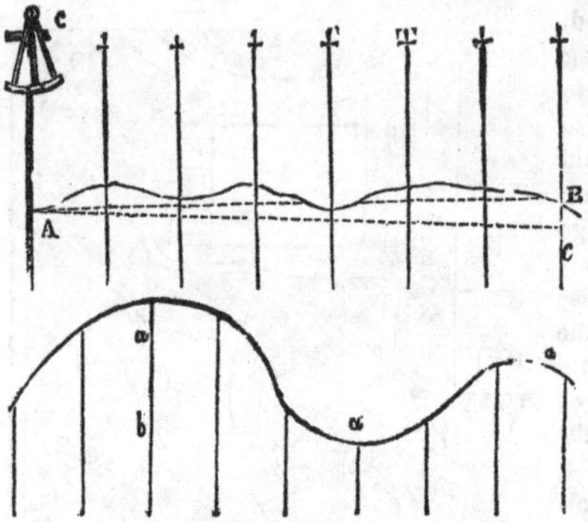

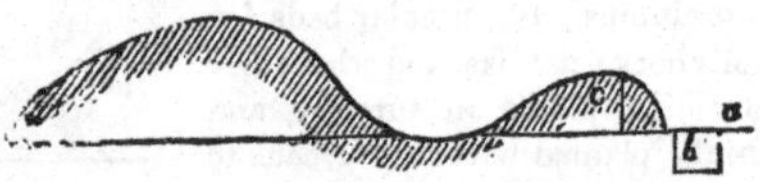

Where hills or mounds are to be thrown up, stakes should be inserted of the desired height, and a line stretched across their tops to show the conformation of the surface, as in the cut. These stakes, in all garden operations, should range from 10 to 20 feet apart, 15 being a good average: they are not only necessary for ascertaining the levels, but enable the men to perform their work with the utmost ease and certainty as to the result.

One of the chief things to be attended to in levelling, is to retain all the best soil for the surface: this increases the labour and expense, but is of the first importance in all garden operations. However, if judgment be exercised in the performance of the work, the surface-soil can generally be passed over on to the new level without the intervention of carts or barrows. This will be obvious from the section above, in which *a* is the desired level, *b* an open trench to get rid of the worthless subsoil, and *c* the section of the next ground to be levelled. Of course, the surface-soil would be thrown from *c* into the trench *b*, up to the level of the line *a*; the subsoil would then be carted or wheeled where it was wanted, and the same process be repeated throughout the entire section. The new level would then be furnished with a depth of from two to three feet of good soil, fit for all cultural purposes.

3.—DRAINAGE.

However high and apparently dry a situation may appear, it is quite possible that it requires to be drained. The object of draining is not only to get rid of superfluous moisture, but also to prevent the little there may be from remaining stagnant. It is quite a common occurrence to find a piece of ground that is never too wet, but which is, nevertheless, sour and unfitted for the cultivation of delicate flowers. It should, therefore, be the first care of the florist to make drains from the highest part of the ground to the lowest, three feet from the surface, dug the shape of a V; and if there be no outlet at the lowest part, to dig a hole, or well, or pond, into which all these should lead, even when there is no apparent means of getting rid of the water. At the bottom of these drains a row of common 2-inch earthen pipes may be placed, end to end, and covered up again with the soil. These are too deep to cause any danger of disturbance in ordinary operations; and the effect is to let air into the soil, if there be no surplus moisture, and to prevent the lodgment of water anywhere. About a rod apart, in parallel lines, will be sufficiently close for the drains, and a larger drain along the bottom, or a ditch, may lead at once to the outlet or the receptacle for the water. Suppose, however, the soil is really surcharged with water, and there is no place but the pond made for the purpose into which this water can pass, and suppose, while we are imagining evils, that this pond or hole fills higher than the bottoms of the drains, it is obvious, in such cases, that the drains cannot empty themselves. Still, even such drains are of use; if they can only discharge all the water in the driest season, immense good is done by them. If the pond be not too large, a garden engine may be set to work to lower the water by throwing it over the surface; and although it may fill as fast as the water is taken away, there is a circulation of water going on in the soil, instead of moisture being stagnant, and the ground made sour.

The rationale of drainage is very well explained by Mr. D. S. Fish. "Drainage," he says, "as popularly understood, means the art of laying land dry. This, however, is a very imperfect definition, both of its theoretical principles and practical results. Paradoxical as it may appear, drainage is almost as useful in keeping land moist as in laying it dry. Its proper function is to maintain the soil in the best possible hygrometrical condition for the development of vegetable life. Drainage has also a powerful influence in altering the texture of soils. It enriches their plant-feeding capabilities, elevates their temperature, and improves the general climate of a whole district, by increasing its temperature, and removing unhealthy exhalations. It lays land dry, by removing superfluous water; it keeps it moist, by increasing its power of resisting the force of evaporation; it alters the texture, by the conduction of water, and by filling the interstices previously occupied by that fluid with atmospheric air; it enriches the soil, by separating carbonic acid gas and ammonia from the atmosphere, and by facilitating the decomposition, absorption, and amalgamation of liquid and solid manures. It heightens the temperature of the earth, by husbanding its heat, and surrounding it with an envelope of comparatively dry air, and by substituting the air for water withdrawn through the interstices of the soil; for while the tendency of excessive moisture in the soil is to bind the whole mass into an almost solid substance, the tendency of air is to separate its particles into atoms, and render it porous: and the more porous a soil is, the greater is its power of resisting evaporation. For this reason porous soils are more moist in hot weather than those of a more tenacious character.

Drainage enriches soils in another way. All rain water is more or less charged with carbonic acid gas and ammonia. Now, the larger the quantity of rain water that passes through the soil, the greater will be the amount of these gases brought in contact with the roots of plants. Nor is this all: solid manures of the richest quality are comparatively useless on wet, heavy soils; for, while a certain amount of moisture is essential to the decomposition of manures, an excess arrests the process, and all the most soluble portions are washed out long before it is sufficiently decomposed to enter into the composition of plants. Judicious drainage, therefore, places the soil in a proper hygrometrical condition for performing its important functions."

1

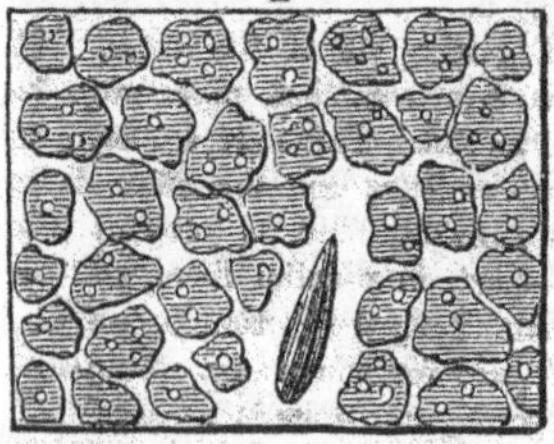

This will be rendered still more obvious by the accompanying diagrams, which prove the beneficial influence exercised by drainage upon the soil, and were exhibited before the Highland Agricultural Society by Dr. Madden, of Penicuik, in 1844. They are highly magnified sections of soil in three different conditions. Under the microscope, soil is seen to be made up of numerous distinct porous particles. Fig. 1 represents it in a perfectly dry state; both the soil and the channels between being quite dry. Fig. 2, on the other hand, represents a soil perfectly wet; the particles themselves are full of water, and so are the channels between them. In fig. 3 the particles are moist, while the passages between them are filled with air. These diagrams show that soil in the condition exhibited in figs. 1 and 2 was totally unfit for the germination of seed. In fig. 1 there is no water; in fig. 2 there is no air; in fig. 3 both are present, in the proportions favourable to the growth of seeds, and these are requisite to ensure the vigorous growth of the plant throughout all its stages; fig. 3, therefore, is the condition of soil desirable for all cultural purposes, and exhibits that congenial admixture of earth, water, and air, that plants delight in, and which efficient drainage only can provide.

2

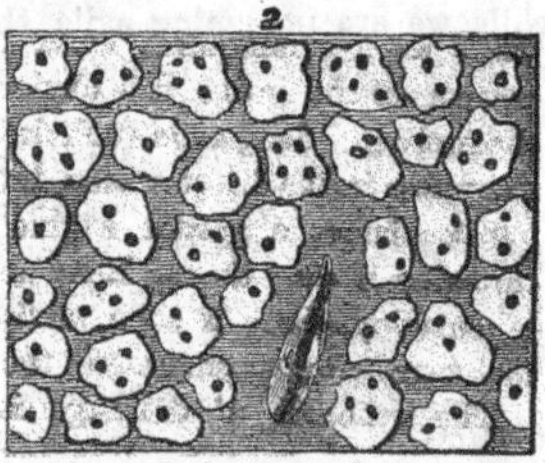

3

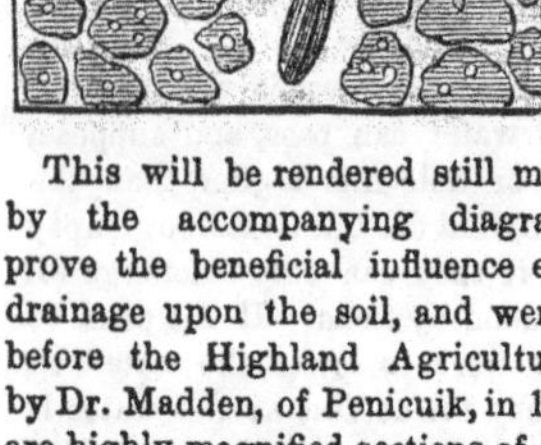

4.—DRAINING MATERIALS.

The materials employed for drains are very varied; brushwood, rubble, stones, bricks, and pipes being all in use. In clay countries it is no unusual thing to form pipes with the clay itself, by inserting an arched framework of wood, and withdrawing it when consolidated. The best and cheapest drains, however, are drain-pipes, which are now obtainable everywhere on moderate terms.

5.—DRAINING IMPLEMENTS.

The implements used in draining are a spade, and in deep draining, and in a clay soil, a series of two or three spades, varying in size and each sloping to the point, and slightly rounded, so as to make a circular cut; a spoonlike implement also is required for lifting the loose soil clear out of the bottom of the trench: and a level, which may easily be formed by fixing three perfectly straight-edged boards in an upright position and in a triangular form, held together by a vertical board in the centre, with an opening at its base for a line and plummet.

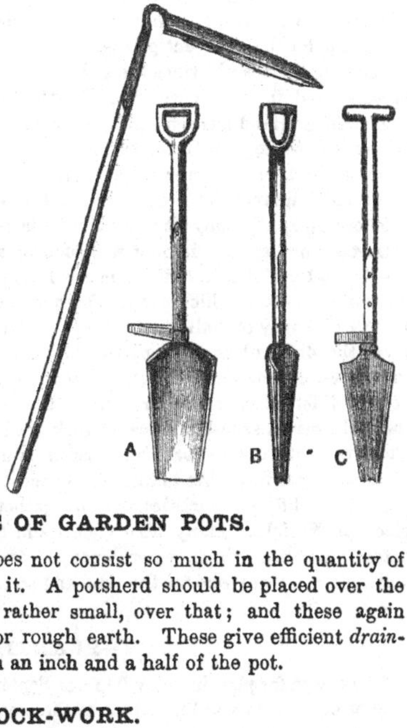

6.—DRAINAGE OF GARDEN POTS.

The effectual drainage of pots does not consist so much in the quantity of drainage, as in the arrangement of it. A potsherd should be placed over the hole; some pieces of pot broken rather small, over that; and these again covered with a layer of peat-fibre or rough earth. These give efficient *drainage*, and need not occupy more than an inch and a half of the pot.

7.—ROCK-WORK.

Few ornaments of a garden have a better effect than rock-work properly disposed; while at the same time it is also very useful. By means of it, not unfrequently, an ugly corner may be turned to very good account, and very many plants will be found to flourish and do well upon rock-work which can hardly be kept alive elsewhere. Sometimes, when the garden or pleasure-ground is very extensive, a piece of rock-work may appear to be needed on its own account, to form a break in the scene; in which case it will be desirable that the work be constructed of the stone of the county, to give to it as natural an appearance as possible; but, in a general way, for rock-work which is intended to be covered with plants, any material that comes most readily to hand may be made use of. The flint stones from the chalk and marl pits, where they can be had, form excellent rock-work; and so, of course, do the different spars of Devonshire and Derbyshire also. As a general rule, rock-work should never be raised on grass, but on gravel, or on a concrete foundation. It is also well placed around a pond or water-tank. In the centre of a square gravelled plot, a tall piece of rock-work is a very pleasing object.

It may be constructed by using the roots of old trees piled one upon another as a basis, which should be well covered with a good coating of fine loam. On this the stones may be built up, in any form that good taste may suggest, interstices, with more or less of surface, being left, which will in this way form beds for the different plants. The spring of the year is the best season for making rock-work, since the soil will have time to settle, and the stones to become fixed in their position before the next winter's frost. Almost every county in England has some material natural to it from which rock-work can be formed—even the larger stones of the gravel pits may be used for this purpose; and, in the absence of anything else, blistered clay from the brick-yards and clinkers from the smith's furnaces are not to be rejected. The seashore also, all along the coast, affords plenty of material, out of which a little taste and good judgment will soon arrange something both agreeable to the eye and useful as a bed for many different classes of plants. On the tall piece of rock-work which has just been described may be planted almost every variety of hardy or half-hardy creepers—lophospermums, *Maurandya, canariensis*, the different sorts of periwinkle, &c., &c.; while lower down, between the stones, cistuses, saxifrages, and sedums may be grown. The wild sedums of our different counties form most interesting collections when placed by themselves in a separate piece of rock-work; and so also do the wild ferns. The writer of these remarks has formed on a piece of rock-work, under a north wall in his garden, what to himself is a most interesting collection of seventeen or eighteen different varieties of ferns, gathered with his own hands in different places in Norfolk. Many other counties in England are much richer in these natural beauties, which, when arranged in rock-work as county collections, will well repay any one for the time and trouble spent in searching for them.

8.—PLANTING.

The season for planting may be from September to March. Many arguments may be brought forward in favour of the month of November, if the weather be open and free from frosts. Spring is always a busy season in the garden; digging, sowing, grafting, and pruning are then in full operation. "And why should planting be added to the number?" asks the Rev. Mr. Lawrence. "It makes part of the wise man's pleasure and diversion to have always something to do, and never too much. Amusements and recreations of all kinds should come to us in regular and orderly succession, and not in a crowd; besides, some intervals of time for meditation between different kinds of work in a garden are very desirable to a good and thoughtful man."

The Pear loves a silicious earth, of considerable depth; Plums flourish in calcareous soils, and the roots seek the surface; the Cherry prefers a light silicious soil; and all cease to be productive in moist, humid soils. The Apple accommodates itself more to clayey soils, but does best in a loamy soil of moderate quality, slightly gravelly. In preparing stations, therefore, suitable soils should be supplied to each. The station is prepared by digging out a pit about three feet square, and the same depth, in ground that has been well drained. In the bottom of this pit lay 10 or 12 inches of brick or lime rubbish, the roughest material at the bottom, and ram it pretty firmly, so as to be impervious to the tap-root: the remainder of the pit is filled in with earth suitable to the requirements of the tree. When the surrounding soil is a tenacious clay, the roots of the young tree should be spread out just under the surface, and rich light mould placed over them, forming a little mound round the roots; but in no case should the crown be more than covered: deep planting is the bane of fruit-trees.

The stations being prepared, the trees require attention; it is necessary to prune the roots, by taking off all the small fibres, and shortening the larger roots to about six inches from the stem; if there be any bruise, the root in which it occurs should be removed entirely, by a clean sharp cut. Two or three spurs are sufficient: but if there be more good ones they may remain, after careful pruning. The roots may be laid in milk-and-water or soap-suds, a few hours before the trees are planted. The process of planting will differ, according as the trees are intended to be dwarf, standard, pyramid, or wall-trees. With dwarf, standard, or espalier, place the tree upright in the centre of the station: spread the roots carefully in a horizontal direction, and cover them with prepared mould to the required height, supporting the young plant with a strong stake, driven firmly into the ground, and tying the stem to it with hay, or something that will not bruise; press the soil gently, but firmly, over the extended roots, having first cut away the tap-root. Then mulch the place. This process, called mulching, consists in spreading a layer of short, half-rotten dung five or six inches thick round the stem, in a radius six inches beyond the extremity of the roots; the mulch spread evenly with the fork, and gently pressed down by the back of the spade, or, if exposed to the wind, pegged down to prevent its being blown away. In the case of a wall-tree, let the root be as far from the wall as convenient, with the stem sloping to it, the roots being extended and covered in the same manner with the soil.

The nature of the soil is to be regarded, and the tree planted at a greater or smaller elevation above the level of the surrounding soil accordingly: where the subsoil is a stiff clay, the mound in which it is planted should rise from nine to twelve inches; in a warm dry soil, a very gentle elevation suffices. The roots should be planted in the richest mould; and various expedients may be used to keep them moist and cool, and free from canker. The mould requires to be pressed gently round the roots with the hand, so that the soil may be closely packed round them: with these precautions, no fear can be entertained of productive fruit-trees.

9.—PROPAGATING.

In a state of nature all plants are propagated from seed, and the multifarious form of the seeds and envelopes with which they are provided is one of the many interesting subjects of investigation to the lover of nature. For our purpose it is sufficient to state that most seeds are covered with a hard shell or envelope, which protects them from external injury; that within this envelope lies the embryo plant. All seeds in this latent state contain an organ, or Germ, which, under favourable circumstances, shoots upwards, and becomes the stem of the plant; another, called the Radicle, which seeks its place in the soil, and becomes the root; and the Seed-lobes, which yield nourishment to the young plant in its first stage of growth. Moisture, heat, and air are necessary conditions for the development of all seeds; and most of them require, in addition, concealment from the light.

It is found, however, that except in the case of annuals (as plants raised from seed, which grow, produce their seeds, and ripen their fruit in one year, are called), much time is lost by following this mode of propagation; it is also found that the seed does not always produce the same identical plant; above all, it is found that none of the double-flowering, and few of the herbaceous-flowering plants, with which our gardens are furnished, ripen their seeds in our climate. This has led to other methods of multiplying; for, besides the roots properly so called, which attach themselves to the soil, and draw from it the principal nourishment of the plant, it is found that each branch conceals under its outward covering a bundle of fibres or tissues, which, under favourable circumstances, develop roots, and become the basis of an independent plant, identical with that from which it sprung. Many plants have also a crown with buds or eyes, each capable of propagating its species. Every plant with roots of this description may be divided into as many portions as there are eyes, taking care that a few fibres are attached to the root, and each will become an independent plant. The potato, and all the bulbous and tuberous plants, are familiar examples of this principle of propagation; so are the Dahlia and Pæony, which grow better when the set is confined to a piece of the tuber with one eye attached, than when planted whole. So conspicuous is this in the potato, that, where it is planted *whole*, all the eyes except one, or at most two, should be scooped out with a sharp knife, because the young plant will thus be supplied with more of its natural pabulum while it is rooting, and will increase its vigour.

Other plants throw off short stems, like the daisy and house-leek, by means of which they are propagated. Others again, like the strawberry, throw off runners, each of which is furnished with its root-fibre already elaborated, only requiring soil in which to root itself. But there are others, where nature requires the aid of art, and this has given rise to the operation which will be found explained under the article *Layering*.

10.—PRUNING.

The pruning of fruit-trees is performed at two seasons,—winter and summer. Winter pruning should be performed while vegetation is entirely at rest,—the period which follows the severest frosts, and which precedes the first movement of vegetation, that is to say, the end of February or the very beginning of March, in ordinary years. If trees are pruned before the strong frosts of winter set in, the cut part is exposed to the influence of the severe weather long before the first movement of the sap takes place, which is so necessary to cicatrize the wound, and the terminal bud is consequently often destroyed. Equally troublesome are the wounds made during frosts: the frozen wood is cut with difficulty; sometimes the cuts are ragged, and do not heal; mortality attacks the bud, and it disappears. To prune after vegetation has commenced, except summer pruning, is not to be thought of; therefore let all chief pruning be done in February, if the frost has disappeared, especially for the peach, whose buds, placed at the base of last year's shoots, are particularly exposed to the action of the ascending sap.

The implements required in pruning are a hand-saw, a pruning-knife, a chisel, and a mallet. For garden trees the knife is the most important: it should be strong and of the best steel, with a considerable curve, so as to take a good hold of the wood. The manner of operating is far from indifferent. The amputation should be made as near as possible to the bud, but without touching it; the cut should begin on the opposite side, and on a level with the lower part, made at an angle of 45°, and terminate just above the bud.

If it be necessary to cut away a branch altogether, a small portion of it should be left on the stem, and the cut should be a smooth and bevelled one,

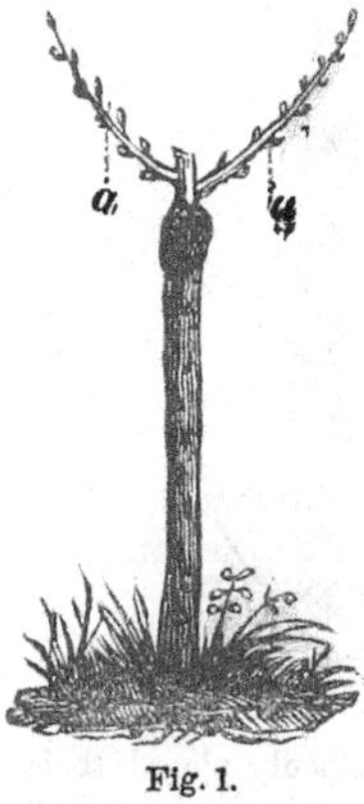

Fig. 1.

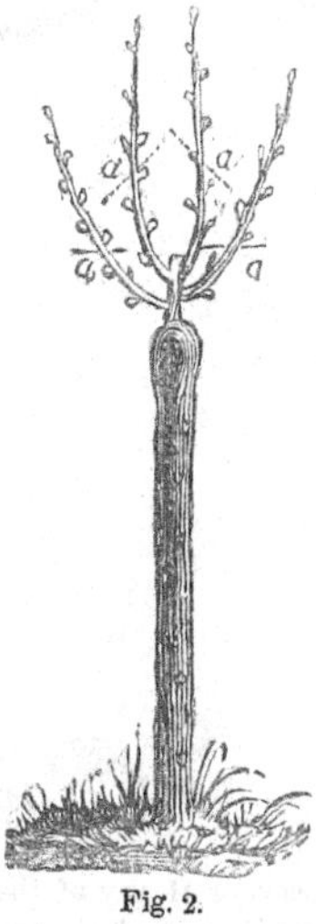

Fig. 2.

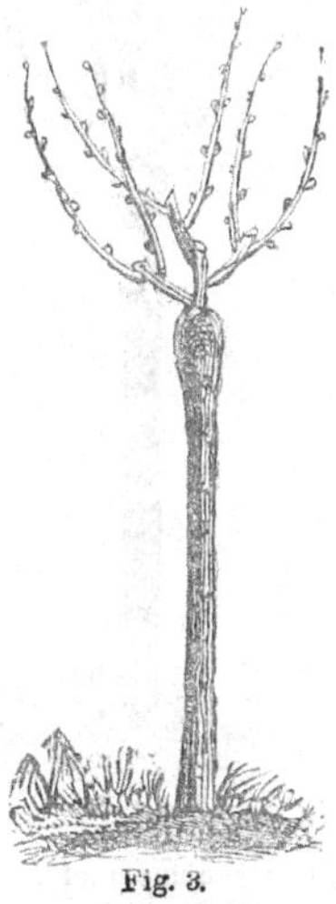

Fig. 3.

presenting the smallest possible extent of wounded surface. If made with the saw, it should be made smooth with the knife or chisel, and covered with grafting-paste.

The first object in pruning a standard tree is the formation of its head. The first pruning must take place at the end of the first season after grafting, when the scion has made its growth, as represented in fig. 1, when two shoots have sprung from the graft. To form a full round head, the two shoots should be pruned in to *a*, *a*. The year after the tree will present the appearance represented in fig. 2; or, if three shoots have been left the first year, and the whole three headed in, in the following year they will appear as in fig. 3, each shoot having thrown out two new branches. The one tree now presents a head of six, and the other four shoots. At the end of the second year both are to be headed back, the one to the shape indicated by the crossing lines, *a*, *a*, *a*, *a*, the other as nearly as possible to the same distance from the graft.

Another year's growth will, in each instance, double the number of main shoots, which are now eight and six respectively, as represented in fig. 5. If a

Fig. 4.

Fig. 5.

greater number of shoots appear, or if any of them seem badly placed, their growth should be prevented by pinching off the tops when young, and pruning them clean off when the tree has shed its leaves. The time for pruning is any of the winter months between November and February before the sap begins to stir. Those trees which have produced six shoots should be pruned exactly like those with eight, to form a compact head, as in fig. 4; and when the standard tree has acquired eight main branches by these various prunings, it has attained its full formation, and with a little care may easily be kept in shape.

When a standard tree has reached its bearing state, the object of the pruner is the production of fruit. If the branches are well placed, let them have free course, and they will throw out bearing-spurs at the extremities. Little more need be said on the subject, except that all unproductive wood, crowded sprays, and decayed branches, that cross each other, should be cut out, the tree kept open in the centre, and the open cup-like form rigorously maintained. These remarks apply chiefly to apples, pears, and other trees which bear their fruit on spurs. These spurs will in time become long and scrubby, with many branches, as in fig. 6, where we see a spur with many branches getting further and further away from the main branch. To bring it back to its proper position, cut away, neatly, the upper shoot at *a*, and the side shoot at *b*, cutting out, also, the central shoot, when the small bud, *c*, will push out and form blossom-buds the following year.

Fig. 6.

When a tree is very vigorous, the buds will break strongly and run into wood too strong to form blossom-buds. The remedy in this case is to break the young shoot near the third bud from the main branch, leaving the broken part hanging down. The time for this operation is about the middle of March. The broken part, while it droops, nevertheless draws up a portion of the wood-sap. The following winter, when the buds are turned into blossom-buds and become fruitful, the hanging shoot should be neatly pruned away, when a fruitful bearing-spur will be formed.

11.—BUDDING.

Grafts of this description present the following characters: they consist in raising an eye or bud with a piece of the bark and wood, and transferring it to another part of the same plant, or any other plant of the same species. Budding is chiefly employed on young shoots or trees from one to five years old, which bear a thin, tender, and smooth bark.

The necessary conditions are, that the operation takes place when trees are in full growth, when the bark of the subject can be easily detached from the *liber*, and it may be performed generally from May to August. The buds adapted for the operation should present well-constituted eyes or gemmæ at the axil of the leaf; if they are not sufficiently so, it is possible to prepare them by pinching the herbaceous extremity of the bud; thus producing a reflux of the sap towards the base; and in about twelve days' time the eyes will have become sufficiently developed: then detach the bud from the parent tree. Suppress all leaves, only reserving a very small portion of the petiole, or leaf-stalk, as *c*.

Having fixed upon the intended stock and bud, take a sharp budding-knife, and with a clean cut remove the bud from its branch, with about a quarter of an inch of the bark above and below: remove all the wood without disturbing the inner bark of the eye; for it is in this liber, or inner bark, that the vitality lies. Now make a cross-cut in the bark of the intended stock, and also a vertical one, T, and shape the upper part of the shield, or bud, *a*, so as to fit it exactly. Having fitted the parts correctly, raise the bark of the stock gently with the budding-knife, and insert the bud: afterwards bandage lightly above and below the eye, bringing the lips of the bark of the stock together again over the bud by means of the ligature, in such manner that no opening remains between them, and, above all, taking care that the base of the eye is in free contact with the bark of the stock.

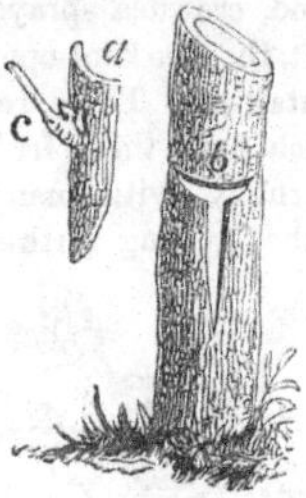

Some weeks after, if the ligatures seem to be too tight, they may be untied and replaced with smaller pressure. When the operation takes place in May, the scion will develop itself as soon as the suture is completed. In order to provide for this, cut the head of the stock down to within an inch of the point of junction immediately after the operation.

When the operation takes place in August, the head is never cut till the following spring, when the scion begins to grow. If the same practice as in earlier budding were followed, the consequence would be, that the bud would develop itself before winter; and, having no time to ripen its new wood, it would perish, or at least suffer greatly When the buds begin to grow, they require to be protected from strong winds; otherwise they would be detached from the stem. This is done by driving a stake, *a*, firmly into the ground, attaching it by a strong cord to the stem of the stock above and below the junction, as in the engraving, and tying the shoot of the young scion firmly to the stake above, protecting it by a bandage of hay or other substance, to prevent the bark being injured.

The weather most suitable for budding is a subject of dispute among practical men. Cloudy weather has generally been preferred; but Mr. Saul, of Bristol—no mean authority—protests strongly in favour of warm sunny weather, provided the stock and buds are in proper condition. "In warm weather," he says, "the sap is more gelatinous, and the bud, on being extracted and inserted in the stock, quickly and properly tied, soon takes. On the contrary, in wet, cloudy weather the sap is more thin and watery, and the bud will not unite so freely besides this, a fall of rain, after the buds are inserted, likely enough, in such weather, will fill up the interstices, and rot the buds before they have time to unite with the stock."

Another question Mr. Saul agitates:—Is it necessary to extract the wood from the eye of the bud? "American writers say, no; but I answer, yes," he says. "It may suit their hot, dry climate, but I must give the preference to our old system of extracting the wood from the bud, not only for roses, but for fruit, ornamental and forest trees. In rose-budding," he adds, "the bud in the shoot should be commenced with, cutting out from it about the eighth of an inch below the bud or eye, to about half an inch above it. Take out the wood without touching the liber or inner bark; next make an incision in the branch on which the bud is to be placed, quite close to the main stem, half an inch long, with a cross-cut at the upper extremity, thus: T. Raise the bark with the end of the budding-knife, without bruising it, and insert the bud, tying it well with worsted thread, giving one turn below, and two, or at most three, above the eye of the bud. Worked in this way, they grow out from the axil of the branch, and look neat and workmanlike; and after a season or two, when headed back and healed over, it presents a fine bushy head, growing apparently out of the main stem, without scars, wounds, or knots."

The shoots selected for budding or grafting, whether for fruit or rose-trees, should be firm and well-ripened: watery shoots, or watery buds, are valueless. For grafting, the branches should be of the preceding year, well ripened under an August sun—*aoûté*, as French fruitists say.

The stock should be in a state of vegetation slightly in advance of the graft; otherwise the flow of the sap is insufficient to supply the wants of the scion. In order to provide for this, the graft may be removed from the parent branch a little before the operation, and buried under a north wall: there it remains stationary, while the stock is advancing to maturity.

12.—BUDDING ROSES.

To the foregoing remarks on budding generally, it may be well to add some special directions respecting roses.

In selecting buds of roses, take those of moderate size; clean off the thorn, cut the leaves off, leaving only about half an inch of the stalk or petiole to hold by: then with a sharp knife take out the bud, beginning half an inch above the eye, and bring the knife about the eighth of an inch below; with the point of the knife separate the wood from the bark, without interfering with the wood which remains in the eye, leaving it so that, when inserted on the stock, the wood left may be in immediate contact with its wood.

Having removed the thorns on the intended stock, open the bark at the most convenient spot for the insertion, by drawing the point of the knife down the centre of the shoot, and by a cross-cut, where the other begins, raise the corners of the bark sufficiently to introduce the lower end of the bud: press it down till it is opposite to the corresponding bud on the stock, and bind it up with a piece of fine bass or worsted thread, leaving the eye so that it is just visible. After three or four weeks it should be examined, and the band loosened a little. In cases where the bud does not separate freely from the bark, the wood may be tied in also; but the operation is both neater and more efficient when all the wood except that in the eye is removed. Cloudy weather is generally recommended for the operation; but Mr. Saul, of the Durdham Downs Nursery, Bristol, says he prefers bright, warm, sunny weather, provided the stocks are in proper condition. Rose-budding may be performed any time from June to September, and even as late as October, August being suitable for the greatest number of roses, the test being of course the maturity of the shoots.

13.—GRAFTING.

Gardening ingenuity has invented many kinds of grafting: we shall describe a few of these processes, in order to explain their principle. Select a suitable stock, whose height will be according to the purpose for which it is intended, also a graft, which should be from an early branch of the previous year's wood which has ripened under an August sun, so that the wood has been thoroughly constituted before the early frosts set in. It should also be selected so that the graft is in the same state of vegetation with the intended stock. Where the texture of the wood is less advanced in the graft than in the stock, the latter intercepts the descent of the pulpy sap, and forms the bulging on the stem which is observable on so many trees; when the case is reversed, the swelling occurs in the branch above the graft; for the principle of the union is, that the pulp from the scion descends to the point of junction, where, being excluded by the ball of grafting-wax, which surrounds it, from the light and air, it forms woody fibre in place of the roots which it would have formed in the soil; in the meanwhile, the sap from the stock rises into the graft, where it is elaborated into pulp by the action of the leaves, and returns again, but in a more consistent state. It is necessary, therefore, where the graft selected is in a more advanced state of vegetation, to detach it from the parent stem, and bury it in the ground, under a north wall, until both are in a similar state: the graft will here remain stationary while the stock is advancing.

In gardening nomenclature, the term "stock," or "subject," is applied to the tree on which the operation is performed; that of "graft," and sometimes "scion," to the portion of the branch which is implanted in it.

Cleft- or Tongue-grafting.—In cleft- or tongue-grafting, the crown of the stock is cut across, and a longitudinal wedge-shaped slit, *c*, is made about four inches long, according to the size and vigour of the intended graft: this cleft is kept open by a wooden wedge until the scion is prepared. The scion is then selected, having a bud, *a*, at its summit; and the lower part of it is shaped with the knife so as to fit the slit in the stock. The double-tongue graft only differs from the first in having two grafts in place of one; and it is preferable, when the size of the stock permits of its use; the wound heals more quickly, and the chances of success are greater than in the single graft.

In placing the graft, it is to be observed that the top, whether single or double, should incline slightly inwards, as at *e*; thus leaving the lower extremity slightly projecting, as at *d*, in order that the inner bark of the graft and stock may be in direct contact with each other. Finally, bind the whole, and cover it over, from the summit of the stock to the bottom of the cleft, with clay or grafting-paste.

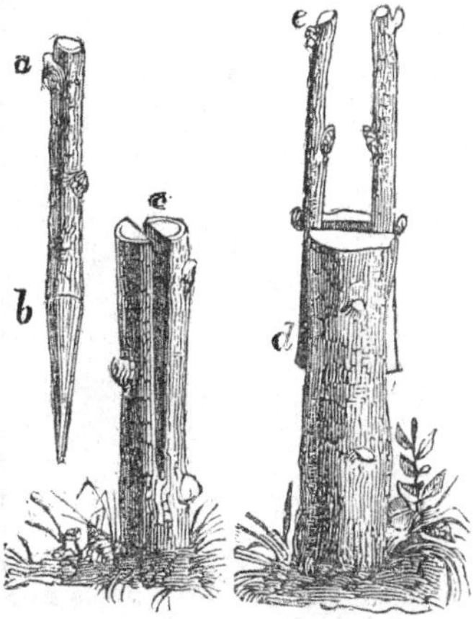

Double Grafting.—In double grafting, where they both take, it is necessary to suppress the least vigorous as soon as the wound is completely closed, especially in the case of standard trees; otherwise the head gets formed of two parts completely estranged from each other. During the first twelve days after the operation, protect the head from the action of the air and the heat of the sun by some kind of shade. A square piece of paper, twisted into the shape of a bag, such as grocers use for small quantities of sugar, answers very well for this purpose, protecting it at the same time from the attacks of insects; and when it begins to grow, protect the graft from being disturbed by the wind, or by birds lighting on it by attaching it to some fixed object. A perch formed of an osier rod, having both ends tied firmly to the stock, and having the young shoot attached to it, as in the engraving, will serve both purposes.

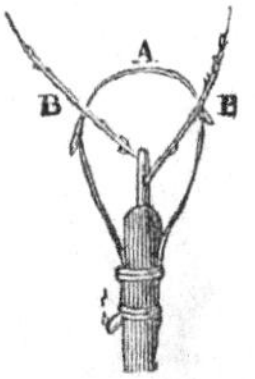

When the young scion begins to grow, it is necessary to suppress all buds which develop themselves on the stem, below it, beginning at the base, and upwards to one or two inches long.

The Bertemboise Graft.—A very neat mode of grafting, called by the French the "Bertemboise graft," is described and figured by M. Breuil. Cut the crown of the stock at a long bevel, leaving only about an inch at the top square, cutting out an angular piece to receive the graft, and operating in all respects as in the former instance. When the stock is not large enough to receive a graft on each side, this mode is preferred, as forming the neatest union, as well as the most rapid; for all the ascending sap is thus drawn to the summit of the bevel on which the graft is placed.

Theophrastes Graft.—A graft honoured with the name of Theophrastes is sometimes practised on trees having healthy roots, where it is desired to improve the fruit. Having cut the stem of the tree itself horizontally, or selected a single branch to be operated upon, about twenty inches from the principal stem, three vertical cuts are made in the bark, at equal distances from each other, about an inch long. Having selected three or more grafts, *a*, and shaped their lower extremities into a tongue somewhat like the mouth-piece of a flageolet, with a neck or shoulder at the upper part, then introduce a graft under the bark of each vertical cut, raising the bark for that purpose with the spatula of the grafting-knife, and placing each graft in such a position that the inner bark of the graft is in immediate contact with the inner bark of the tree. When neatly arranged, bandage the whole, and cover with the grafting-paste.

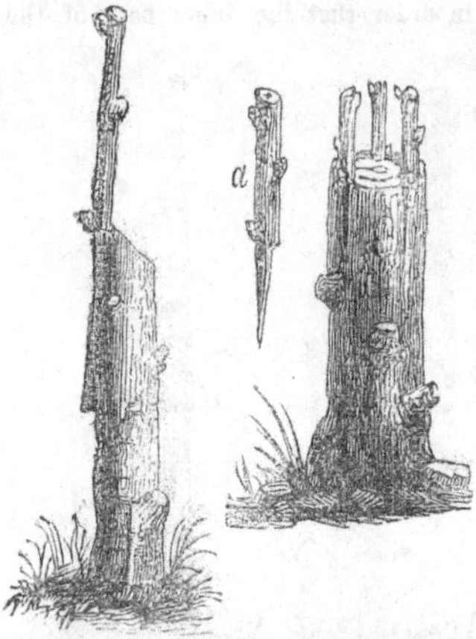

Slit-Grafting.—In place of the vertical cut through the whole of the stem, in this process a triangular cut is made in the side of the stock, as in the next engraving; the lower end of the graft is then cut so as to fit exactly into the gap made, so that the inner bark, or liber, meets in contact at all points: this done, it is covered with clay or grafting-paste, and bound up until amalgamation takes place.

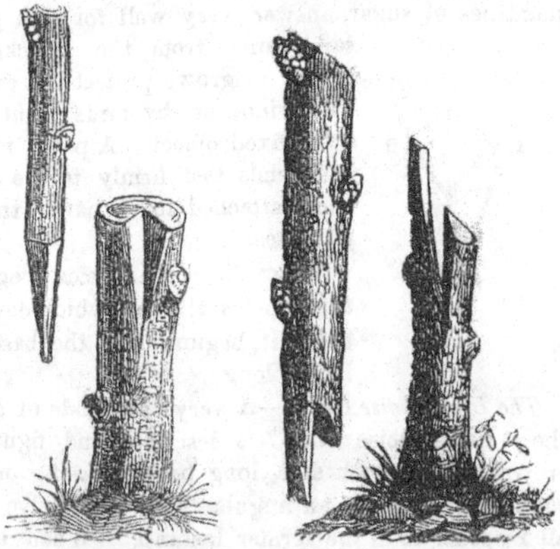

A strong and efficient mode of grafting is represented in the second cut. Make an elongated bevelled cut in the proposed stock from left to right; make another vertical wedge-shaped cut, three inches long, from left to right, leaving a narrow shoulder at the top on the left side, and terminating in the centre of the stock, so as to resemble that in the engraving. Take the intended graft, of the same diameter as the stock, and shape its lower extremity so as to fit into the cleft thus made; bind up in the usual manner, and cover the joint with grafting-paste. This forms a very strong and very useful graft in species which unite slowly.

Herbaceous Grafting, as initiated by Baron de Tschudy, consists in choosing branches still in active growth. Pines, walnut-trees, oaks, and other trees which are multiplied with difficulty by other processes, are easily produced by this one. The mode of operating differs slightly, according to the species. In the case of pines and resinous trees, when the terminal bud of the subject, *a*, has attained two-thirds of its growth, make an horizontal cut at *d*; then make a slit downwards to the point where it begins to lose its herbaceous character in the ligneous consistence of the tree; stripping a part of its leaves, and leaving only a bud or two at the top to attract the sap. The graft, *b*, is now prepared, having a cluster of young buds at its summit, and its lower extremity is shaped to fit into the slit, where it is so placed that the upper part projects over the cut in the stock. It is now covered with grafting-clay and bound, beginning at the top, below the bunch of leaves left on the stock, so as to avoid disturbing the leaves, and working downwards. This done, break off, an inch or so from their axils, the branches *c c* of the stock below the graft. When operating on delicate species, it may be desirable to envelop the graft in a covering of paper, to preserve it from the over-dry atmosphere or the heat of the sun, for ten or twelve days after the operation. Five or six weeks after grafting, the union will be complete, and the bandage may be removed, or at least relaxed; and when the suture is perfect, the leaves at *d* may be removed, otherwise they will originate buds and branches from the old tree.

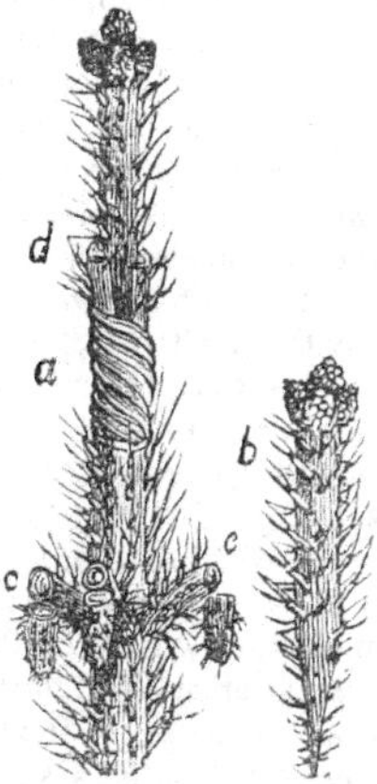

In other species proceed as follows:—Towards the end of May, when the terminal bud of the tree is in a state of active vegetation, make an incision, crossing the insertion of the petiole of the third, fourth, or fifth leaf, as at *b*, penetrating half the diameter of the stem; the choice of the particular leaf depends upon its state of vegetation as compared with the proposed scion. If the axis of the leaf, *a*, is examined, it will be observed that it has three eyes, or *gemmæ*, the centre one being most developed: it is between the axis of the central eye, at *b*, and one of the lateral ones, that the oblique cut is to be made, stopping in the centre about half an inch below the axis of the leaf. The graft, *c*, consists of the fragment of a branch of the same diameter as the stock, and in the same state of vegetation; it is cut to the same length as the prolongation of the stem, *d*; it is wedge-shaped, fitted to the slit into which it is inserted, bound, and covered with some grafting-paste. The leaf, *a*, is left on the stock to draw the sap upwards for the nourishment of the graft. The leaf of the graft, *e*, assists in

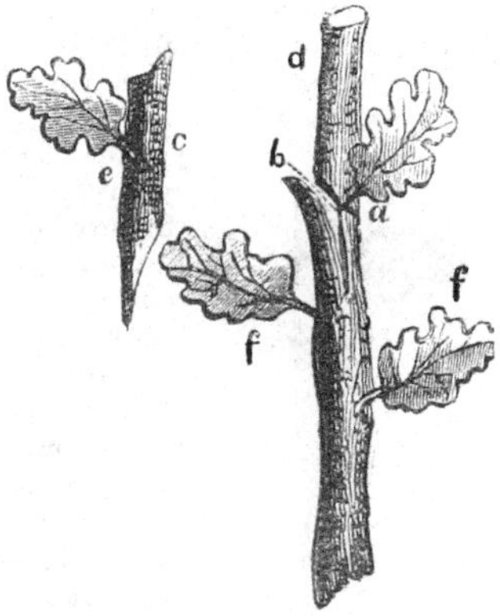

the process by absorbing it to the profit of the young scion. The fifth day after the operation, the central eye, *a*, is suppressed; five days later, cut the disk of the leaf at *f*, reserving only the medial nervure, rubbing off at the same time the eyes at the axil of these leaves, repeating the same suppression ten days later. At this time, also—that is, twenty days after the operation—cut the disk of the terminal leaf, *a*: these several suppressions will force the sap progressively from the roots into the graft. Towards the thirtieth day the graft enters on its growth: at this time remove or relax the bandage, protecting it by a paper coronet from extreme drought and the sun.

Side-grafting.—In side-grafting it is not essential, as in other cases, to amputate the head of the stock, the graft being attached to the side, as its name indicates. Having made a cross-cut into the bark of a tree, as at B, and a vertical incision in the bark from its centre, thus marking a cut in the form of a T, each cut penetrating to the liber or inner bark; having also prepared the scion A by a longitudinal sloping cut of the same length, as B C, and raised the bark with the spatula of the grafting knife, the graft is introduced, and the whole bandaged in the usual manner. This kind of graft is particularly useful in replacing branches on fruit-trees which are necessary to complete the symmetry of the tree for horizontal training.

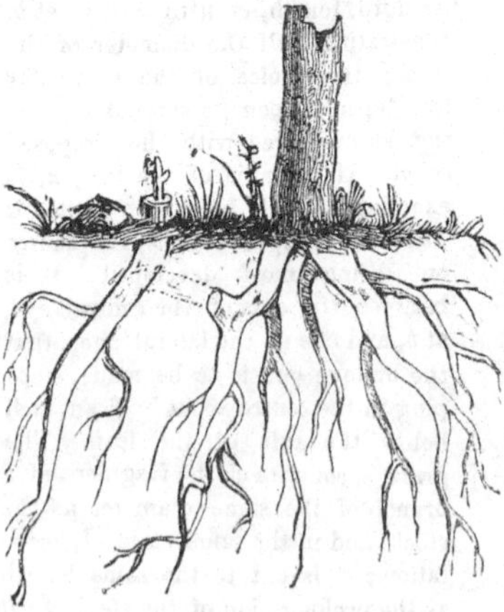

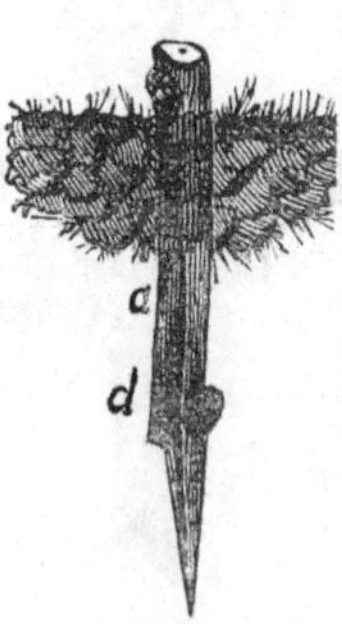

Root-grafting.—In root-grafting the roots are operated on as the stems have hitherto been. Although it is by no means in common use, this mode of grafting is very convenient on some occasions. Having laid bare the roots to be operated on, shape the graft, *a*, by cutting its lower extremity into a shape resembling the mouth-piece of a flageolet, with a tooth or shoulder, *d*, in its upper part. Cut the root across as at the dotted lines, and make a vertical cut in the separated part to receive the tongue of the scion, with an opening also corresponding to the tooth in the scion. Bring the scion and vertical cut together, so that all the parts cut, meet, and cover each other, meeting just below the last bud on the scion. This root, being already fixed in the soil, will serve to multiply plants which do not even belong to the same species.

Shield-root-grafting.—Shield-grafting is also usefully practised on the root in some cases even where the stock and scion are not of the same species. To discover the larger nd best roots, trace them with the finger, and graft upon it in the spring, leaving the spot *a* (opposite) occupied by the cushion uncovered. In the following spring, when the graft has pushed forward, separate the root from the parent tree. We thus obtain a new individual.

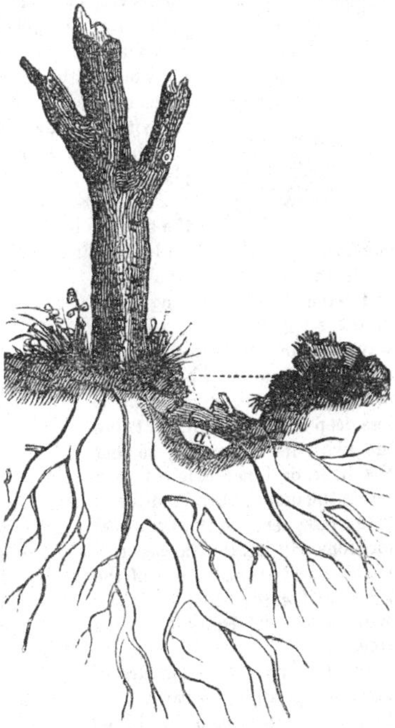

Circle-grafting.—These grafts are composed of one or many eyes or buds, carried by a ring of bark including the liber. They are applied generally to the multiplication of certain large trees—as the walnut, chestnut, oak, and mulberry. Towards the decline of autumn, as the sap returns to the roots, choose a mild day, free from rain. From the tree to be operated on, select a branch of the same size as the scion, having well-formed eyes. Upon this branch raise a ring of bark, *a*, without detaching the branch from the tree, making two circular incisions all round it, and making another vertical incision afterwards on one of its sides, and remove it gently.

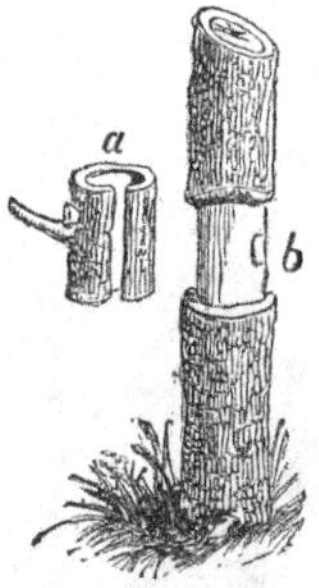

Detach from the intended stock another ring of bark of the same size, and place the ring of the graft in its place at *b*, and the ring of the stock on the place whence the scion was taken; bind up and cover the joinings with grafting-paste. In the following spring, if the graft has taken, cut the head of the stock immediately above the rings, which will favour the development of the buds which they carry.

Another application of this mode is practised. When the spring sap is about to rise, cut the head from the tree to be operated on, and remove a ring *b* from the top *a*. Choose a tree of exactly the same size, on which the operation is to be performed; detach from it a ring furnished with two or three eyes, as *b*, and of the same length. Adjust this cylinder in the place of the ring detached, making it coincide exactly at its base with the old bark, and cover the whole with grafting-paste. Of all these grafts this is the most solid, and least subject to be disturbed by the wind; but even this requires protection, so that it is not shaken in its place till complete suture has taken place.

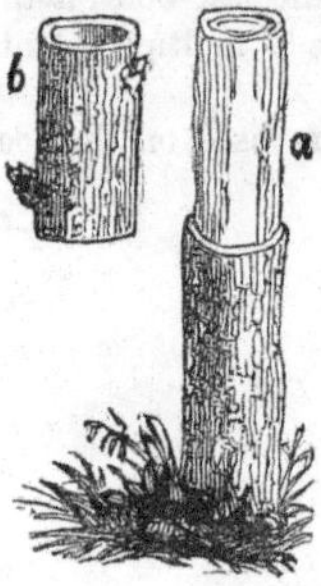

Approach-grafting.—In approach-grafting, supposing the stock to be planted, and the scion in a pot, as in the engraving below, make a longitudinal cut in the stock, of such extent as to reach the medullary canal at *a*, and leave a corresponding notch in the scion at *b*, but in such a way that in the scion it is less deep at the base, *b*; while, on the contrary, the cut in the stock is less deep at the summit, *c*. Bring the two cuts in contact, so that the liber, or inner skin, of each meets the other; then bind them. The consequence of these unequal incisions will be, that in separating the head at the point *d* of the graft, and *a* in the stock, there will be less deformity left in the tree.

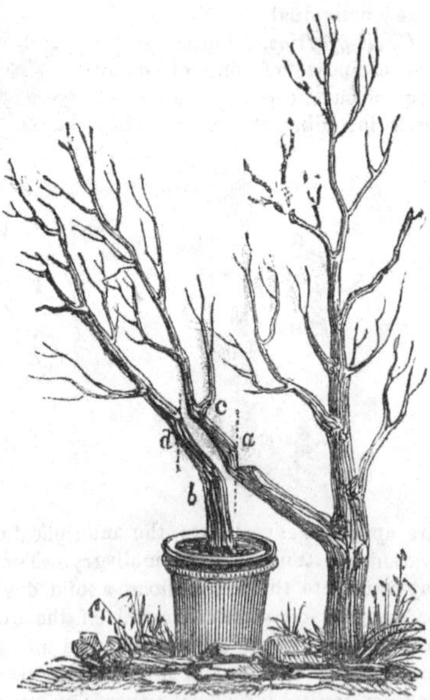

In the preceding examples of approach-grafting, the parts of the branch operated upon should be of the previous year's growth at least. It is sometimes desirable, however, to apply the principle to branches of the same year's growth. Accident may deprive a tree of the branches necessary to its symmetry and a year's growth be saved by applying an herbaceous or green graft to supply the deficiency, if there happens to be a lower branch of the same tree available for the purpose. Let us suppose that a void exists at A A A, on an otherwise healthy peach-tree, and that side branches, or fruiting spurs, are required at these points to balance the tree and restore its symmetry, and that a lower branch from B is available to supply them any time between June and August. Supposing the shoot to have attained sufficient length, an incision is made in the branch, about a quarter of an inch long, with a cross cut

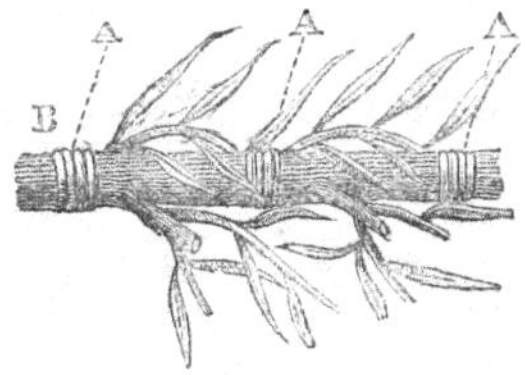

at each extremity, deep enough to penetrate to the inner bark; the bark is raised from the wood on each side of the longitudinal cut by means of the spatula at the end of the budding-knife. A thin slice is now cut out of the shoot B, on the lower side, and opposite to a leaf-bud, corresponding in length with the incision on the branch. The parts thus laid bare are brought together, the lips of raised bark brought over the shoots, and the parts are again bound together. The process is continued as often as is deemed necessary, or the length of the shoots will permit, taking care that in each case a leaf-bud is left above the point of union, and that it is left uninjured by the ligature, but leaving eight or ten days between each operation.

In the following spring the union will be complete; but it is better not to separate the grafts till the second spring. At this time cut each shoot which has furnished the graft immediately below the ligature, and submit each of the new shoots to the usual training.

14.—GRAFTING WAX (COBBETT'S).

Take of pitch and resin four parts each; beeswax two parts; tallow one part; melt and mix these ingredients, and use them for grafting when just warm.

15.—GRAFTING IMPLEMENTS.

The implements necessary for the operation are, a hand-saw, sometimes made with a folding blade, the peculiarity of which is, that the blade should be thin at the back, with very open teeth, A B; a grafting-knife, and a chisel and mallet bevelled on both sides, used where the graft is too large to be cut by the knife; also a supply of small quoins, or wedges of hard wood, to keep the slit open while the graft is preparing. The grafting-knife is furnished with a smooth spatula, of hard wood or bone, at its lower end. A bundle of coarse hemp, or worsted thread, or of willow bark which has been softened and rendered pliable by being soaked in water, and some composition which shall protect the graft from the atmosphere and from rain; these complete the appliances necessary in grafting.

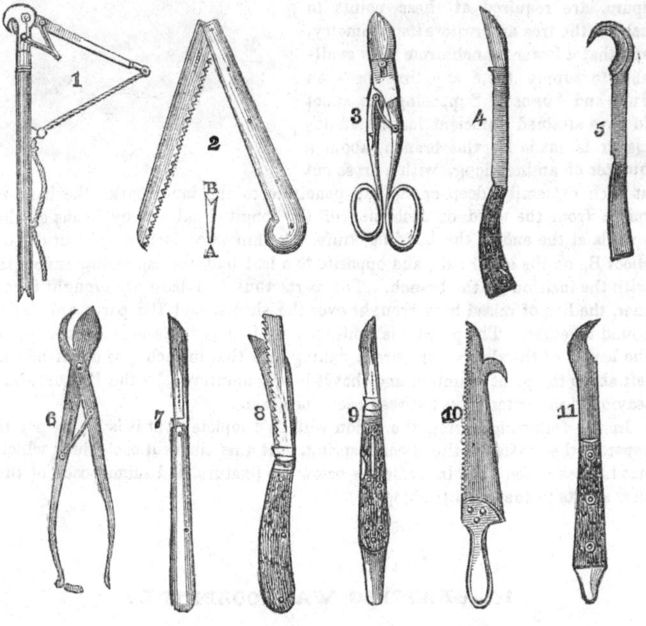

GRAFTING IMPLEMENTS.

1. Averancators, six feet long. 2. Folding Pruning Hand-saw, with tooth, A B. 3. Bow-slide Pruning-shears. 4 and 5. Gooseberry Pruning-knife, straight and hooked blade. 6. Hand-sliding Pruning-shears. 7. Pruning-knife, with straight blade and smooth spatula. 8. Pruning-knife and Saw. 9. Budding-knife, buckhorn handle, with ivory spatula added. 10. Gentleman's improved Pruning-saw, with Billhook. 11. Grafting-knife, with strong curved blade serving as a chisel, with spatula added.

16.—LAYERING.

Layering, which consists in arresting the circulation of the sap on its return to the roots, is one of the most useful methods of propagation. In this operation an upward slit is made half across a joint; and by fixing the part so cut in favourable soil, the latent fibre expands into a root, and the branch becomes an independent plant. The process is well adapted for pinks, carnations (the example illustrated in the engraving in the following page), roses, and many other plants. It is, indeed, a very important operation in gardening, and should be neatly executed. Choosing, the suitable branch of the carnation, for instance, which is first stripped of all branches below the joint selected, and being furnished with a very sharp knife, the operator begins his incision a quarter of an inch below a joint, passing the blade through it in an oblique direction, to a quarter of an inch above, taking care that the cut terminates as nearly as possible in the centre of the stem: the tip of the tongue thus made is cut off with a clean sharp cut, and the layer pegged down in a little fine rich mould, but not more than an inch under the soil. In the case of carnations, the plant is in a fit state for the operation as soon as the flowering season is over; and no stem which has already produced flowers should be employed for the purpose.

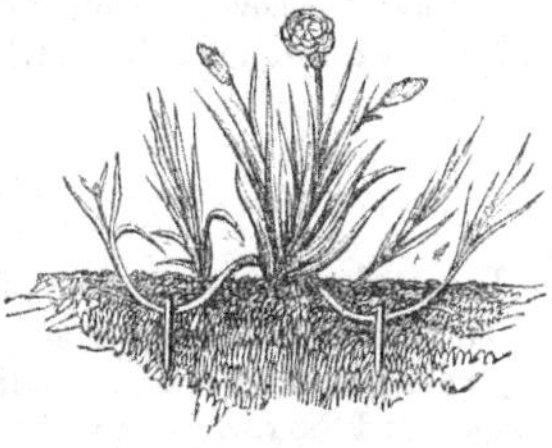

In the case of roses, and other shrubby plants, all that is required is to run the knife sufficiently through a joint to make an opening or crack near it, and peg down the wounded part three inches below the surface of the soil, pressing the soil lightly round it, and leaving that part of the branch above the soil as erect as possible. The roots will soon form, and then the new plant may be separated from the parent tree.

17.—ROOT PRUNING.

This is an important operation in gardening: it is performed by laying bare the roots three feet from the stem of the tree; then with a sharp axe, or chisel and mallet, cutting through a portion of the strongest roots, according to the requirements of the tree. If the tree is extremely vigorous, without producing fruit, two thirds of the stronger roots cut through in this manner will probably restore it to a state of perfect bearing; the trench being filled up with fresh virgin mould, and the tree left at rest for a year. The proper season for root-pruning is the autumn, when the roots will send forth small fibrous spongioles, which elaborate the sap, and form blossom-buds. Should this operation fail to check the superfluous vigour of the tree, the roots may be again laid bare in the following autumn, and the remaining large roots then cut away, avoiding, as much as possible, all injury to the smaller fibres which have pushed out from the previous operation. Should the tree still present an over-vigorous growth, it must be taken up entirely, and all the strong roots pruned in, then re-planted, taking care that in re-planting, the tree is raised considerably above its former level,—a severe operation, but certain to be successful in reducing a tree to a fruitful state.

18.—DISBUDDING.

The object of disbudding is to remove all useless sprays not required for next year's branches, which would not, from their position, ripen into desirable fruit-bearing wood; it is, in fact, to relieve the tree from nursing wood that must be cut away in the autumn-pruning; but it must be the care of the operator to avoid removing shoots well placed for future branches, or which would expose the tree to too severe a trial of its vital power.

19.—WIRE-TRAINING.

This mode of training is practised in many places with considerable success; at Trentham, for instance, there are five lines set out for the culture of choice apples and pears on the wire-training system; they are alternately cup or vase-shaped, and the reverse, or umbrella-fashion, which has an excellent effect, and appears to be very fruitful.

In the gardens of the Luxembourg, the quarters in the fruit-garden are surrounded with borders planted with cherry, plum, and apricot trees as standards, while others are trained in the vase-shape, on 4-foot stems. The head is formed hollow, the shoots being annually tied down to hoops of wood of the circumference required, and all shoots which spring up in the centre of the vase pinched off. Supposing a tree to have six shoots, they are tied to a hoop at equal distances all round, and shortened to a few inches above the hoop, so as to leave them about a foot in length. From each of these, two other shoots are trained towards a wider hoop, parallel with the first, in the following season; and this is continued annually, until the desired form is attained, and the head of the tree completely balanced in its cup-like form. At Trentham and elsewhere this form is attained by means of a light iron wire frame, the umbrella form being produced in the same manner, the wires radiating from the centre corresponding with the number of shoots issuing from the stem, with hoops of thin wire on which the lateral shoots are trained.

Another arrangement of the cup-trained tree is recommended by M Du Breuil. Starting with four primary lateral shoots, these are suffered to extend themselves horizontally for the first year. When they are tied to the hoop and stopped, each of these will throw out two shoots the second year, being kept at an equal degree of vigour by pinching. At the end of the second season, these eight shoots are cut back to a foot from their base, with two buds on each; those issuing on the left side of the shoot being trained on a loop towards the left, at an angle of about 30°; those on the right, at the same angle, in the contrary direction; each thus crossing another at the same angle, and consequently at the same distance from each other. During the third year sixteen shoots are suffered to grow, the circle widening with each year, or rising at the same circumference, according to the taste of the trainer; and this special training is continued until it is seen fit to stop its further growth, each of the branches being grafted by approach to that which it crosses.

The advantage presented by this mode of training is said to be a more equal distribution of the sap in the whole extent of its branches, and consequently a more equal distribution of fruit; the fruit on spurs being obtained in the same manner as in the pyramid trees.

20.—TRAINING FRUIT-TREES.

Various modes of training fruit-trees are in use among gardeners, but none are more graceful than the pyramidal form; and it is profitable as well as graceful, inasmuch as double the number of trees may be planted in the same space without crowding. This mode of training is extensively adopted in continental orchards, chiefly with pear-trees, but it is equally applicable for apples, cherries, and plums. The form is, of course, the result of pruning as well as training; and the commencement is from a young tree with a single strong leader, which may be obtained at any of the nurseries; though the best and surest way is to plant stocks where the trees are to stand, and graft them with suitable varieties for the purpose, taking care that one shoot only is allowed to spring from the graft. If they are procured from the nursery, plant them in properly prepared stations, supported by a strong stake driven firmly into the soil, and leave them for a year, in order that the roots may have a secure hold of the soil, and send up plenty of sap when growth commences, to push the buds strongly. We will assume that the young trees have plenty of buds nearly down to the graft; then in the following autumn, cut off the top of the shoot at *a*, fig. 1, with a clean cut. At the end of the second year it will have made several shoots, and will probably, in many respects, resemble fig. 2; but as we still require vigorous growth, it will be necessary to cut in again severely at *a* and *b*, *b*. The summer following, the side-shoots will spring forth with great vigour, spreading on all sides; and now the first foundation of the pyramidal form is laid, by extending the shoots horizontally, and tying them firmly to stakes so placed that the range of branches forming the bottom of the pyramid should project from the tree at nearly right angles, and at equal distances from each other. If they are too numerous, the superfluous shoots should be cut off. The third summer if it continues in a healthy state, the tree will present the appearance of fig. 3, with this exception, that the lower branches will be more horizontal than they are here represented, in consequence of being tied to the stakes. If some of the branches have grown more vigorously than others during the summer, such shoots should be pruned in to where the lines cross the branches. On the other hand, should others develop themselves feebly, they should be left at their full length, so that the descending sap, elaborated by the leaves, should deposit a larger amount of cambium. Strong shoots may also have their vigour modified by making an incision immediately below their junction with the stem, just before the sa-

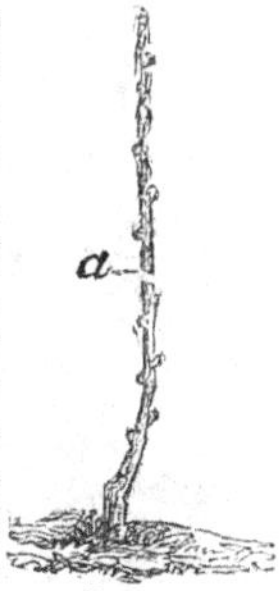

FIG. 1.

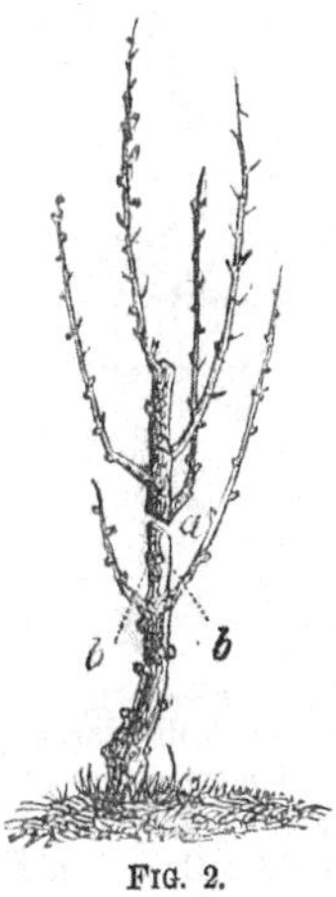

FIG. 2.

rises in the stem; and if a desirable bud remain dormant, it may be forced into growth by making an incision just above it. Where a large vacancy occurs between the branches, then a side-graft should be inserted to fill up the space. The branches should again be cut at *a*, and the fourth year will present the appearance of fig. 4. It will then most likely begin to throw out fruiting-spurs; these should be carefully encouraged, for it depends on the number of spurs which a branch exhibits whether the tree is to bear a good show of fruit or not.

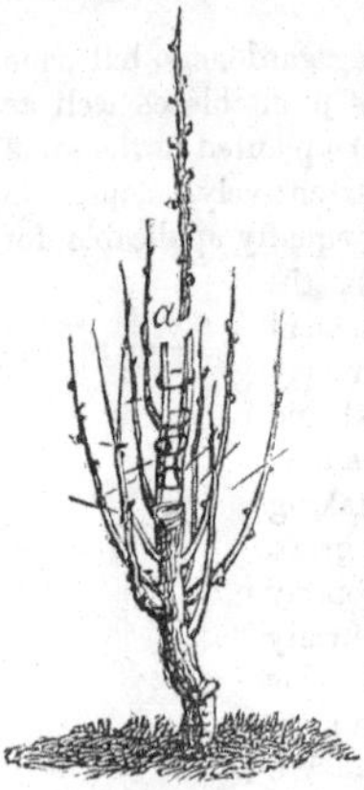

FIG. 3.

The tiers of branches, as they advance in height, should be regulated, so that every side is furnished with an equal number of branches. In the autumn of the year the tree will resemble fig. 5. The pruning is now confined to shortening the leading shoots and the laterals where the lines cross (fig. 5) the branches. The spurs should be carefully examined, and if any of them get long and branching, prune them in, as described and illustrated in fig. 6. If any of them promise to be unfruitful, follow the method described in figs. 1 and 2.

The fifth year the tree, continuing its progressive growth, presents the appearance of fig. 6. It is now a tree of considerable size, and requires, besides the regular annual pruning of the leading shoots and spurs, that the lateral branches should be cut in a line as nearly as possible to that indicated between *a* and *b* in fig. 6. We see in the figure some short lateral shoots crowding towards the centre; all these, if present, should be pruned away. After this, careful pruning is all the tree requires, taking care that the lower branches are not shaded by the upper ones, which is attained by pruning them at greater length than those above; for it is one of the great principles on which this mode of training has been advocated, that the trees should be so managed that the advancing tier of branches shall not interfere with the swelling and ripening of the fruit on the lower tier by overshading them. During every summer, all superfluous shoots should be rubbed off as they appear, and all strong shoots in the spurs should also be stopped during that season, in order to insure vigorous action in the remaining buds, while the base of the pyramid is to be extended as far as is consistent with the development of fruit-bearing habits; and this will probably be best attained by making it a rule, that as soon as a shoot has extended from eight to ten inches, the point should be cut. By this practice the more powerful shoots are checked, and the

FIG. 4.

weaker shoots encouraged. The advantages derived from this system of training may be stated as follows:—

FIG. 5.

FIG. 6.

1. An increased number of trees in the same space.
2. The trainer has his trees more directly under control.
3. Increase of crops.
4. Ornamental and uniform appearance.

Some years ago, M. Cappe, a gentleman of a very great experience in the management of fruit-trees, and curator of the gardens belonging to the Museum of Natural History in Paris, devised a modification of the pyramidal form, which secures a more complete diffusion of light and air in the interior of the trees, causing the centre of the tree to be more fruitful, and the fruit to ripen and colour more perfectly. He termed it the winged pyramid, and certainly a tree so managed exhibits the highest skill in pruning and training, and is, independently of its fruit-bearing qualities, a most beautiful object. In adopting this system of training, a long pole of oak, or some other imperishable wood, some thirty feet in length, and charred at the lower extremity for about four feet, is employed. The charred end is driven firmly into the soil, close to the stem of the tree. At the upper extremity of the pole is fixed a strong iron ring or hoop, perforated with five holes equidistant from each other ; five strong stumps of oak, charred like the pole, are driven into the earth, at spots corresponding with the holes in the hoop, the tops being four inches above the ground and having a strong staple attached to them. Five iron rods with hooks are attached to the staples at one end, and to the holes in the hoops at the other; this is the framework of the winged pyramidal tree. These preparations being completed, the tree is planted in the soil, and on a station previously prepared for it, and pruned annually, so as to produce lateral branches in the manner already described. The branches are to be trained in right lines, slightly rising at the points towards the iron rods. It is evident that there will be regulated lines or openings between each of the five rods, into which light and air penetrate without obstruction, the openings, also enabling the operator to reach every part of the tree with great facility; for the radiating branches should not

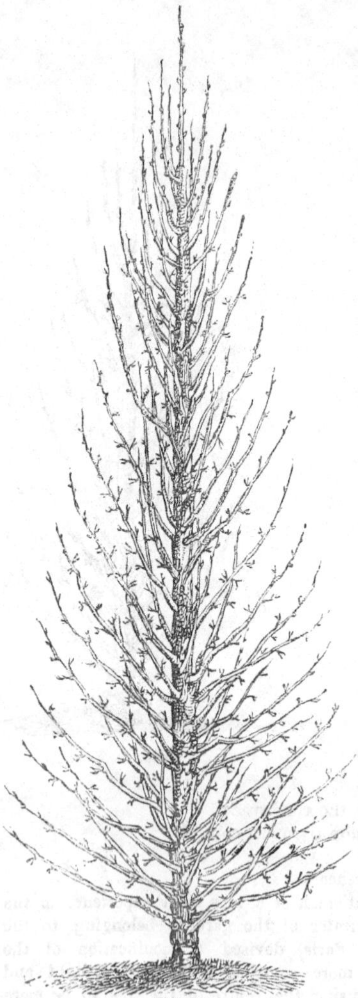

FIG. 7.

be too close together in trees trained in this manner: 20 inches would be a proper distance, though that would greatly depend upon the habit of the tree. In one of vigorous growth, that distance would not be too great; but for one of delicate growth 16 inches would probably be better.

Any large and straight tree that has been allowed to grow in a wild manner, may, by grafting, be converted into the pyramidal form, like that illustrated in fig. 7. By a process of this kind, following the directions already given for side grafting, fine new varieties of fruit may be raised in a comparatively short period, and a comparatively lifeless tree converted into an object of great beauty.

FIG. 8.

Another method of training the pear tree, which has obtained some reputation with our French neighbours, was first practised by M. Verrier, chief gardener at Saulsaye, by whose name it is known. The tree is subjected to pruning when it has attained a central stem and two lateral branches, as in fig. 8. In the autumn or winter pruning of the following year, the two side-branches are trained horizontally, as in fig. 9, and pruned back to about two-thirds of their length, with a bud immediately below the cut. The stem itself is pruned back to about 18 inches above the side branches, taking care that there are three buds immediately below the cut,—one on each side, well placed, and a third in front to continne the stem. With the fall of the leaf in the following year the tree will be as represented in fig. 10, with two horizontal shoots, a central stem, and two other untrained side-shoots. When the pruning season arrives, the same process of cutting back takes place, each of the new side-shoots being cut back to two-thirds of its length, the two lower branches to two-thirds of the year's growth, and the stem to within 18 inches of the second pair of laterals, leaving three well-placed buds immediately below, as before, to continue a third pair of side-branches and the stem. With the fifth year's growth the lower side-branches will have attained as much horizontal extension on the wall or espalier as it is intended to give them. Having, therefore, nailed or tied them to the trellis, give the end of the shoot a gentle curve upwards.

FIG. 9.

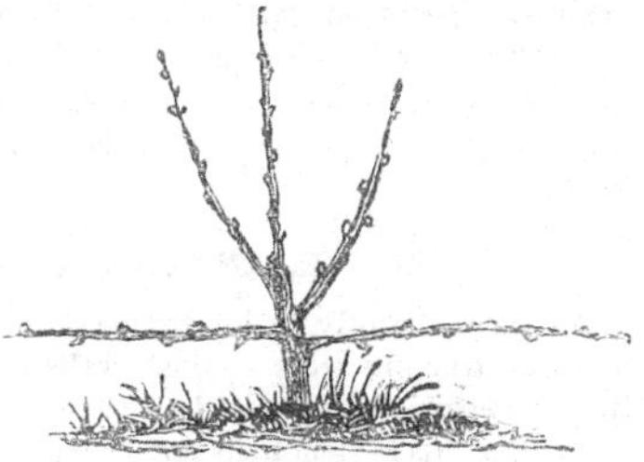

FIG. 10.

Continuing this annual process of cutting back after each year's growth, in some eleven years from the graft, the tree will have covered a wall 12 or 14 feet high, and 6 feet on each side of the stem;

FIG. 11.

each side-shoot, when it is within 18 inches of the one immediately below it, receiving an upward direction, until the tree is as here shown. The stem, as well as the side-shoots, having reached the top of the wall, the extremities of the branches are pruned back every year to about 18 inches below the coping, in order to leave room for the development of the terminal bud, which is necessary to draw the sap upwards for the nourishment of the fruit. After sixteen or eighteen years, a healthy tree, properly trained on this system, presents a surface of upwards of sixty square feet of young fruit-bearing wood. The symmetry of the tree is pleasant to look at, and it is said to be admirably balanced for vegetation, and consequently for fruit-bearing.

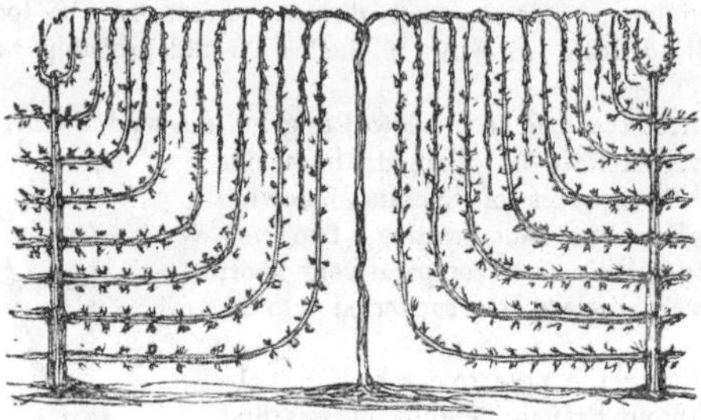

FIG. 12.

One objection to this mode of training is, that the buds do not always occur at the right spot for projecting new side-shoots. When this is the case, the process of shield-budding is had recourse to in August. In other respects, the same principle of pruning is adopted as in pyramid-trained trees, the only modification being the removal of the spurs thrown out between the tree and the wall. Another objection to the system is the time which must elapse before the wall is covered; but this is inseparable from any mode of growing apples and pears on walls, and may be met by planting vines between each, running a central rod of the vine to the top of the wall; stopping it there for the first year, and carrying a shoot on each side under the coping, with descending rods at intervals, calculated not to interfere with the side-shoots of the pear tree.

21.—TRAINING ON WIRES (Novel Mode).

We have recently had introduced to our notice a very neat and efficient mode of training trees against walls; also of training all sorts of espalier trees, raspberries, and indeed all bush fruit, by means of stretched galvanized wire. By this system nails and shreds are entirely dispensed with, the walls are not injured, and no harbour is afforded for insects, and the tying of the branches is effected in much less time than nailing. The wire, together with a very ingenious contrivance for stretching and tightening it, and full instructions both for wall-trees and espaliers, may be had of Messrs. Barnard and Bishop.

22.—FAN-TRAINING.

Numerous fanciful modes of training wall-fruit have been recommended; but trees to be healthy should be made to assume their natural position as near as possible. For peaches and nectarines the fan system is the best, because it is the most natural form that trees so artificially placed can be made to assume. Much injury is done to trained fruit-trees merely to give them an artistic appearance.

23.—TRAINING FLOWERS.

When training is required, it should be done neatly and tastefully, using thin and pointed sticks, and very fine fibres of matting or soft twine; avoid anything like stiffness or formality, which is the opposite extreme to the graceful habit of plants.

24.—TRANSPLANTING.

In transplanting any tree or shrub, especially evergreens, be careful to preserve the same aspect; that is, keep the same sides to the north, south, east, and west, as before. This will greatly facilitate the speedy establishment of the plant in its new situation.

Transplanting is an important operation, and in a general way November is the best month for it. To save time, it is frequently desirable to transplant large trees and shrubs. The effect of ten or twenty years' growth is gained on any given spot at once. This is of immense importance in the lifetime of a man, and the practise of transplanting large trees is therefore popular and highly to be commended; neither is there much risk of failure, with proper caution and skill, and it is not so expensive as many imagine. With the aid of McGlashin's patent transplanting machines, trees of almost any size may be safely and expeditiously removed: in fact, these machines forcibly remove earth and roots and all, with the minimum risk of failure. But even very large trees and shrubs may be safely moved with no other machinery than a few strong planks nailed on a harrow sledge. In this mode of transplanting, a trench is dug round the plant at a distance from the bole of two-thirds the diameter of the top, and to a depth of two to five feet, according to the age and size of the tree, character of the soil, depth of roots, &c., leaving a space of from two to three feet at the back of the tree untouched. At the same time, the front, or part where the tree is intended to come out, should be approached at an easy angle of inclination, extending from two to three feet beyond the circumference of the trench already begun. The earth is rapidly removed from the trench; the roots carefully preserved as you proceed. The size of the ball in the centre must be determined by the nature of the soil and size of the plant. Its size is of less consequence than the preservation of the roots.

As the removal of the earth proceeds, a fork must be used to separate the roots from the soil, and they should be carefully bent back and covered over until the work is finished. After excavating from one to three feet beyond the line of the bole of the tree or shrub, according to its size, introduce into the vacant space a sledge or low truck; cut through the solid part at the back line, and the tree will rest on the machine. This should be furnished with four rings at the corners, through which ropes or cords should be fastened and firmly fixed to the bole of the tree. Of course, some soft substance, such as hay or moss, will be introduced between the bole and the cords, to prevent them chafing the bark. The tree is then ready for removal; the necessary horse or manual power can be applied : the plant will slide gently up the inclined plane, and may be conveyed any distance desired with facility. Sometimes it may be impossible to fix the cord through the back rings until the tree is out of the hole. In that, and indeed in any case, cords had better be attached to the top, and carefully held by men, lest a too strong vibration of the top should upset the machine, or topple the tree over. The hole destined to receive the tree should have an inclined plane on each side to enable the horses to walk through. When the ball arrives towards the centre of the hole, the horses stop. If the tree is not too heavy, the truck or sledge is prised up by manual strength, and the plant gradually slid off. If very heavy, a strong chain is passed under the ball, attached to a couple of strong crowbars; the horses are applied to the other end of the truck, and the tree drops off into its place. The roots are carefully undone, and spread throughout the whole mass of soil as the process of filling-up goes on ; three strong posts are driven in to form a triangle, and rails securely fixed to them across the ball to keep it immovable; the top reduced in proportion to the mutilation the roots may have suffered ; the whole thoroughly drenched and puddled in with water and covered over with four inches of litter to ward off cold and drought, and the operation is complete. If this operation is well performed, the loss will not average more than from five to eight per cent. The principle involved in all planting is the same, and only of secondary importance to securing as many healthy roots as possible. The stability or immovability of both root and top comes next ; for, if not attended to, every breeze that blows is analogous to a fresh removal. No sooner do the roots grasp hold of the soil than they are forcibly wrenched out of it again, and the plant lives, if at all, as by a miracle. The planting of young trees and small shrubs is so simple as scarcely to require instructions. Always make the hole considerably larger than the space required by the roots, whether few or many, so that they may find soft recently-moved soil to grow in; and yet the soil must not be left too loose. If so moist as not to need watering, which will moisten and also consolidate the soil, it may be gently trodden down round the roots.

In reference to the proper distance at which shrubs should be planted, much depends upon the object in view. A safe rule, however, is to plant thick and thin quickly: from three to four feet is a good average for small shrubs and trees. In three years, two out of three plants should be removed; and in planting it is well to introduce rapid-growing common things amongst choice plants, to nurse them up; only the nursing must not continue too long.

25.—REGENERATING FRUIT-TREES.

It frequently happens that grafted fruit-trees, some at one period of their age and some at another, cease to assimilate themselves with the stocks upon which they have been worked. This is to be seen by a thickening of the tree just above the place where it had been worked. This thickening, which in some parts of the country is called a burr, is always to be regarded as an effort of nature to throw out new roots and preserve life, and should be treated accordingly. If the tree has originally been worked, and the burr consequently shows itself at some distance above the ground, a large box should be provided, and placed round the burr, in such a way that it may contain a quantity of soil, into which the tree can strike out its new roots. This soil should be a light loam, and always kept moist. In the second or third year, new roots will have been formed, and the tree may safely be separated by a saw from the old stock, and let down into the earth beneath. When the tree has been worked close to the surface, a place about a yard square may easily be built up with bricks or tiles, and filled with light soil a few inches over the burr, to receive the new roots. In this way the writer has preserved two small trees of the Sturmer pippin, which he found fast dwindling away, the stocks on which they were worked not having power to sustain them. By a somewhat similar process, the healthy branch, *b*, of a favourite tree may be preserved by layering it in a box or pot, *a*, as in the engraving.

Another operation, the object of which is to utilize the roots and stem of the old tree, is connected with the foregoing.

The final cause of the languishing state of some trees being the absence of vigorous young shoots and the imperfect organization of the cambium and liber, and, finally, the abortion of root-fibres in consequence, these trees can only be restored to health by the production of more healthy and vigorous organs; and this may be done by concentrating the whole energy of the trees on certain points. Amputate the principal branches, *A* (*see* the engraving given in illustration), about seven or eight inches from their base at *c*, the branches *b* being left entire for the present, the amputations being so made that the branches left are not required to carry out the new system of training to be adopted, passing, in all cases, the four largest branches. These branches are retained for the present, it being doubtful if the tree has strength to develop upon the old bark the new buds necessary to fulfil the functions of the roots; for if the buds perish, and there is no outlet for the rising sap, the tree dies. By preserving these branches, their leaves and shoots provide against such accidents.

To facilitate the issue of buds on the tree, the hard dry bark should be removed by a plane, and its place covered by a coating of chalk and water,

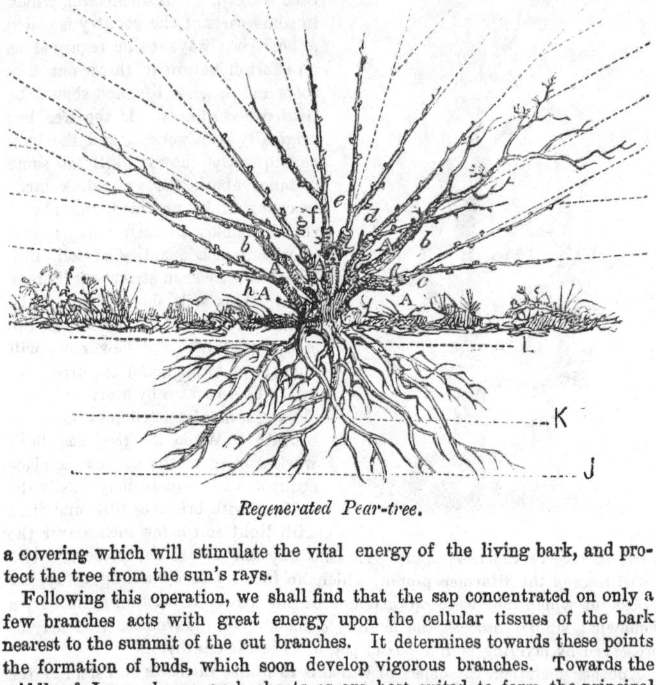

Regenerated Pear-tree.

a covering which will stimulate the vital energy of the living bark, and protect the tree from the sun's rays.

Following this operation, we shall find that the sap concentrated on only a few branches acts with great energy upon the cellular tissues of the bark nearest to the summit of the cut branches. It determines towards these points the formation of buds, which soon develop vigorous branches. Towards the middle of June, choose such shoots as are best suited to form the principal branches for horizontal training; such would be *c d e f g h* in the accompanying engraving. The others must be cut towards the middle of their length.

The year following, in the spring, train the principal branches according to the plan laid down; for example, in the fan shape, as in the engraving, break the tender branches close to their junction with the stem or main branch, and, during the summer, pinch the leading shoots off, so as to convert into fruit-spurs the shoots not intended to form main branches.

In the following spring the tree will be as represented above. At this time, the branches *b*, left for precaution, may be entirely suppressed, the several cuts being covered with grafting-paste. These new suppressions increasing the energy of the young branches, they will henceforth grow with great vigour, and will soon replace the ancient tree.

In the same proportion in which the stem is operated upon, so must the roots be. As soon as buds begin to appear upon the portion of the branches left, the leaves which are developed send towards the roots a quantity of ligneous fibre and corticle, or wood-fibre. In its course towards the roots, this sap meets with beds or layers of cambium and liber, through which they extend themselves in a languishing state (since it is now deprived of the fluid which facilitated its passage), taking their natural direction, and penetrating the cells in the bark upon the roots, they give place to new organs, at once more nourishing, more healthy, and more vigorous than the old roots. If, after three or four years, a tree operated upon as we have indicated is transplanted, it will be observed that the lower half of the old roots, comprised between the lines J and K, is decayed, and that young roots, comprising those between K and L, have been thrown out. The tree has reached the state represented in the engraving, and is supplied with young, healthy, and vigorous roots, as well as more vigorous branches, with new layers of cambium and liber. It is, in reality, a new tree, which has taken the place of the prematurely old one, whose organs have ceased to live.

Analogous treatment to that which we have indicated for espalier trees may be followed with standards and pyramid trees; removing the objectionable branches eight or ten inches from the stem and placing a crown or cleft graft on each, if it is considered necessary, but taking care to leave a fourth of the old branches, till the branches cut down have thrown out young shoots. In the second year the remaining branches may be removed altogether, the extremities of the severed cuts, when made perfectly smooth, being covered over with grafting-paste.

By these processes, it is possible, except in cases of complete decay, to restore the tree to its first vigour, especially in the case of pip fruit, as the apple and pear. In stone fruit, the success is less assured; above all, it is doubtful in the peach, which scarcely ever produces buds on the old wood; and the application of grafting is had recourse to when it is desired to regenerate this tree, or to substitute an improved variety on an old but healthy stock. In this case, crown-grafting is adopted, and a graft placed at the extremity of each branch, which is cut down in the manner already described, favouring, in the meanwhile, the development of young wood at the base of the tree, by short pruning, and pinching off the buds at the summit.

26.—SOILS AND COMPOSTS (Harvesting of).

All who have examined this question admit that the value of manures is in proportion to the nitrogen or phosphates which they contain, more especially the former; for nitrogen is almost synonymous with ammonia, that being the chief source of nitrogen for plants. "Let the cultivator," says Dr. Scoffern, "take care of his ammonia; let him take care of his phosphates; let him prevent the loss of all soluble matters from his compost-heap."

The first and most important source of these elements is farm-yard manure, which, in its fresh state, consists of the refuse of straw, of green vegetable matter, and the excreta of domestic animals. Horse-dung varies in its composition according to the food of the animal; it is most valuable when they are fed upon grain, being then firm in consistence and rich in phosphates. Sheep-litter is a very active manure, and rich in sulphur and nitrogen; for if a slip of white paper, previously dipped in a solution of lead, be exposed to the fumes of fresh sheep-dung, the paper will be blackened; a sure test of the presence of sulphur.

Cow-litter is cooler, and less rich in nitrogenous or azotized matter; but it is rich in salts of potash and soda, and thus better adapted for delicate and deep-rooted plants. Swine's dung is still less azotized and more watery, and full of vegetable matter; but the most important of all manures is the urine from the stables and drainings of the dung-heap, which is wasted daily to an enormous extent. "The urine of carnivorous animals," says the authority we have already quoted, "is rich in the principles urea and uric acid. In herbivora, hippuric acid takes its place; but in all cases it is rich in nitrogen, and, when allowed to putrefy, ammonia is evolved. Urine is thus one of the most important constituents of farmyard manure."

The composition of manure is a very heterogeneous mixture. It may be broadly viewed as a mixture of humic acid bodies fixed in alkaline salts, and nitrogenous bodies capable of yielding ammonia; and it becomes an important question how is its strength best economized. Some advocate the practice of allowing the compost-heap to be entirely decomposed into an earthy mass—thus permitting the whole of the ammonia to escape; others have gone so far as not to permit of any fermentation at all, stopping all action by continual turning. "It must be a bad practice," says Dr. Scoffern, "to allow so valuable an agent as ammonia to go to waste; and this is the inevitable result of permitting manure to undergo its last degree of fermentation. In the second place, it is doubtful whether the full and immediate virtues of the manure can be brought into play if it has not been submitted to incipient decomposition." There are means, however, of fixing the ammonia and retaining it in all its strength, while the manure containing it is reduced to a state suited for assimilation as food for plants. It may be absorbed by gypsum or sulphate of lime, which, being cheap, is often mixed with the compost-heap for the purpose; the ammoniacal salts thus formed being afterwards decomposed by the vegetable organism, or by its agency combined with atmospheric influences.

Collecting and preparing manure, and transporting it where it is wanted, are operations that should be attended to when other operations become impossible. The waste, not only of liquid, but solid manure, in this country, is enormous. Everything that has ever been endowed with life, and all the excrements proceeding from them, are available for manure. Their nature, qualities, influence, and the mode of their application, may be endlessly varied; but all alike possess a power of enriching the earth. The hard texture of bone or wood fibre may render it desirable to subject them to chemical action, or the influence of fire, to render them more speedily available to the wants of plants; but these hard substances possess the elements of plant-food in common with the soft constituents of plants and animals. The influence of sulphuric acid upon bones is well known. When fire is used to break down or soften woody fibre, it should be applied so as to char, and not to burn it. Charring is effected by covering the heap of wood to be operated upon with turf or earth, so as almost entirely to exclude the air, and thus insure slow combustion. Almost any vegetable refuse, including roots of weeds, can be charred; and this charcoal, saturated with urine, is one of the best fertilizers. It may be usefully drilled in with seeds, in a dry state. The scourings of ditches, scrapings of roads, decayed short grass and weeds, half-rotten leaves, soot, and every bit of solid manure that can be got, should be collected and thoroughly mixed together. The excrements of most animals are too rank and strong for flower-garden purposes, applied in a pure state; by mixing, however, the bulk of the manure may be quadrupled; it will be sooner available, and much more valuable. There are some very useful fertilizes sold under the name of Portable Garden Manures, in canisters at 2*s*. 6*d*. each.

Genuine Peruvian Guano.
Rape Dust.
Sulphate of Ammonia.
Cubic Petre.

Genuine Peruvian Guano, 4*d*. per lb.
Bone Dust, 3*d*. per lb.
Wood Charcoal, 3*d*. per lb.

27.—HOTBED (Making of a).

To make a hotbed, let a quantity of stable-dung be got together, proportioned to the size of the frame: two double loads for a three-light frame are usually allowed for the body of the beds; but it is as well to add an additional load if the bed is required to last some time.

The dung should be turned over four or five times during a fortnight, and wetted, if dry. This preparation is most important; the inexperienced operator, unless he would run the risk of destroying his plants at the beginning, should follow it to the letter; for, unless the material has been well worked before the bed is made, it is apt to heat too violently, and burn the roots of plants. An equal quantity of leaves mixed with stable-dung may be used with advantage; the leaves give a sweeter and more moderate, as well as more lasting heat.

When the material is ready, measure the frame, length and breadth, and mark out the bed, allowing a foot or 18 inches more each way than the length and breadth of the frame. At each corner of the bed drive a stake firmly into the ground, and perfectly upright, to serve as a guide to build the bed by; then proceed to build up the bed, shaking up the dung well and beating it down with the fork. The whole should be equally firm and compact, so that it is not likely to settle more in one part than in another. The frame and lights may now be placed in the centre, but the lights left off, so that the rank steam which always rises from a newly-made hotbed may escape.

This is the usual way of making a hotbed; there is, however, another which is very effective, and greatly economizes the manure. The trimmings and prunings of trees may be tied up into fagots, and with these the walls of a pit should be built, the exact size of the frame: on this the frame rests. The fagots are fixed by means of stakes driven through them into the ground, the walls being about 4 feet high. After the frame is put on, the mixture of dung and leaves is thrown in and well beaten down. The dung is piled nearly up to the glass to allow for sinking; in other respects, the management is the same as for an ordinary bed. The advantage of this plan is, first, it requires a trifle less manure; secondly, the heat from the linings penetrates through the fagots under the bed, and is found more effective.

When the bed is made, the frame and lights put on, and the rank steam passed off, which generally takes five or six days, let a barrowful of good loamy soil be placed under each light; by the next day this will be warmed to the temperature of the hotbed, which will now be ready for use.

The heat of the manure is not lasting; consequently the bed will require watching. It is advisable to have a thermometer in the frame, and as soon as the heat gets below 70°, apply a lining of fresh dung, which has been prepared as the body of the frame, to the front and one side of the bed; and when this again declines, add another to the back and the other side. The bed can be kept at a growing heat for any length of time by this means, removing the old linings, and replacing them by fresh.

In covering the lights, during frosts or rough winds, it is advisable to avoid letting the mats, or what not, hang over the sides, as there is often danger of conducting rank steam from the linings into the frame. Straw hurdles which exactly fit the lights are better than mats. The coverings should be used just sufficiently to protect the plants from frost or cutting winds, without keeping them dark and close.

28.—INOCULATING.

This term in gardening is usually confined to a peculiar process of creating grass lawns by distributing over the surface of the ground small pieces of turf, rolling them in, and leaving them to take root and get together. The process, if properly carried out, is a very good one. The pieces of turf should be free from weeds and the surface made level to receive them.

29.—POTTING.

This is one of the most important of gardening operations. Adapt the pots to the size of the plants as near as possible—or rather, to what the plant is expected to be—as allowance must be made for growth of the root as well as the plant. Let the pots be perfectly clean. Effectual drainage of the pots does not consist so much in the quantity of drainage, as in the arrangement of it. A potsherd should be placed over the hole: some pieces of pot, broken rather small over that; and these again covered with a layer of peat-fibre or rough earth. This gives efficient *drainage*, and need not occupy more than an inch and a half of the pot. Hard-wooded plants should be potted rather firmly; soft-wooded should be left rather loose and free.

30.—WATERING.

In watering fresh-potted plants, it is important that the whole of the soil be effectually moistened, which can only be accomplished by filling up two or three times with water. No fear need be entertained of over-watering; if the plants have been rightly potted, all surplus water, beyond what the soil can conveniently retain, will drain away. Irregular watering is frequently the cause of failure in plant-culture, even with experienced growers. A certain amount of tact is necessary in giving plants, which have been so neglected, just as much water as they should have, and no more. In watering, much depends on the weather, and also on the season: plants require less in winter than in summer. The proper time to water them in winter is when they are in bloom, or growing rapidly—in summer as soon as the least dryness appears; but a little practice will be more useful than a lengthy description. In giving air, it may be observed that all plants which are not tender, that is, all plants which are natives of temperate climes, may be exposed to the air at all times when the thermometer indicates a temperature above 40°, except in case of rough winds or heavy rains. Hardy plants may be exposed at any temperature above 32°; for, although frost will not kill them, it may spoil their appearance for a time. Plants in bloom should never be kept close, or exposed to wet or wind: the flowers last longest in a soft, mild atmosphere, free from draught. Plants should never be wetted overhead in cold weather, or, rather, while they are in a cold atmosphere; and never, except to wash off dust, should those having a soft or woolly foliage be so treated; but some plants, as the Camellia, Myrtles, Heaths, and others with hard leaves, may be plentifully syringed, or watered overhead from a fine rose, in warm weather, especially when in full growth.

31.—CUTTINGS (The taking of).

Cuttings in general may be considered as of two kinds,—matured wood and young green shoots. The former, whatever they may be, strike readily, and with very little care. An American plan, which is very successful, is to lay them in slightly-damped moss, or to drop them lightly into a wide-mouthed bottle, having a piece of damp sponge at the bottom and a covering of muslin over the top. In either of these methods a callus is soon formed, and the cuttings readily throw out roots. Cuttings of young green shoots, however, require a very different treatment: they must be so managed as never to be allowed to flag, and the following appears to be the best method that can be pursued. Put silver-sand about an inch deep into shallow pans (common saucers answer every purpose), and in these plant the cuttings. Then pour carefully upon the sand enough water to make a thin sheet about it. The lower leaves of the cuttings are to be removed before planting, and the stalk fixed firmly into the sand before the water is poured on. These tender young green shoots, or cuttings, will be better for a little shade and heat. A piece of thin muslin or tissue-paper will provide the former, and heat may be had by placing the pan of cuttings over a basin of hot water, re-filled twice a day. These cuttings will be rooted and nearly ready for potting off before the water in which they are grown has dried up.

Cuttings of all sorts of geraniums for bedding the following year should be struck early: from the last week in July to the end of the first week in August is a very good time. They should be taken in dry weather, when the parent plant has had no water for some days, and they should be kept to dry twenty-four hours after they have been prepared for potting. The more succulent sorts, and any that appear difficult to strike, may with advantage be touched at the end with a small paint-brush dipped in collodion, which will serve to hasten the callus which the cutting must form before it will throw out roots. They may be potted four or six in a pot, according to sizes. It is essential that the pots be well fitted with drainers, that the soil be light and sandy, and that it be pressed tight round the joint of the cuttings, which should be buried in it as fleet as possible. When filled, the pots may be sunk in the ground on a south border, and well watered in the evening, when the sun is off. They will require no shading, except the sun be very scorching; and, in this case, they must not be kept from the light, but merely screened from the scorching rays of the sun. They may flag a little; but this is of no importance; in two or three days they will recover, and put forth roots. If they grow too freely before it is time to take them in for the winter, the top shoots should be broken off, and in this way they will make strong bushy plants.

To preserve cuttings from frost where there is no greenhouse, dig a pit about four feet deep, strew the bottom well with ashes, and sink the pots in the same. Over it place a common garden-frame, bank up the outsides with straw and a coating of earth. In such a pit, verbenas, calceolarias, fuchsias, &c., &c., may be preserved during the severest winters, provided the pots be kept in the dark by being well covered with matting during frost.

32.—PROTECTING BEDDING-PLANTS.

It is always desirable to get bedding-plants out as early as possible, and yet there is much danger both from wind and frost in so doing. It is an excellent plan to stick sprays of evergreens, Scotch and spruce firs, in different parts of the bed as a protection. By this means the force of the wind is broken, and the plants take hold of the ground sooner: the tender leaves also are saved, which otherwise not unfrequently turn brown and fall off, retarding the growth of the plant.

33.—PEGGING BEDDING-PLANTS.

Various expedients are resorted to by gardeners to peg down the different sorts of bedding-plants—verbenas, petunias, &c., &c. Some use ladies' hair-pins, and some use small pegs made of hazel or other wood; but the neatest, the cheapest, and most efficient pegs which have come under the writer's notice, are cut from the brake, a wild fern which grows freely in every lane and on almost every common in England. Many a poor boy might earn an honest penny by cutting these in autumn, when the wood is tough, and selling them in bundles for next summer's use. Galvanized wire pegs are very durable, and may be bought at 3/. per thousand.

34.—NAILING.

This is a difficult operation, for nailing is no ornament, and the less it shows itself the better. The gardener's skill must be exerted to conceal his nails and shreds as much as possible. List is generally used; but strips of leather or black tape are preferable. They not only have a neater appearance, but afford less harbour for insects. Fruit-trees should be nailed close on to the wall; ornamental shrubs, &c., merely fastened in for the sake of support.

35.—MULCHING.

This operation consists in spreading a layer of stable-dung over the roots of trees and plants, and in times of drought watering through it. After a time the dung may be forked into the soil.

36.—CLIPPING HEDGES, &c.

All evergreens and hedges, especially evergreen hedges, should be cut to a point pyramidically; for if the top be allowed to overhang the bottom, the lower shoots will invariably die off. With hollies and laurels use the knife in pruning, to avoid the rusty appearance of the withering of half-cut leaves. Privet and thorn may be clipped with the garden shears.

37.—FRUIT GATHERING.

Fruit-gathering is one of the most cheerful and agreeable employments. It usually enlists every hand in its service, and in an abundant year finds all hands plenty to do. To the following plain and simple directions, those who are intrusted with the superintendence of fruit-gathering will do well at all times to attend. It is important, in the first place, to remark, that no fruit should be gathered for storing before it has arrived at maturity. By this we are to understand not necessarily its full flavour and ripeness, but the completion of its growth or size; and as all fruit, even upon the same tree, does not come to maturity at the same period, it will frequently be found the safest and most economical plan to make the gathering at two or three different times. It is very easy to ascertain when any particular fruit is ready; for ripe fruit always leaves the tree upon a gentle touch,—the fruit-stalk parts from the twig on which it grows without any signs of rending or violence. In a general way, with both apples and pears, several of the most forward fruit will have fallen before the general crop is in a fit state to be gathered; and this fallen or bruised fruit should never be mixed with that which is intended to be stored; all unsound fruit which may be found upon the trees at the time of gathering should also be rejected. Fruit, in fact, which ripens in summer and autumn, should be gathered a little before it is absolutely ripe: thus gathered, it is better in quality and higher flavoured than when absolutely ripe. But this must not be carried too far. A single day before they are perfectly ripe suffices for peaches and other delicate stone fruit; a week for apples and pears; but cherries are only gathered when completely ripe. Apples and pears, which arrive at complete maturity in winter, are best gathered at the moment when the leaves begin to fall, and the sap to withdraw from the branches in October. All gathering should take place in dry weather, and the fruit should not be handled or pulled about more than is absolutely necessary. The middle and afternoon of the day will usually be found the best time for gathering, as autumn mornings, even in the finest weather, are always more or less humid; and to avoid any risk in keeping, all fruit should be quite dry before it is taken from the tree. The most convenient baskets for fruit gathering are peck and half-bushel baskets, with cross-handles. These should be provided with a line and a hook, by means of which they may hung to the branches of the trees, and thus allow the gatherer the liberty of using both his hands: by the line, the baskets, when full of fruit, can be lowered and emptied, and drawn up again.

With regard to the choicer sorts of pears, especially those growing on trees against walls, or on dwarf trees, it will well repay the little extra time and trouble it may cause, to gather these by their stalks without touching them with the hand, and to remove them at once to the fruit-room on the trays or in the drawers in which they are to be stored. There is on the skin of all fruit a secretion more or less marked, known commonly by the name of bloom. This, though less conspicuous on apples and pears than on plums and peaches, is nevertheless present, and its use is to protect the skin of the fruit from the ill effects of excessive moisture. While this bloom can be preserved, the fruit will never require wiping, and will retain its full flavour and freshness.

38.—FRUIT-STORING.

The following statement appears to embrace the best methods, and those that are most generally adopted, for the storing and preservation of fruit. It must be borne in mind that they are not arranged in order with any reference to their respective merits; some of them are decidedly objectionable; but the good and bad of each will be noticed as we proceed :—

1. Apples and pears may be sweated, *i.e.*, laid in heaps and left to heat, and then stored away in an apple-room, on dressers, or in a dry dark vault in heaps, uncovered, except during frost.

This plan, though very generally adopted where apples are kept in large quantities for sale, is always open to the objection of being more or less injurious to the quality and flavour of the fruit.

2. Fruit may be stored on open shelves, and on the floor of a fruit room, spread out upon straw, and covered, when necessary, with the same material.
3. In the same way, put upon dried fern leaves, and with fern leaves for a covering.
4. In baskets or hampers, lined with straw or fern leaves, but without any material between the layers of fruit.

If the fruit be dry when placed in the basket or hampers, and the storeroom of an even temperature, it keeps very well in this manner. However, for the reason assigned above, fern-leaves are preferable to straw.

5. In boxes or casks, with sawdust.
6. In boxes or casks, with bran.
7. In boxes or casks, with wheat-chaff, or with oat-flights.

Sawdust is decidedly objectionable, even though taken from the hardest and most inodorous wood, for it is almost certain, after long keeping, to become musty and unpleasant: and so also does bran, which is naturally a fermenting substance, and soon heats if put together in any quantity, especially with fruit among it. In shallow trays bran will answer for a time very well, but it will require attention. For packing fruit for conveyance, both bran and sawdust also may be used with good effect. Wheat-chaff, as well as oat-flights, is liable to produce the same mischief. It is quite impossible to be certain that the fruit will not become tainted by means of them, more particularly in closed boxes, and where there is no ventilation.

8. In boxes, with dry sand between the fruit.
9. In boxes, with powdered charcoal in the same way.

By adopting either of these methods, fruit may be preserved for a long period; but though sand and charcoal are good materials for keeping fruit sound, they are both open to the great objection of making the skin gritty and unpleasant.

10. In jars, without any material intervening between the fruit: the jar, when covered with a piece of slate or tile, to be buried in dry sand of a depth sufficient to exclude all air and to ensure preservation from frost.

This plan will, undoubtedly, answer its purpose, as far as preservation is concerned; but it is attended with much greater trouble and inconvenience than most persons would deem desirable.

11. In deep drawers one upon another, without any substance between them.
12. In deep drawers, with sheets of paper or dried fern-leaves placed between the layers of fruit.

Both these plans are good, and if the fruit be stored sound and dry, there will be little need of any intervening material.

13. In single layers in shallow trays or drawers resting upon fern-leaves, and to be covered when necessary with the same.

This is the plan which we should recommend as the safest and best, under ordinary circumstances, to be adopted.

14. In heaps in dark, dry, well-aired vaults.

In this way both apples and pears, in large quantities, may be well and easily kept: and if the vaults be thoroughly dry and sufficiently beneath the surface to exclude frost, the fruit will require no further protection, and give but little trouble.

Filberts and walnuts, to be stored for winter use, should be gathered when full ripe, and on a dry day. The latter must be cleared of their husks. They may then be packed in glazed earthern jars, tied down with coarse brown paper, and kept in a damp cellar. Filberts keep best in this manner without their husks; but if the husks are to be preserved, the fruit must be left to stand for a night in open baskets, and be well shaken to get rid of earwigs. Many persons shake a little salt over the last layer of nuts before the jars are tied down.

All drawers, shelves, boxes, or jars containing fruit, should be labelled every year as soon as the fruit is stored, so that the different sorts may be easily known.

39.—VEGETABLE STORING.

There are several sorts of vegetables which require storing for winter; potatoes, carrots, beet, and onions are the chief of them. Potatoes do best when harvested in clumps in the open ground, care being taken to protect them from rain and frost. A long ridge is the best form. The ground should be dry and thoroughly drained. The potatoes should be heaped on a ridge, tapering from a base of three feet to a foot and a half, or less, at the top, separating the different sorts by divisions in the ridge. It is usual to cover this ridge with a thatch of wheat-straw, and then with six or eight inches of mould; but some authorities highly disapprove of this, McIntosh recommends the tubers being covered with turf, and afterwards with soil; and, in the absence of these, laying on the soil at once without any litter. After having laid on nine or ten inches of soil, thatch the whole over an inch and a half thick, with straw, fern leaves, or any similar non-conducting material; "the object being, he says, "first to exclude frost and wet, and, secondly, to exclude heat; for which purpose earth is not sufficiently a non-conductor of heat and cold."

If the weather is fine when the tubers are taken up, and the potatoes are required for early use, much of this labour may be dispensed with; but if for spring and early summer use, the precautions will be found necessary.

Carrots, beet, and other similar root-crops should be taken up before the frosts set in: they may either be stored in a dry cellar, covered with dry sand, or after the manner of the potato. The London market-gardeners winter their beet and carrots in large sheds, in moderately damp mould, and banked up with straw; "for," says Mr. Cuthill, "it is a mistake to pack them in dry sand or earth for the winter; and the same may be said in regard to carrots, parsnips, salsafy, scorzonera, and other similar roots; and by this means," he goes on to say, "the roots retain their natural sap, and the colour is preserved."

It is probably unnecessary to add that in roots and tubers, as with fruit, all cut or bruised ones should be thrown aside: when the skin is cut, or a bruise exists, the elements of decay are soon introduced, and all others within reach contaminated. A dry day should be chosen for lifting them, and they should be exposed a few hours before collecting into heaps, that the soil adhering to them may dry.

Onions should be lifted a little before they have altogether ceased to grow: the leaf turning yellow and beginning to fade will be the sign. As they are taken up they should be placed in a dry airy place, but without being exposed to the sun. If they are thinly spread out on a dry floor or shelf covered with sand, or on a gravel-walk partially shaded in fine weather, they will do very well. As they dry, the roughest leaves should be removed, when dry, they should be removed to a warm dry loft, where they can ripen more thoroughly. When in a proper state for storing they should be gone carefully over and separated, the smallest ones for pickling, the ripest picked out, as likely to keep longest: those with portions of leaves to them are best stored by stringing and suspending them from the ceiling of the room, which promotes ripening. The stringing is done by twisting a strong piece of matting or twine round the tails of each in succession, so that they may hang as close together as possible without forming a cluster, until the string is about a yard long.

40.—BED FORMING.

Divisions of a flower-garden formed in different-shaped figures. The following are some of the most pleasing varieties:—

Basket-beds of Ivy on grass plots: these have a pretty appearance, and may be made round or oval, according to fancy. A frame of wicker-work should be made, the shape of the bed, about 1 foot or 1½ foot high, around which, on the outside, should be planted, quite thick, either the large Russian, or the small-leaved and variegated ivy. In a year or two, with little care and attention, the wicker work will be quite covered, when the ivy must be kept well cut in, and the earth in the basket may be raised or not at pleasure. With a little trouble the ivy may be made to trail over wands, and form a handle to the basket.

Cross-shaped Beds.—The various sorts of crosses also form very ornamental beds, and have this advantage over fanciful figures, that they may easily be designated by their particular names. Nothing is more brilliant than a Maltese-cross bed, filled in each separate compartment with different shades of verbenas, or in the opposite compartments with the small dark blue lobelia and *Gazania splendens.* The St. Andrew's cross also forms a nice bed, and so do the different forms of upright crosses, when the stem and the transverse are filled with flowers of such shades and colour as contrast well with each other.

Leaf-shaped Beds.—Some of the prettiest beds for lawns may be made by cutting them out into the natural form of the leaves of trees, shrubs, and plants. The form of the common ivy-leaf makes a very pretty bed, so does the heart-shaped ivy; also the oak, the maple-leaf, the horse-shoe geranium, and an endless variety of others. Beds so formed have this advantage, that they can be called by the name of the different trees, shrubs, and plants from which they have been taken.

Oak and Holly Beds.—Acorns sown very thick round a bed in a drill about two or three inches wide, in the course of a year or two form a very pretty edging; and, owing to the thickness with which they stand, with an occasional clipping, the small oaks may be kept four or five inches high, and in this manner have a very good effect. Hollies also may be used in the same way; but in this case it is better to raise the plants on a seed-bed, and transplant them to the bed for which they are required as an edging when about two or three inches high. They may be kept dwarf by cutting, and will not become too large for their position for some years.

Tile-beds for Grass-plots form very pretty and very ornamental objects made on grass borders, or on lawns of kitchen or mixed gardens. The tiles, or pipes, as they are called in some parts of the country, should be of bright red clay, twelve inches long and about three inches in diameter, and all carefully formed in the same mould. These should be placed upright in a circle, or any other figure, buried, according to taste, about four to six inches in the ground; the earth and the beds being raised to the level of the outstanding part of the tiles. A very effective centre bed can be made with these tiles in three tiers, the edges of each tier being built in scallops, and a border left about one and a half or two feet in diameter. These three borders have a beautiful effect when filled with different plants. Take, for instance, *Calceolaria aurea floribunda* for the top department; Tom Thumb's, or Frogmore's, for the middle border; and Mangleis variegated geranium for the lower. These beds have an agreeable appearance even in winter when cleared, on account of the contrast between the bright red tiles and the grass; and in spring they may be made very gay with hyacinths, crocuses, and other bulbs.

Pincushion Beds.—So named from having plants of different coloured flowers and different habits dotted on all over them. The old *Verbena venosa* dotted among Mangle's geranium forms a pretty pincushion bed. Other pincushion beds may be formed by mixing plant for plant, of Dandy geranium and *Lobelia speciosa*, and Lady Plymouth with Purple King verbena, or the old scarlet *Verbena Melindres*. The Variegated Alyssum or Cerastium, mixed with Lobelia, also makes a soft chaste pincushion bed. Golden Chain geranium, dotted among white verbenas and edged with *Lobelia speciosa*, is also, beautiful. Mrs. Pollock and Bijou, alternate on a ground of Lobelia or Purple King verbena, have a rich effect.

Tent Beds.—These are formed by the aid of chains, flexible wires, or ropes. Drive a tall stake of the desired height, say 10 feet high, in the centre; a circle with a radius of, say 8 feet. Insert six or eight stakes at equal distances on this line, say six feet high; join the centre stake to each side, one with a chain or wire, and the frame of a tent bed is formed.

41.—TRELLIS-WORK (Making of).

This may frequently be introduced with good effect in the mixed flower and kitchen-garden, to shut out buildings or unsightly objects. Small oak stands, or small larch poles, about five or six feet apart, and having the intervals filled with thin iron wires crossing each other, form the most durable trellis-work. Against the walls of a house a very nice trellis-work may be made with a lacing of copper wire over nails of the same. This may be worked in any pattern and carried in any direction; to this wire the creepers may be tied when necessary; and in this way the walls of houses may be covered with flowers or evergreens, without any injury to the brick-work from continual nailing.

10

13

16

11

14

17

12

15

18

Pl. II.

OUR COLOURED PLATE.—No. 2.

10. *Carnation.* — What valuable flowers are carnations, pinks, and Picotees, which all belong to the same class. Our illustration is the Carnation, which differs from the Picotee chiefly in its mode of colouring. In the Carnation, as is shown in the plate, the colour is disposed in unequal stripes from the centre to the outer edge, while in the Picotee the colour is uniformly disposed along the outer edge of the petals and radiates inwards.

11. *Cloth of Gold Rose.*—This is one of the most lovely of our roses, whether we regard the bud or the full-blown flower. Unfortunately, however, it is a shy bloomer, and somewhat difficult to cultivate. It requires a wall, and a full south aspect, a rich soil, and close priming, which should never be attempted before the end of March, when there is no longer danger of severe frost.

12. *Cineraria.*—A great improvement has been made in the growth of Cinerarias since they were first brought into notice. Mr. Glenny tells us that to make a truly fine Cineraria, we must have a white ground, which renders any colour a good contrast, the most striking being crimsons and blues. The edging should be even, forming an even band of colour alike all round, and having a well-defined circle of white, surrounding a disk of some determinate colour. We mention this in order that any of our readers who like to amuse themselves by raising seedlings, may know what flowers to reject, and what are not worth saving.

13. *Coreopsis Tinctoria.*—This is one of our most popular annuals, producing a very agreeable variety when massed with other things in a large bed. Plants should be raised from seed in pots in a cold frame in the spring, and set out with other bedding plants early in June.

14. *Cyclamen.*—Few flowers are more admired than the Cyclamen. We have met with the pink variety growing wild in the woods of Bavaria and Austria, where the air is literally scented with them. These and the white

variety, C. Europeum, are quite hardy, and will thrive and blossom freely in our gardens, provided they find a soil and situation suited to them. It is the Persian variety which is generally grown in pots in greenhouses—this is not hardy.

15. *Dentzia.*—Some few years ago these pretty little deciduous shrubs were only used as pot plants under glass. They have, however, proved themselves to be quite hardy, and they now produce their graceful flowers quite as freely in the open ground as in a confined greenhouse.

17. *Euphorbia Splendens.*—The variety here illustrated is one of the best of the greenhouse Spurges or Euphorbiæ. It has large brown thorns along its stems, and if the plant is not suffered to grow too rapidly, it produces an abundance of flowers, which will hang on during the greater part of the year.

18. *Epacris.*—All the Epacris are very pretty, and many of them, from their compact growth and abundance of blossom, make useful plants for the dinner table. When these plants have done blooming, they should be cut in freely. In this state they may be kept in a pit until they begin to make fresh growth, when they should be forced on in a little heat. The young shoots, when about six inches long, should be lopped—the weaker once, but the stronger will bear this operation twice. After this, the wood must be ripened by exposure to light and air. When the sprays show blossom, the plants may be brought into the greenhouse.

42.—SUMMER HOUSES.

Summer-houses and seats add much to the comfort, as well as the ornament, of pleasure-grounds and gardens. Almost any clever carpenter can put up a rustic arbour, at any rate with the assistance of a few hints. Arbours, however, as well as seats, can be bought ready-made. Very neat buildings may be formed with young oak stands, ornamented with pieces of oak billet and thatched with reed; also of Scotch fir poles split, showing the bark on the outside. Such summer-houses as these may be boarded inside and lined with matting, or made more ornamental by a panelling of split hazel worked into different patterns. The flooring can be of brick or stone. More substantial houses can be built wholly of flint or stone, and fitted up accordingly. Of garden seats the variety is infinite. In the wilder portions of the plantations and shrubberies, the more simple and rustic these seats are the better. The stump of a tree, or the stem placed lengthways, may be fitted up for the purpose.

43.—RIBBON-PLANTING.

The ribbon style in border-planting is very effective. As an illustration of it take the following arrangement:—Supposing there be room for five or six rows, each row a foot or 18 inches wide,—a double row of *Lobelia speciosa* next the edging, followed by a row of verbenas,—Mrs. Holford's Snowflake, or any other white sort; these, again, followed by *Calceolaria aurea*, this by Tom Thumb geranium. If there be room for more rows, the above may be followed by *Salvia patens* (blue), *Coreopsis lanceolata* (yellow), a row of white phlox, and a back row of dahlias. These should graduate in height and colour. This is merely given as a sample of what may be done. There are many plants that may be used in the same way, as *Kœnigia variegata*, isotomas, *Phlox Drummondii*, which are all dwarf, and suitable for front row; petunias, heliotropes, lantanas, &c., might form a second; ageratums, galardias, salvias, a third. Again, mirabilis, still taller; then dahlias and hollyhocks, tallest of all. Ribbons are also very pretty planted with annuals, as *Phlox Drummondii*, stocks, asters, zinnias, xeranthemums, and sweet peas, all which graduate in height, and vary in colour. These may be raised in frames in March and planted out in May, or sown in the open ground in May. Hardy annuals may be sown early in spring, and be allowed to flower, and then followed by bedding-plants or by biennials, which are best sown in May and planted out. Hardy herbaceous plants alone may keep a border perpetually gay, but are not well suited for massing. They should, however, be arranged with regard to height and colour. Pansies, daisies, primroses, silenes, &c., being dwarf; pinks, cloves, carnations, veronicas, &c., taller; phloxes, various sorts of campanulas, chrysanthemums, &c.; starworts, rudbeckkias, &c., being tallest of all. Plants of this class flower at various times of the year, from early spring to late in the autumn. Where spring-flowering bulbs are mixed up with them, it is not advisable to plant them near the edge of the beds. Plant them far back; as they flower when the borders are comparatively bare, they are sure to be seen to advantage; and the long grassy leaves do not disfigure the borders after they have flowered. Late bulbs, as gladiolus and lilies, being tall, should be placed far enough back to correspond with the other plants. A very good effect may be produced by planting a ribbon border or clump with plants of ornamental foliage. These look better than most people would imagine. The very commonest and cheapest of plants may be made use of; for instance, a front row of variegated arabis, which is a very common, hardy, herbaceous plant; second row, Henderson's beet, treated as an annual. This is a dwarf, and very bright crimson-coloured sort, and grows about 8 or 10 inches high; third row, antenaria, or variegated mint; fourth row, *Perilla nankinensis*—annual; fifth row, ribbon-grass; sixth row, purple arack, *Atriplex rubra*. These graduate in height and colour, have a very pretty effect, and last the whole summer and autumn.

44.—HINTS ON SEWING SEEDS.

There are two points in connection with seed-sowing which are of paramount importance to the success and vigour of germination, and the regularity, strength, and luxuriance of the crop, besides that of having good and perfect seed. These are the proper mechanical condition of the soil, and the regular and uniform depth at which the seed is sown. The presence of air, moisture, and a certain degree of warmth, is essential to the germination of seeds. In the absence of these agents, the process of germination will not go on. The soil is the medium by which a supply of air, moisture, and warmth is kept up; but, unless the soil be in a proper condition, it cannot supply these. If it be very dry, it contains too much air and too little moisture. The proper condition of the soil is when it is neither very dry nor very wet; it is then moist, but not wet; it has the appearance of having been watered, and is easily crumbled to pieces in the hand, with its particles adhering together.

A state of too much dryness seldom occurs in this country; but the presence of too much water is not uncommon: it is, however, remedied by drainage. The grand point is to get the soil thoroughly well pulverized, by means of which, with proper drainage, it will be in a condition favourable to germination of seeds. Temperature exercises a powerful influence over the time required for germination; and, within certain limits, the higher the temperature is, the more rapidly does germination go on. The soil receives its heat through the medium of the air; consequently the surface-soil is more quickly heated than that lower down. Whenever the air is warmer than the soil, the surface will be warmer than that below; when, on the other hand, the air is cooler, the surface will, by contact, cool much more rapidly than that below the surface. From this it follows, the more rapid germination will occur at about 1 inch below the surface, to which depth the heat will soon penetrate, and which, nevertheless, will not be so readily cooled during the night. Seeds near the surface will generally grow most rapidly, and the germination of others will occupy more time as the distance from the surface is increased.

It is owing to this fact that seeds too deeply sown do not grow at all, the temperature not being sufficiently elevated, and the supply of air being too limited to set the chemical process at work, which is essential to germination.

45.—STRAGGLING PLANTS.

To preserve a neat appearance in the flower-borders, all perennials that have a tendency to run about or stray,—and there are many of which this is the habit,—should have their roots confined under the surface with tiles. Old chimneys and seakale-pots are very good for the purpose. By this means they may be kept within due bounds, but, of necessity, will require every few years to be transplanted into fresh soil.

46.—NORTH BORDERS.

A north border under a good wall in a garden is generally much undervalued. In the flower-garden, a north wall, if it happens to exist, is frequently looked upon as a nuisance, and covered with ivy; in the kitchen-garden it is only more profitably occupied by Morello cherries and red currants, while, in both cases, the border is kept as shallow as possible, and turned to little or no account. Many plants and shrubs, however, will flourish upon a north border and against a north wall, and show themselves hardy there, which in any other situation would not outlive a winter's frost. In the flower-garden let the north wall have a good deep border of bog, and against the wall all the hardy sorts of camellias will flourish and blossom freely. The green and the black tea-plant also, not having their bark exposed to the scorching sun of summer, will survive our severest winters in such a situation. Rhododendrons will also do well, and so will chrysanthemums. All our hardy indigenous ferns do better upon a north border than under any other aspect. Those persons who wish to acclimatize any tender plants should, by all means, make their first experiments upon a north border or against a north wall. The shady side of a wall is decidedly the best position for all cuttings during spring and summer, to enable them to stand the severity of winter.

THE FLOWER GARDEN.

47.—FLORISTS' FLOWER-GARDEN.

The florist's flower-garden comprises the Dahlia, Hollyhock, Chrysanthemum, Tulip, Polyanthus, Auricula, Heartsease or Pansy, Ranunculus, the Anemone, Carnation, Pink, Picotee, Hyacinth, &c., &c., for florists have largely increased the objects of their care; but probably the present list includes all which can properly be called florist's flowers.

A florist's flower-garden is usually planted in formal and rather stiff-looking beds, the flowers in right lines, those of dwarf habit occupying the outsides of the beds, with the taller sorts in the centre. The garden itself should be very near to the dwelling of any one who has charge of it, for no plants require greater attention to grow them properly; they need all the air that can possibly be given to them, while a slight frost coming suddenly on after a warm April day—above all, heavy storms of rain, a hailstorm, or even a boisterous wind—will be destructive to many of them.

Heat is nearly as injurious. They should only meet the morning and evening sun: a shade of light calico for an hour before and after noon, during the floweringseason, will much prolong their bloom.

The bed for the reception of auriculas, polyanthuses, carnations, and flowers of similar habit grown in pots, should be placed in an open airy situation, where they can readily receive shelter from rain and from a too ardent sun.

A useful bed or stage for the reception of these may be prepared by laying down about six inches of coal-ashes upon the natural soil, over which a platform should be made by a flooring of square tiles, closely fitting into each other. Over this are to be laid seven rows of bricks, equidistant from each other; and on these, at regulated distances, the pots may be ranged after the operation of potting has been performed in May.

Shelter and shade can be provided in the following manner:—a row of strong stakes, sufficiently close together to support a top rail without interfering materially with the pots which occupy the centre of the bed. These stakes should be six feet high, and five inches by four in size, and should be driven at least twenty inches into the solid soil; the top rail being sufficiently strong to receive and support the shutters. A similar row of stakes, with top rail, must be driven on each side of the bed, and about two feet and a half from it. Three inches on each side of the bed there should be another row of six stakes of equal strength, but only eighteen inches above the ground; the top being notched in the form of the letter V.—The use of these stakes will be obvious; they are to receive and support the lower end of the shutters—the central top rail when closed; and the outside rails are to receive and support them when it is desired to throw them open.

The shutters, each four yards long and three feet wide, when closed, form a span-roofed pit open at the sides, three posts supporting each shutter. They are made with feather-edged inch deal, forming a solid frame; the centre may be deal, felt, or any material impervious to heavy rains; if of glass, there should be an arrangement for shading with calico, or some other material; and the frames should lap one over the other at the sides, and meet at a proper angle at the top, so as to form a ridge.

Florists' Flowers, as we see them in their present state of cultivation, prove how immense is the field which Nature lays open to reward the industry and intelligence of man. Who can place the different flowers which have passed under the florist's hand for cultivation side by side with their wild originals without being struck with wonder at the almost marvellous results which follow from the ingrafting of nature and art? Compare the pansies of some of our recent prize-shows with the wild heartsease of the woods, and it is hardly possible to realise the idea that the two stand to each other in any sort of genealogical relationship; and the same is true of pinks, and hyacinths, and anemones. Nor, indeed, is the contrast yet at its height, for every year fresh progress is being made in symmetry, in richness and variety of colouring, or in size. Look at the dahlias and chrysanthemums of the present day, and think what were considered good flowers, and actually called forth admiration, some twenty years ago. The different varieties which come under the head of florists' flowers, are so rich in beauty that most persons take delight in them. Indeed it is quite impossible for a garden to be really gay without its share of them, and with them any garden may be gay at all seasons, except in the depth of winter. Even as the year declines—late in the autumn months, though tulips, and carnations, and auriculas, and ranunculus, and hyacinths, &c., &c., are at rest, chrysanthemums and dahlias are in their glory; and these continue to enliven our gardens till an envying frost cuts them off. At all seasons of the year there is something to be done with florists' flowers, and the reader will find notices of them under each head. Instructions also for the management of particular flowers are to be found in the calendar of each month.

48.—DAHL'IA (nat. ord. Compos'itæ).

The Dahlia, so named after Dahl, the Swedish botanist, belongs to the same family, and is a native of the same country, as the patato—namely, the mountains of Mexico. There it was found in the sandy plains 5,000 feet above the level of the sea, and was sent to Europe in 1789 by Cervantes, the Spanish director of the Mexican Botanic Gardens, who named it *Dahlia coccinea*. Under the impression that sandy soil was its proper compost, it lingered in our gardens, a miserable scraggy plant, till 1815, when a fresh and improved stock was introduced from France, and it was taken up by the florists. Under the influence of cultivation, it has been so much improved in form as to become one of the finest flowers of the garden, while the shades of colour are so numerous, so diverse, and so opposite, that it would be difficult to find another plant at once so hardy and so showy. Probably its importer never dreamed that the naked stem and imperfect flower of *D. coccinea* would, by the efforts of cultivation, become so ornamental in European gardens; nevertheless, such it has become; and few gardens are now without their collection of dahlias, while the nursery lists of named varieties swell into hundreds, of every shade and colour, except the much-prized blue, which was for some years the object of tho florists's pursuit.

Dahlias may be multiplied by seeds and by dividing the tuber, every eye, when separated with a portion of the tuber, making a plant. They will also grow from cuttings off the young shoots under the lower leaves, which may be struck in small pots filled with sandy soil. Experiments have even been made to ascertain how far grafting would succeed with the Dahlia.

Seedlings are procured by sowing the seeds in shallow pans and plunging them into a hotbed, or by sowing on hotbeds prepared for the purpose, in March. The soil should be light and sandy, with a mixture of peat-mould. The seed should be chosen from the best varieties only; it should be lightly covered with soil. A few days will bring them up, when they require all the air which can be given them safely. In April they will be ready for potting off either singly in the smallest-sized, or round the edge of 6-inch pots: this strengthens them for final planting out. Towards the middle or end of August, if successfully treated, they will begin to bloom: at this time they should be examined daily, all single and demi-single blooms thrown away, unless they present some new colour or show some peculiar habit of growth, in which case they may be improved by further cultivation and crossing. Caution in this respect is necessary, as it is a habit of the dahlia to improve under a second year's cultivation, and some of our finest varieties have come up with indifferent flowers as seedlings. When done flowering, the young bulbs are taken up and treated as old tubers.

Cuttings are taken as follows:—In February or March, and even as late as the first week in April, the tuber, which has been carefully wintered in a dry place, is put into soil placed over a hotbed, and in a very short time as many shoots as there are eyes in the tuber make their appearance. As soon as these are two inches long, they are taken off just below the leaves, struck singly in small pots, and again placed in the hotbed. Some prefer cutting up the tuber as soon as the eyes are distinguishable, and replacing them either in the soil of the hotbed or in pots. Mr. T. Barnes, an undoubted authority, tells us that to obtain short-jointed, stout and healthy plants, "they should be rooted from cuttings taken off in April, and struck in a gentle hotbed." As soon as rooted, they should be potted in 5-inch pots, and again placed in a gentle heat, but with plenty of air. "Cuttings struck at this time," he tells us, "are more healthy than those struck at an earlier period," and consequently form better flowering plants. A week after they are potted they should receive a watering of liquid manure made from guano and powdered charcoal, well mixed with rain-water, repeating this occasionally till the time of planting out; fumigating the frame with tobacco, should there be any appearance of the green-fly.

Early in May beds are prepared for the reception of dahlias, if they are to be grown in massed beds. The form of the beds will depend on the general design of the garden: if a portion of the garden is devoted to them, either for the plants or the flowers, they will be best displayed in beds three feet wide, with alleys between. The beds being marked by stakes placed at each corner, four inches of the surface-soil is removed, and four inches of thoroughly-rotted manure put in its place, and the whole deeply dug, and the manure thoroughly mixed with the soil in digging. In the beds thus prepared the plants are placed, the collars, as they have grown in the pots, being on the surface of the beds. The 3-foot beds will receive each a row; the stakes are firmly fixed 4, 5, or 6 feet apart, according to the size of the plants; the plants themselves are set 4 inches deep, so that the crown of the plant is just above the surface. As the plants increase in growth, tying up commences; at the same time a diligent search should be for made slugs, earwigs, and other pests of the garden. These must be rooted out, or they will root out the dahlias, or at least destroy their flower. Where any of the plants show a weak and drooping growth, time will be saved by re-striking the top: although these will bloom later, the flowers will be stronger than they would be from a plant which has received a check.

During June and July dahlias require careful attention in watering and stirring the soil about the roots. As the lateral shoots attain sufficient length, tie them up so as to prevent their breaking, placing other stakes for the purpose, should that be necessary. This prevents their clinging too near to the stem, and permits of a free circulation of air round the plants. When they are intended either for exhibition or for highly developed flowers, only one bud should be left on a shoot, shading the flower both from the sun and rain by tin sconces, oil-skin caps, or inverted flower-pots, placed over the top of the stake to which it is tied, while all superfluous and useless shoots are removed, and the growth of the plant encouraged by every possible means. As autumn approaches, the swelling shoots render it necessary to examine those tied up, slacking the strings where necessary, to prevent them from galling.

Where dahlias are merely to fill a place in the general arrangement of the garden and shrubbery, care should be taken to supply them with suitable soil. Peat-mould, mixed with sand, is useful in developing stripes and spots on the flower. As the plants progress, the lateral shoots, as well as the central stem, require support by tying up, and the roots should be assisted by stirring the soil with a fork every two or three weeks, and by copious watering, removing all dead or straggling shoots, and keeping the plant trim and well-staked.

Dividing the Roots.—Another and more common practice in gardens is to place the whole tuber in some warm place in March, and, when the eyes show themselves, cut up the tubers, and in May plant them at once six inches below the surface, in the place where they are to bloom, staking them and leaving them to nature until they are sufficiently grown to compel attention; but even for common bedding out purposes, and for filling up gaps, the plant is worthy of greater care than this amounts to. Light-coloured flowers are confirmed in their beauty by seclusion from sun and air while they are developing their bloom; darker flowers, on the contrary, lose much of their brilliancy if too much shaded: these should, therefore, only be shaded partially from the direct rays of the meridian sun.

In October the dahlia begins to fail. This then is the time to revise the names, and see that they are all correct, and to take care that seed from such as it is desired to propagate from is secured before they are injured by the frost. Provide also against severe weather coming on suddenly, by drawing the earth round the stems in a conical form, which will protect the roots from frost while they are yet in a growing state, as well as diminish the moisture, which encourages growth. Even in November, in mild seasons, the dahlia will remain fresh and gay if the weather is open and clear; but in general the earlier flowers will have passed away: their time of rest is come. When the frost turns their foliage brown or black, take them up, cut off the roots, leaving six inches or so of stem attached, and plunge them into a box of ashes, chaff, or sand, or any other method of preserving them from damp, frost, and heat as during the winter.

As to giving a list of dahlias. when it is considered that the eight or ten winners of prizes at the national dahlia-shows must exhibit upwards of 160 varieties, it will be obviously vain to attempt it. We may state, however, that Mr. Turner of the Royal Nursery, Slough, has repeatedly carried off the great prizes for fifty dissimilar varieties, as well as for fancy varieties: it is pretty clear, therefore, that he can supply the very best selection possible.

49.—HOLLYHOCK.

There is no finer ornament of the autumnal flower-garden than the Hollyhock: its noble tapering spike-like stem and rich rosettes of flowers clustering round the footstalks of the leaves, its pinnacled head and luxuriant massive leaves, render it the most effective occupant of a gap in the shrubbery, or in the back row of an herbaceous border. In rows in the flower garden, or in beds by themselves, their variety of colour renders them also most attractive objects. Hollyhocks may be propagated by seed and by cuttings.

The seed of the hollyhock should be gathered only from the most perfect plants, in which the flowers have been round, the florets thick and smooth on the edge, the colour dense and decided, and the flowers close to each other on the stem. About the middle of March, or not later than the first week in April, the seed-bed should be prepared, four feet wide, with an alley on each side. The soil should be rich and in good heart: such soil as would suit a cabbage will grow the hollyhock in tolerable perfection. Trench the bed two feet deep, throwing the top spit to the bottom, and bringing the second spit to the surface, breaking up the surface thoroughly. On this bed, raked smooth, sow the seed so thickly as to come up an inch apart, and sift over the seeds some rich dry soil, so as to cover them for about an inch. When they come up and begin to grow, the weeds must be kept down, and vigorous growth encouraged by watering in dry weather. In June the plants will bear removal to a nursery-bed, prepared in the same manner as the seed-bed. If the seedlings have been growing vigorously, the roots will be strong, and must not be broken in taking up: this may be prevented by soaking the bed thoroughly the night previous to removal, and lifting the plants cautiously with a fork inserted under them, as in lifting potatoes. Plant them in the new bed six inches apart each way, using a dibber, making a hole large enough to receive the roots, and pressing the earth round them by making another hole on each side with the point of the dibber, watering the bed thoroughly when planted. When dry and somewhat settled, rake the bed smooth, giving the same care as to weeding and watering when dry, as well as to destroying slugs, earwigs, and insects. In the autumn the plants will be strong and fit to put out where they are to bloom. If they are intended to bloom in rows where they stand, every other plant must now be removed, so as to leave them one foot apart all over the bed; here they may be supported by strong stakes placed at both ends of each row, and a strong cord carried from one to the other, to which the plants are to be tied.

As they come into bloom, in the second year, all single flowers which do not exhibit some desirable character of habit or colour, should be thrown away before they begin to ripen seed: the majority will be in this category. Those selected for further experiment should be cut down to within three inches of the ground, the earth round them stirred with a fork, to loosen the soil and let in the air, having previously named or numbered them in your book, and described the qualities for which they were selected.

In propagating by cuttings, as soon as the first flowers of an old plant open sufficiently to judge of the flowering, the superfluous side-branches having no flower-buds should be taken off, with two or three joints and leaves. Cut the shoot through with a clean cut just under the lower joint, leaving the leaf entire; cut it also at about two inches above the joint either joint will do, provided they have growing eyes, with a leaf and a piece of ripened wood to support the bud until roots are formed. These cuttings, planted in a light sandy soil, placed under a hand-glass, and watered occasionally, and shaded from the sun, will require little further care except keeping clear of weeds and dead leaves. When rooted, pot them off in 60-sized pots, and put them in a cold frame where they can remain during the winter. In spring plant them out in the open ground, where they are to flower, taking care to furnish the roots with the proper soil.

The old plants in autumn furnish another source for new plants. When the flowers are becoming shabby, cut the plants down, and, beginning at the bottom joints, continue to make cuttings, as described above, until the fibre gets too soft for the purpose—each joint having eyes will furnish a plant: these struck under a hand-glass, on a very slight hot-bed, will grow vigorously, the soil being gritty sand, loam, and leaf-mould, in equal proportions. Sprinkle the cuttings slightly with water every day in fine weather.

Mr. W. Paul, of Cheshunt, who has made the hollyhock his peculiar study, finds that the season of flowering may be greatly extended by striking and transplanting at different seasons. "There is," he says, "a difference of six weeks in the period of flowering between plants removed early in autumn and late in spring; and of this we may avail ourselves to lengthen the succession. Early-rooted cuttings and old plants may be induced to bloom in July, and late-rooted cuttings and spring-sown seedlings in November: hence there is no difficulty in obtaining flowers for four successive months."

Mr. Paul attaches much importance to a free supply of water in the spring months, when the plant has just been turned out of the pots, and when it is most desirable to have rapid and vigorous growth.

Three flowering-spikes should only be allowed to the strongest plants—to weakly ones only one. When a foot and a half high, stake them, placing two to each plant, one of these stakes being driven in on each side of the plant, the stakes being five feet long and 2½ feet into the ground. "Pass the bass round the stem of the plant," says Mr. Paul, "drawing it first to one of the stakes, and tie it;" then perform the same operation on the other side a few inches higher up, tying it in the opposite direction, "until the plant is rendered quite secure. When there are two stems to one root, three stakes will be necessary, placed in a triangle, tying the stems alternately as before;" care, however, should be taken that the stakes are covered with the foliage of the plant as much as possible.

The following selection from Messrs. Paul's numerous list of plants is particularly worth notice, towering, as many of them do, with spikes 10 feet high:—

White and blush-colour,—Vista; Celestial; Lady Tarlton.
Buff, fawn, and salmon,—Empress; Mr. Oakes; Queen of the Buffs.
Lemon,—Walden; Masterpiece.
Pink,—Lady Franklin; Perfection.
Rosy carmine, — Beauty of Chesthunt; Beauty of Walden.

50.—CHRYSANTHEMUM (nat. ord. Compositæ).

Unusual importance attaches to the cultivation of Chrysanthemums, from the facility with which they may be grown in the very heart of large towns, as has been proved by the efforts of Mr. Broome in the Temple Gardens, where he has grown all the best varieties in a manner which has astonished many who have examined his collection. The flower is of easy culture, and cuttings may be struck almost up to the time of flowering. Nothing is finer than the display of these flowers in October and November, ranging as they do from pure white to a deep orange, from a pale blue to deep red and crimson; but, like the dahlia, the first frost sadly spoils them. Where the collection is a choice one, chrysanthemums are best trained against a wall, or in beds, where protection against sudden frost can be easily applied. By means of pot-culture, which is now extensively used, a splendid show of flowers may be preserved even up to Christmas, with comparatively little trouble.

Cuttings of chrysanthemums should be potted pretty thickly together in sandy soil, and the pots plunged to the rim in a gentle hotbed. If the number is not large, put them separately in thumb-pots, so that there may be no derangement of the plants when they are repotted. Nothing roots more certainly than the chrysanthemum; and if rooted pieces of the old plant are taken instead of cuttings, propagation will go on without any trouble at all. In order to give the plants every advantage, they must never be either root-bound or allowed to flag from drought; these evils are to be avoided by frequent repottings and constant attention to watering. The blooming-pots should generally be 10 inches deep and 8 in diameter at the top; and between the thumb-pots and these final ones there should be at least three shiftings. Fresh rich soil must be used to fill up the space in the larger pot. Plenty of fresh air and sunlight must be afforded all through the growth of the chrysanthemum, and as the plant absorbs rapidly, it requires some care to prevent the leaves from flagging. In hot sunny weather, half a dozen waterings in a day would scarcely accomplish this, and prevention must be tried by burying the pots up to the rim in the moist soil of the garden: thus treated, one or two good waterings in a day will be sufficient. Care must be taken that worms do not get into the pot, by placing them on bricks, slates, or coal-ashes. They should be turned round twice a week to prevent their roots striking into the material beneath the pots. Liquid manure may be supplied rather plentifully as the flower-buds begin to expand: when judiciously applied, it produces finer growth and deeper colour. The best compost consists of two parts light loam to one part of well-decomposed dung, freely mixed with sand. The chrysanthemum may also be raised from seed. Messrs. Barr and Sugden say, "Really valuable varieties have been raised from the seed saved by our Sardinian correspondent. The seed we offer has been saved from the very finest of the new varieties, and is warranted to produce 75 per cent. of double flowers."

The chrysanthemum, like other plants producing terminal flowers, has a tendency to send up one leading stem, which, if not interfered with, would produce a bunch of flowers at the top. This tendency is counteracted by stopping the terminal shoot, which produces a compact shrubby growth, and a great many more flowers. As a rule for the large-flowering kinds, stopping should cease in July; while with pompones it may be extended to August. The general law is, that letting the plants run up is favourable to fine flowers, and stopping, to a more plentiful supply. It is the practice of some growers to stop the plants at every third eye until the middle of August, watering freely with manure-water three or four times a week, and sprinkling the plants overhead with water every morning. Early in September the best plants are selected, and repotted into 12-sized pots, using the compost as before, giving ample drainage, and placing them under a south wall; the smaller plants being transplanted at the same time into 24-sized pots, and placed under an east wall. By the middle of October the earlier plants will be showing flower, and should be placed in a cold green-house, or cold-pit, where they can receive plenty of air, leaving those intended for late flowering under the east wall as long as the weather will permit. By the middle of November all should be housed, or at least provided with shelter, and a good supply of bloom for the next two months should be the result.

The number for selection is well-nigh endless, and they have latterly been divided into Large, Pompone, and Anemone-flowered varieties. The following are good sorts, that cannot fail to give satisfaction. Those marked thus * are the most expensive; but they can, with a few exceptions, be supplied by most nurserymen at 6*s.*, 9*s.*, 12*s.*, and 18*s.* per dozen. The time to purchase them is when they are in bloom; and then you can please yourself both in colour, substance, shape, and size of flower, as well as in the habit of the plant:—

Large Flowering Varieties.

Alfred Salter,—fine form; light pink; first-rate.
Annie Salter,—fine golden yellow.
Aimée Ferière.—white, tipped rose-pink; incurved; fine.
Adriane,—cream, buff-tinted.
Aurora,—very full; ochre-yellow.
*Bacchus,—rosy fawn; large; incurved.
Beauté du Nord,—violet-carmine.
Boadicea,—creamy rose.
Cassy,—orange and rose.
Chevalier Domage,—fine gold.
Cloth of Gold,—fine; large.
Defiance,—fine; incurved; white.
*Dido,—white sulphur; incurved; dwarf habit.
*Draco, — incurved; dark red; free bloomer.
Dragon,—puce; light centre.
Etoile Polaire,—deep yellow; incurved.
Excelsior,—bright crimson; very full.
*Finstriatum,—primrose; fringed.
Formosum,—finely incurved; pale sulphur.
Golden Christine,—golden buff; large.
Golden Queen of England, — canary; fine.
*Guelder Rose,—pure white; finely incurved.
*Imbricatum album,—incurved; white.
King,—light peach; incurved.
La Reine d'Or,—golden yellow; fine.

*Lady St. Clair,—pure white; finely incurved.
Le Prophète,—golden fawn; a splendid flower.
Lord of the Isles,—creamy yellow and orange.
Marceau,—rose; bordered white.
Novelty,—large; blush.
Picturatum roseum,—salmon; incurved.
Prince Albert,—bright crimson.
Queen of England,—fine blush; incurved.
Queen of Lilacs,—fine; incurved.
Queen of the Isles,—white; large.
*Snowball,—fine white.
*Snowflake,—very dwarf.
Versailles Defiance,—rose and lilac; fine form.
Vesta,—ivory-white; one of the oldest and best.
Vulcan,— bright red crimson.

Pompone Varieties.—These are beautiful for front shelves in conservatories, or beds or borders out of doors; generally of compact, close growth, and the flowers about the size of large daisies, and rivalling the large ones in colour; and they are at once the neatest and most ornamental plants for furnishing:—

Adonis,—rose and white.
Aigle d'Or,—canary-yellow.
*Apollo,—fine; incurved; chestnut-yellow.
Bob,—fine dark brown.
Brilliant,—crimson-scarlet.
Christina,—canary-yellow; brown tips.
Diana,—white; fine form.
*Fairy,—light lemon.
Fennella,—bright orange.
Hendersonii,—yellow; early flowering.
Ida,—pale clear yellow.
*Indian Prince,—carmine-red and gold; incurved.
*Lucinda,—rosy lilac and blush; full.
Madame Fould,—cream; one of the best.
Madame Mielly,—violet-amaranth.
Marabout.—white-fringed; fine.
Mrs. Salford,—white; fine.
Mrs. Modell,—fine white.
Nemesis,—bronzed orange.
President Ducaisne,—rosy carmine; scarlet.
Queen of Lilliput,—rosy blush.
Sacramento,—golden yellow.
Scarlet Germ,—crimson-scarlet.
Stella,—light-shaded pink.
The Little Pet,—blush-white; incurved.

Large Anemone-flowered Varieties.

Almost every one has seen and knows the Golden Orange-coloured Gluck; it is a perfect representative of this class:—

George Sands,—red, with gold centre.
Fleur de Marie,—large white.
Juno,—white.
King of Anemones,—crimson.
Margaret of Norway,—red and gold.
Margaret of Versailles,—blush.
Louis Bonamy,—lilac; high centre.
*Lady Margaret,—pure white, with row of ground petals.

Pompone Anemone-flowered Varieties:—

Reine des Anémones,—fine white.
Perle,—rose; lilac centre.
President Morel,—cinnamon; gold centre.
M. Astre,—golden yellow.
Madame Peuter,—pure white.
Golden Cedo Nulli,—golden canary.
Cedo Nulli,—white, with brown spots.
Boule de Neige,—pure white.
Adriane Amaranth,—gold centre; fine.

51.—TULIPS (nat. ord. Tulipa'ceæ).

There is a peculiarity belonging to tulips which does not, so far as we are aware, belong to any other flower. The seedlings, in their first bloom, generally produce flowers without any stripes or markings, all the upright portions of the petal being self-coloured, flowering for years without any such variegations, when they are called breeders. After some years they break out into stripes: if these are liked, they are named; but they have multiplied in the breeder state, and may have been distributed in all directions, each person possessed of one which has broken using the privilege of naming it; hence many, with different synonyms, are one and the same thing. It is another peculiarity, that of twenty of a sort in the same bed, scarcely two may come up alike, although good judges can recognise them. These peculiarities, interfere with their cultivation, though, as Mr. Glenny surmises, they may be one of the charms of tulip-cultivation.

The perfection of soil for tulip-culture would be three inches of the top of a rich loamy pasture, the turf of which, cleared of wire-worm, grub, and insect, has lain by till thoroughly rotted, and which has been repeatedly turned and picked: the decayed vegetable matter will suffice without other dressing. The tulip-bed should run north and south, with drainage perfect, but without stones or rubbish at the bottom. The bed may be dug out 4 feet wide and 2 feet 6 inches deep, and the compost previously prepared filled in till it is a few inches above the path, the centre being 2 inches higher than the sides. All water must be withdrawn from the bottom of the bed: it is not enough to give drainage, unless an outlet is found, so as to avoid stagnant water. Giving a few days for the bed to settle, rake all smooth, leaving the bed 3 inches above the path. On this the tulips are placed in seven rows across the bed, and six inches apart in the rows. They are pressed in a little; soil is then placed upon them, three inches above the crown of the bulbs, so that the bed being raised in the centre, the middle row will be covered 4 or 5 inches. The bulbs are planted, of course, according to their height and colour—those growing 15 or 18 inches occupying the outside rows; the second rows on each side are those growing 2 feet, and those growing 2 feet 6 inches occupy the three centre rows. When planted and covered, they may be left until the leaf-buds begin to peep through the ground. Of course the sides of the bed must be protected by edgings either of wood or tiles.

As frost approaches, while giving as much air as possible, they should be protected against it by mats or other shelter, but not longer than is necessary; otherwise they get drawn up, and weakly. In February they begin to appear, when the ground should be stirred, all lumps broken, and pressed close round the stems. As the spikes begin to open, they form a receptacle for the wet, and the frost must not then be allowed to reach them. When the colours begin to show, in order to protect their bloom, a top-cloth must be provided to shelter them from the sun, taking care that no more air than is absolutely necessary is excluded, the cloth being let down on the sunny side only, and that only when the sun is powerful.

By the end of June the stems will have turned brown or yellow. As soon as the leaves begin to decay, the bulbs may be taken up, dried, and stored away in drawers provided for them, which are usually marked in seven compartments, so that each row in the bed occupies a similar place in the drawers.

The best time for planting is the last fortnight in October, or early in November. Tulip-seed may be sown either in spring or autumn, and in the soil already described; it should be saved from the best flowers only, and those grown by themselves, where no inferior pollen can reach them. The small offsets should be planted by themselves and labelled, in similar soil to that already described. Breeders such as we have described may be grown in any soil.

Tulips are divided into Roses, Byblomens, and Bizarres. Roses have a white ground, and crimson, pink, or scarlet markings.

Byblomens are those having a white ground, and purple, lilac, or black markings.

Bizarres have a yellow ground, with any coloured marks that present themselves.

Self-tulips are those which are of one colour, such as white or yellow, showing no inclination to sport into other colours.

Early Tulips.—For the purpose of winter and spring gardening, early-flowering tulips, double and single, are even more indispensable than the hyacinth, narcissus, and crocus. Their rich, brilliant, and diversified hues constitute, for the sitting-room, conservatory, and winter garden, a charm of no ordinary character, while in the beds and borders of the spring flower-garden their picturesque effect and glowing combination of colours exceed those afforded by any other section of garden favourites. Their extreme hardiness, certainty of blooming, and the absence of all difficulty in their cultivation, distinctly entitle them to a preference in the choice of occupants for the spring flower-garden.

For the development of spring gardening, where tulips are made to form a leading and prominent feature, the country is deeply indebted to her Grace the Dowager Duchess of Sutherland; and no small credit is due to Mr. Fleming, her gardener. The spring flower-garden at Cliveden, with its annual display of tulips, &c., and the attention given to these bulbs at the Crystal Palace and the Royal Botanic Gardens, Regent's Park, have done much to stimulate the extended cultivation of spring flowers. Their influence has been felt not only around London, but in the most remote parts of the island; so that grounds which at one time possessed little or no attraction till the bedding plants or spring-sown seeds commenced blooming, are now as rich with floral beauty during April and May as in August and September.

Tulips do well in the shade, and there they remain long in bloom.

Early Single Tulips.—No section of tulips displays so great a variety of delicate, striking, and attractive colours. Of Selfs, there are beautiful scarlets, crimsons, whites, and yellows; of parti-colours, snow-white grounds, striped and feathered with purple, violet, crimson, rose, puce, and cerise, and yellow grounds, with crimson, scarlet, and red flakes and feathers; so that only those who have cultivated the varieties of early single tulips systematically can form any just idea of their beauty, either as regards the shape of the flowers, the brilliancy and variety of their colours, or their general value for decorative purposes.

"For the encouragement of those who possess small town gardens," say Messrs. Barr & Sugden, "we may just mention that in 1860 we had a collection in a little London garden, with a due north aspect, and there the colours came out with as much brilliancy as if the plants had been cultivated under infinitely more favourable circumstances."

Culture in ornamental vases and other elegant contrivances should be the same in every respect as that recommended for the hyacinth.

Culture in pots is also the same as that of the hyacinth, except that three bulbs should be planted in a 4 or 5-inch, and five in a 6-inch pot. To produce an effective display, they should be grown close to the glass, and during fine days have abundance of air. Those intended for early blooming should be forced as soon as the shoot appears.

Culture out of doors precisely that of the hyacinth, except that the bulbs should not be planted more than 4 to 6 inches apart, when an effective display is desired, though many persons plant them 6 to 8 inches from each other.

In beds they can be advantageously arranged in distinct colours, either in circles, diamonds, or with the colours intermixed: we have seen charming beds where the tulips have been planted in geometrical figures, the double tournesol, which flowers simultaneously with single varieties, being used to define the figures. These suggestions are merely thrown out to show how capable the tulip is of being used for fanciful designs, &c. The crown of the bulb should be 3 inches under the surface, and should the weather be severe, a few branches placed on the bed will be found ample protection.

Time of Planting for outdoor Decoration.—The early part of November, or as soon after that period as convenient. We have planted the tulip as late as January, and have had a splendid display. We do not, however, recommend keeping the bulbs so long out of the ground.

Early Double Tulips.—The massive form, brilliant, diversified, and beautiful colours, which are leading features in double tulips, admirably adapt them for beds on the lawn, terrace, or flower-garden, and for édgings to rhododendron, azalea, and rose beds, or for planting in the flower and shrubbery borders in groups of three or more.

In pots the varieties of this section are very attractive, but as a rule they are better adapted for out-door than in-door decoration.

Parrot Tulips.—The parrot tulip has a singularly picturesque appearance; the flowers are large and the colours brilliant, so that when planted in flower-borders and the front of shrubberies they produce a most striking effect. When grown in hanging baskets, and so planted as to cause their large gay flowers to droop over the side, the effect is remarkable and unique.

The following are the best varieties, price 1*s.* 3*d.* per dozen:—

Admiral de Constantinople,—red, streaked with orange.
Coffee-colour,—yellow, green, and brown, mottled.
Large Yellow,—slightly striped with red.
Markgraaf Van Baden,—crimson striped yellow.
Monster Rouge,—crimson.
Perfecta,—yellow, scarlet, and green.
Fine mixed, 1*s.* per dozen.

52.—VANTHOL'S TULIPS (nat. ord. Tulipa'ceæ).

These are dwarf varieties which flower in early spring. There are two sorts, double and single; the latter are the most beautiful. Both, however, are very effective as edgings for beds. The cultivation they require is that of tulips generally.

53.—POLYANTHUS.

Divide the roots of the best plants intended for preservation. This operation must be performed every year, or the flowers will soon degenerate. Fresh soil and continual division is the only plan with all florists' flowers which give out offsets. The single varieties of polyanthus alone are looked upon as florists' flowers; and as these seed freely, an infinite variety of polyanthuses may be obtained by those who will take the trouble to select and sow seed. Late in autumn is the time for sowing; for moderate sunlight only is required to bring up the seed, and the young plants will not stand the scorching sun of summer. Sow in boxes, or pans well drained, filled with light rich mould. The seed must be very lightly covered—indeed, it may almost lie upon the surface. The boxes should be placed under glass, and sparingly watered. They require no artificial heat.

54.—AURIC'ULAS (nat. ord. Primula'cæ).

Florists' flowers of great beauty, well deserving of cultivation. They may be propagated by offsets at any time during autumn, but the earlier the better. New auricula borders may be made in October, and old ones should then be carefully gone over and renovated. Let it be remembered that the auricula delights in shade, and will not bear excessive moisture. In planting offsets, be careful that the soil is well pressed round the roots. In this respect the young plants will require attention for some time. The more delicate sorts of auriculas will not flourish so well as border plants; but the harder varieties—the common purple, yellow, and green—form very effective edgings to beds, and even when out of bloom they have a pleasing appearance from the richness of their foliage. Of course offsets will only perpetuate the same varieties. New sorts are to be obtained from seed, which must be raised in a gentle heat.

Fine seed, which may be relied upon, may be had from Messrs. Barr and Sugden, as under:—

Auricula, from finest stage flowers, various colours, half-hardy perennial, ½ ft., 1*s*. per packet.

Auricula, from finest mixed border varieties, hardy perennial, ½ ft., 6*d*. per packet.

55.—HEARTSEASE.

The common *Viola lutea*, with *Viola grandiflora*, *Viola amœna*, are the joint parents of the many beautiful flowers known to us in these days under the general name of Heartsease.

About the first week in October is a good time to make a selection of plants for potting. These should be vigorous, healthy plants. The bed for their reception may be prepared by digging out the soil for about 18 inches, and filling it up, after providing proper drainage, with compost properly mixed: a better plan is to make a raised bed for the purpose. This may be done by placing a row of bricks, 18 inches high and four feet wide, or wood, supported by stakes at each end, of the same height, and of a length suited to the number of plants required, This bed should be filled with compost, consisting of well-decomposed turfy mould, leaf-mould, and thoroughly-decomposed cow-dung: or, failing that, stable-manure, in the proportions of a bushel and a half of the first to half a bushel each of the two latter: where the loam is stiff, a little well-washed river-sand should be added; where it is light and sandy, equal parts of earth should be added.

56.—RANUNCULUS.

Next to tulips these are the most beautiful of all bulbs or tubers, if, indeed, their claw-like roots deserve these names. They rival the tulip in brilliancy of colour, and many prefer the beautifully-arranged balls of the ranunculus to the stiff formal cups of the tulip; both, however, have their distinctive features of beauty, and deserve a place in every garden. They may be planted from October to the end of March, some preferring one period and some another: perhaps no better time could be chosen for planting than the beginning or middle of February. As soon as the beds are in a fit state, lose no time in planting if the weather be favourable; waiting a day, or even a week, is nothing in comparison with placing the roots in soil in an unfit state to receive them. They are best cultivated in 4-feet beds of rich loam mixed with one-fourth part of decomposed cow-dung. The soil should be dug from 2 to 2½ feet deep, and if the situation is moist and partially sheltered, so much the better. A constant supply of moisture is essential to their beauty and growth, although an excess of water would destroy the tubers during the cold of winter and early spring: after their blossom-buds are formed, however, the surface of the beds must never be allowed to become dry; a daily soaking of water will then be necessary in dry weather, not only for the sake of the flowers, but to preserve the roots from injury, these being very near the surface. The roots should be planted about two inches deep and six inches apart; their claw-like extremities should be pressed firmly into the earth, and the crowns be covered with an inch of sand previous to another inch of soil being spread over them; the beds may then be covered with a layer of spruce branches, straw litter, or leaf-mould, to protect them from the frost: this will, of course, be removed before the appearance of the plants above ground.

Ranunculuses are increased by offsets, dividing the tubers, and seed. Offsets is the usual mode of increase, and they are generally sufficiently strong to flower the first year. Choice sorts may also be divided into several plants; every little knot that appears on the top of a tuber will form a plant if carefully divided, so as to insure an accompanying claw. Unless, however, for choice sorts, this mode of increase is not desirable: by seed is the most rapid mode of increase, as well as the only way of securing new varieties. It is said by some persons that ranunculuses never come true from seed, so that variety is certain. Perhaps the best time for sowing is the month of January, and the best place a cold frame. Sow either in the frame or in pots or boxes, on a smooth hardish surface, and barely cover the seed with soil. Exclude the frost, and keep the frame close until the plants show two seed-leaves; then gradually inure them to more air, until the light may be entirely removed in May. The little tubers may be taken up when the foliage is quite ripened off; they will require the same, or even more attention, in watering than the old roots.

Some prefer sowing the seed on beds out of doors in the autumn or spring months. Generally ranunculuses will have died down, and be fit for taking up

5

and storing, by the end of June or beginning of July. The place for storing should be dry; a drawer with a bed of sand being the most convenient. The following are good named varieties:—

Abbé d'Elugne,—dark brown.
Amazon,—light purple.
Argus,—velvety brown.
Belle Forme,—white, red-spotted.
Bergère Blanche,—white, rose-spotted.
Black Prince,—black.
Blanche Amiable,—white.
Ceres,—purple variegated.
Clothilde,—rose variegated.
Comte d'Artois,—white.
Comte de Gloire,—blood-red.
Comtesse de Pompadour,—deep yellow.
Dictator,—scarlet.
El Dorado,—yellow and brown.
Eleonora,—rose.
Fireball,—fire-red.
Formosum,—brown and yellow.
Genevra,—citron.
Gloria Florum,—brown.
Habet Electoral,—golden yellow.
Hasdrubal,—deep rose.
Herculaneum,—purple-spotted.
Horatius,—dark grey.
La Céleste,—violet.
La Charmante,—light rose.
La Pucelle de Paris,—yellow.
Mathilde,—purplish blue.
Mont Blanc,—white.
Nosegay,—yellow-brown spotted, sweet-scented.
Œil-noir,—true black.
Penelope,—rose and yellow.
Polydore,—velvety blackish purple.
Princesse d'Orange,—orange.
Quintinianus,—rose-spotted.
Reine des Violettes,—velvet.
Rinaldo,—velvety black.
Rose sans égal,—white, rose-feathered.
Temple van Apollo,—dark purple.
Valois,—dark brown.
Vocabule,—white and rose.

57.—ANEMONES or WIND-FLOWERS (nat. ord. Ranuncula'ceæ).

Culture.—The Anemone delights in a light, rich, loamy soil, but generally succeeds in any which is well drained. Sea-sand, or a little salt mixed with the soil, is a good preventive of mildew; in other respects, the culture and after-management should be precisely such as we shall recommend for the ranunculus.

They are usually regarded as spring flowers, and most undoubtedly always flower best at that season. In certain soils and situations, however the tendency to growth and flowering in anemones is such that they have no sooner died down after spring-flowering, than they throw out fresh leaves and flower again in autumn. This, however, is not desirable, for it weakens the tubers, and the flowers soon degenerate. Autumn-flowering may generally be prevented by excluding light and air from the beds, by means of heavy top-dressings of well-rotted manure during the summer months. Many persons take up anemone tubers as soon as the leaf has died down; but this is not necessary, nor is it a good plan, unless the soil of the bed requires renovation, for the tubers will not keep many days out of the ground. The finest flowers are generally produced the first spring after a new sowing; but soil and situation have always a great effect upon them.

58.—CARNATION.

This flower is the cultivated *Dianthus Caryophyllus,* found wild in many parts of England, although it is supposed to have reached this country in its cultivated state from Italy or Germany. Gerarde, writing in 1597, makes mention of it as received from Poland; and it is a remarkable instance of the effects of cultivation, that the named varieties of flakes, bizarres, picotees,—pink, purple, scarlet, and crimson, bear scarcely any resemblance to the original. They are propagated by *seeds*, which, however, do not ripen well with us, and are obtained chiefly from South Germany; by *pipings*, by *cuttings*, and by *layers*. The seed should be sown in May in pots of light rich soil, placing the pots in an airy, sheltered part of the garden. When the plants are up, and show five or six leaves, plant them out in beds in the same rich soil, and 10 inches or so asunder; protecting them during winter in a cold frame, or by means of matting. Many of them will bloom the following summer.

For layers, the propagating season is July and August. Having selected the shoots to be layered, and prepared pegs for pegging them down, and soil for their reception, add a little grey sand where the layers are to be placed. Prepare the shoot by trimming off all the leaves with a sharp knife, except five or six at the top; then with a thin-bladed knife make an incision half through the shoot, with an upward cut, beginning below a joint, and passing through it for an inch or so; bend the layer down into the sandy soil prepared for it, pegging it down in that situation in such a manner as to keep the slit or tongue open, and cover it over with rich light compost. Two days afterwards, when the wound is healed, a gentle watering will be beneficial.

Cuttings are made by taking off shoots which cannot be conveniently layered, cutting them right through a joint with an oblique angular cut, and planting them in pots or beds prepared with mixed compost and sand.

Piping consists in drawing out the young shoots from the joints, and inserting them into a light sandy soil, where they take root—a process more generally applicable to pinks than carnations.

In preparing compost for carnations, take two-thirds good staple loamy soil, the turfy top-spit in preference; add to this one-third of thoroughly rotted cow or stable dung, and one measure of driftsand or other sharp grit, to ten measures of the compost. The alluvial deposit from watercourses, like a mill-head, is an excellent substitute for the maiden loam. In preparing the bed for carnations, having filled the bottom with sufficient drainage material, and secured an outfall for the water, fill in the compost till nearly full. On this surface spread out the roots horizontally, and fill up with fresh compost, pressing the whole firmly, but gently down, in that position.

The layers of carnations and picotees should be taken off as they begin to form fibre, and either potted or planted in a nursery-bed till October, in either case keeping them in a close frame till they have rooted. If potted, re-pot in October, and prune to a clean stem, leaving the lower pair of leaves half an inch from the soil, and removing all laterals over an inch long: most of them will strike in the beginning of October, and bloom strongly the second year, if they do not bloom the first.

The chief distinction between the carnation and picotee is, that the colour of the former is disposed in unequal stripes, going from the centre to the outer edge; that of the picotee is disposed on the outer edges of the petals, radiating inwards, and uniformly disposed. Flakes are carnations of two colours only, with large stripes going quite through their petals. Bizarres have their colours in variegated irregular spots or stripes. A perfect carnation should be not less than two inches and a half across; the lower petals six in number—broad, thick, and smooth, and lying over each other so as to form a circle; each row of petals smaller than that immediately under it, rising in the centre so as to form half a ball. The colour should be clear and perfectly defined; and where there are two colours, the contrast should be bold and decided.

59.—CARNATIONS FOR EXHIBITION.

The preparation of carnations and picotees for exhibition is quite an art, and to some a mystery. The base of the petals, which are mere threads issuing from the calyx, supports broad, heavy blades, which form the expanded blossoms; the largest-sized, which should be outside, being frequently in the centre. If the flower were left to itself, the calyx would probably be split all the way down one side, the other side not opening at all. To counteract this tendency, it is necessary to tie the calyx round the middle when the bud is nearly full-grown, and before it splits; and when the bud begins to open at the top, to pull back the five pieces which form the outer leaves of the calyx down to the point where it is tied: this enables the petals to develop themselves properly.

It is necessary, moreover, that these petals should be "dressed," to make them presentable. This is done by bringing all the petals into their proper places, passing the threads of the broadest petals outside those of the smaller, and guiding the others to the centre according to their size. This is obviously a very delicate operation, and should be commenced as soon as the petals begin to develop themselves. The larger petals are to be placed out-side, and should form a complete circle; the next largest follow, making an inner circle, each petal lapping over the centre of that on which it rests; the third row being placed on the joinings of the second, and a fourth row, should there be one, on the joints of the third; while the whole continues to grow and expand, giving a natural effect to this artificial operation of dressing the flower.

The mode of operation is to take hold of the broadest part of the petal with a pair of smooth flat tweezers of ivory or bone, and by a gentle twist to bring the base round into the position it is to assume; the whole being usually supported by a card, in which a hole is cut large enough to let the calyx about halfway through, while the petals fall back upon, and are supported by, the card, which is circular in shape, to correspond with the expanded flower.

60.—PINKS (nat. ord. Caryophylla'ceæ).

Pinks are closely allied to picotees and carnations, and admit of very similar cultivation. New varieties may be obtained from seed, and old plants may be increased by pipings.

Pipings, as the grass is called when it is pulled out of the joint in the parent stem, should be struck under a hand-glass, and when well rooted should be planted in a bed, in rows six inches apart, and three inches between the plants: here they should remain till September, when they may be planted in a bed or pots, in a compost, consisting of two-thirds of loam from decayed turf, and one-third well-decomposed cow-dung. If in pots, let them be 48's, having a few crocks in the bottom, and filled with compost. Lift the plants carefully, without breaking the fibres, adjusting the soil so as to place the plant in its proper position, spreading out the roots on the soil, and filling up the pot to the surface. The roots must not be sunk too deep, but the soil on the top must be on a level with the collar of the plant. When gently watered, the pots may be placed in a common garden-frame, and the glass closed for four-and-twenty hours. Throughout winter the plants give very little trouble, seldom requiring water, but all the air which can be given them. In March they should be repotted in the pots in which they are to bloom, which should be 24's, with an inch at least of crocks for drainage; the soil as before.

The soil best suited to receive the young plants is a mixture of good hazel-loam, with well-rotted manure from old cucumber or melon-frames. This mixture should be made some months before it is required for use, and at the time of planting or potting the layers, a little white sea-sand should be added. Where layers of carnations and picotees are potted, the best plan appears to be to place them singly into small pots for the winter months. In this way they can be packed closely under common frames in old tan or cinder-ashes. Let the newly-potted layers have all the air possible in fine weather; but if the winter prove severe, it will be necessary to cover the glass with mats, straw, or pea-haum.

Pink pipings properly rooted should be planted out in October; avoid the old system of shortening the grass. Where seed is required, the decaying petals should be picked off.

61.—PICOTEES (nat. ord. Caryophylla'ceæ).

These are a kind of carnation, distinguished by a narrow dark-coloured edging to the petals, or by the petals being covered with very small coloured dots. The cultivation is in every respect the same as the carnation.

62.—HY'ACINTHS.

The Hyacinth is a most delightful and valuable flower. In the conservatory or sitting-room it is equally at home, and its well-being is less dependent upon the mysteries of the gardening art and the pure atmosphere of the country than that of almost any other exotic in cultivation. In the most confined streets of London the Hyacinth may be seen blooming as magnificently as if surrounded by all the advantages of the open country, and displaying ungrudgingly for the delight of its city cultivators charms which most other plants, even though indigenous to our own soil, cannot be induced to reveal.

Time of planting.—The first selection of hyacinths, as of most other bulbous roots, arrives in London from the 20th of August to the 5th of September; orders should therefore be given for them as soon after that time as possible. If it be inconvenient to plant them immediately, they may be placed on a cool dry shelf till wanted.

Hyacinths in Moss.—In the new rustic jardinets "Alexandra," "Prince of Wales," and "Rustic Robin," introduced by Messrs. Barr and Sugden, china bowls, glass dishes, jardinets, and any other elegant contrivance having no means of drainage, moss, or better far, amalgamated cocoa-fibre and charcoal, can be used with great advantage: this latter is always sweet, and bulbs strike their roots freely into it, and are less subject to decay than if planted in any other medium with which we are acquainted.

If clean moss is used, at the bottom of the vessel lay a handful of charcoal, on which place the moss and firmly press it down; on this plant the hyacinth bulbs, and cover them with nice green flat moss, which is to be found in most localities. If the amalgamated cocoa-fibre and charcoal be used, press it down moderately tight, and cover the bulbs over with the same material; this again may either be covered with flat moss or any variety of Lycopodium. Water with tepid water overhead two or three times a week, through a fine rose, till the plants are in bloom; thus the freshness of the moss will be maintained, and the plants freed from dust, which, if allowed to remain, would prove injurious.

Hyacinths out of Doors.—If the soil be light or medium, it simply requires to be deeply dug and well worked; if heavy, besides deep digging and well working, the bulbs should be surrounded with sand, or, better still, two good handfuls of cocoa-fibre; if wet or subject to occasional floodings, drain the ground with a series of drains, three feet deep and ten feet apart, or raise the bed six inches above the general level. When manure is added, thoroughly rotted cow-dung or leaf-soil is best; and when winter protection is given, long straw laid loosely on the bed, and hooped down to prevent its littering the garden, or cocoa-fibre,—both are unequalled; but they should be removed as soon as the plants begin to show.

In planting, the crown of the bulb should be four inches under the surface, and to produce a very effective display the bulbs should be planted six inches apart, but many persons plant them eight or ten inches from each other.

Where the beds are required for the summer occupants before the bulbs have matured themselves, they may either be removed to the reserve garden, and carefully planted in a north border to ripen, or the following plan adopted:—Take as many 4-inch pots as you have bulbs, plant one in each, using the best compost you can possibly command, and place the bulbs on the top of the soil. Prepare the ground as already recommended, then bury the pots as thickly as you please, keeping the top at least three or four inches under the surface; as soon as the blossom fades, remove the flower-spike, dig up the pots, and replant them in the reserve garden, or place them on a north border, surrounding them with moss, cocoa-fibre, or whatever will assist in keeping the roots cool.

Hyacinths in Pots.—To cultivate the Hyacinth successfully in pots, a free porous soil is indispensable, and one composed of equal parts of turfy loam, rotted cow-dung, and leaf-soil, adding about one-eighth part of silver-sand, and thoroughly incorporating the whole and passing it through a rough sieve, is undoubtedly the best compost for the production of handsome flowers. As this, however, cannot always be commanded, use, instead, road-scrapings a year or two old, or good garden loam (not of a retentive nature) mixed with silver-sand. Amalgamated cocoa-fibre and charcoal, mixed with rotted cow-dung and loam, makes a fine mixture, if used in equal parts.

The size of the pot must be regulated by the accommodation and requirements of the cultivator; for one bulb a 4- or 4½-inch pot will grow the Hyacinth well, for three bulbs a 5½-inch pot will be sufficient (and here we would remark, hyacinths cultivated in groups are much more effective than grown singly). At the bottom of the pot place over the hole a piece of potsherd and some charcoal, and on this some rough pieces of turfy loam to insure good drainage; then fill the pots with the prepared soil to within an inch of the top, placing the bulb in the centre, or if three, at equal distances apart, pressing them well into the soil, and filling up, leaving only the crown of the bulbs uncovered; moderately water, and place them anywhere out of doors, on coal-ashes or anything that will secure good drainage and at the same time be objectionable to worms; then with coal-ashes, leaf-soil, or old tan, or better still, common cocoa-fibre, fill up between the pots, and cover over two or three inches. In five or six weeks the pots will be full of roots, and may then be removed at pleasure. For a few early blooms some may be removed at the end of three or four weeks and placed in a gentle hotbed, warm green-house, forcing-pit, or vinery, but they must be kept close to the glass, to prevent them from growing tall and unsightly. At first they should be forced very gently.

Hyacinths in Sand.—Besides the advantage which sand offers in being independent of drainage, it may be procured in various colours. The different shades may be arranged either in geometrical figures or fanciful designs, with very charming effect. As soon, however, as the plants appear, the surface should be covered with green moss, the best for this purpose being found near the roots of old trees.

To insure an effective display, it is necessary to plant thickly. Push the bulbs into the dry sand, leaving only the top visible, and to fix the sand, the vessel should be immersed in a pail of water; also, to prevent any subsequent displacement of the sand, and to secure for the plants a sufficient supply of moisture, this operation must be repeated once a week, or oftener if required, a bath of two or three minutes' duration being sufficient; and if the water used be tepid, it will be all the better, as it encourages the development of the flower. An occasional watering of tepid water overhead, through a fine rose, will free the plants from dust, and keep them healthy and vigorous. The amalgamated cocoa-fibre and charcoal is a much better medium than sand.

Hyacinths in Water.—The hints we offer for the cultivation in water are simple but ample, and if followed cannot fail to insure success, provided the roots be sound, of the proper age, and the right varieties for glass-culture. Never use spring water if you can get clean rain-water. Place the bulb on the glass, and let the water just touch its base; for three or four weeks keep it in a dark, cool situation, but avoid a damp, close atmosphere.

When the roots have grown two inches remove the water half an inch from the base of the bulb. All disturbance of the roots should be avoided; therefore never change the water while it remains sweet; as a purifier, place a piece of charcoal in the glass. Avoid a close, hot room, or you will have long stems and small flower-spikes. Choose an airy situaton, and place the glass or Ne Plus Ultra in the lightest and sunniest position, turning it once a day. Never allow dust to remain either on the bulb, leaves, or flower; once a day, or oftener, remove it with a camel-hair brush and water.

Carefully guard against changes of temperature, especially from heat to cold, and never remove the plants from a hot room to a cold one; and when the water is changed or the glasses filled up, the chill must always be taken off the water. Never use a support till the plant requires it.

A preference is generally given to dark-coloured glasses over clear ones, on the supposition that the Hyacinth grows better in them; but it is our opinion that the Hyacinth will do as well in the one as in the other; therefore those who delight in watching the growth of the roots, as well as the development of the flower, should buy the clear glasses.

63.—NARCIS'SUS (nat. ord. Amaryllida'ceæ).

This genus is a very extensive one, embracing Jonquils, the Polyanthus, Narcissus, the Hooped Petticoat, the Poet's Narcissus, and many others.

The Double Roman Narcissus, planted early in September, would bloom in-doors before Christmas, while the Paper-white, combined with the other varieties for in-door culture, if planted in succession from the 1st September to the 31st December, would maintain a rich floral display till the end of April.

Culture in-doors is similar to that recommended for the Hyacinth. The bulbs of the Polyanthus Narcissus being large, a 5-inch pot will be needed for one bulb, and a 6-inch pot for three; a group of six in an 8-inch pot will produce an exceedingly beautiful effect.

Culture out of doors is exactly the same as that for the Hyacinth, except that the crown of the bulb should be at least five inches under the surface, and for winter protection should be covered with about an inch of newly-dropped leaves, or three inches of cocoa-fibre.

September is the time for sowing seed, in order to obtain new varieties. For the mode of proceeding we cannot do better than quote the words of Mr. Leeds, of Manchester, who has been one of the most successful amateur cultivators of the Narcissus. Mr. Leeds says: "To obtain good varieties, it is needful, the previous season, to plant the roots of some of each kind in pots, and to bring them into the greenhouse in spring to flower, so as to obtain pollen of the late-flowering kinds to cross with those which otherwise would have passed away before these were in flower. With me the plants always seed best in the open ground. When the seed-vessels begin to swell, the flower-stems should be carefully tied up, and watched until the seeds turn black. I do not wait until the seed-vessels burst, as many seeds in that case fall to the ground, and are lost, but take them off when mature, with a portion of the stem, which I insert in the earth in a seed-pot, or pan, provided for their reception. I place them in a north aspect, and the seeds, in due season, are shed, as it were, naturally, into the pot of earth. I allow the seeds to harden for a month on the surface before covering them half an inch deep with sandy soil. The soil should be two-thirds pure loam and one-third sharp sand; the drainage composed of rough turfy soil. In October, I plunge the seed-pots in a cold frame facing the south, and the young plants begin to appear in December and throughout the winter, according to their kinds and the mildness of the weather. It is needful, in their earliest stage, to look well after slugs and snails."

64.—ROSE CULTURE.

The Rose is propagated by seeds, by cuttings, by layers, by budding, and by grafting, and new varieties are produced by hybridization, that is, by transfusing the male pollen of one flower into the stigma of another, with the object of producing seed which shall reproduce the best properties of both the parent plants.

The great obstacle to hybridizing in our climate is the difficulty found in ripening the seeds. Tea-scented and Chinese roses must be grown under glass to do so. Many other varieties, however, ripen their seeds sufficiently out of doors, and Mr. Paul gives a list of twenty kinds which ripen their seeds perfectly in this country in ordinary seasons. These are :

Hybrid Chinese,—Chénédole; Marshal Soult; Duke of Devonshire; General Allard.
Hybrid Bourbons,—Athelin; Great Western; Charles Duval.
Moss,—Du Luxembourg; Celina.
Hybrid Perpetuals,—Madame Laffay; Comtesse Duchatel; William Jesse.
Austrian Brier,—Harrisonii.
Bourbons,—Bouquet de Flore; Malvina; Ceres.
Rosomène,—Gloire de Rosomène.
French,—Rosa Mundi.
Multiflora,—Russelliana.
Ayrshire,—Splendens.

Having selected the plant to be impregnated, as well as that from which the pollen is to be taken, and having removed the stigma of the female flower with a pair of round-pointed scissors, just as the flowers begin to expand, both flowers being in a fit state for the operation, the flower of the male parent is brought just over the seed-bearing tree. If the weather is calm, the petals of the former are cut away, the operator holds a finger over the flower of the latter, while the tree deprived of its petals is struck a sharp blow. This sudden shock drives the pollen into the other flower, and the work is done. When the wind is high, it is better to collect the pollen on a camel-hair pencil, and convey it to the styles of the female flower. It is of much importance that both plants are grown in the same temperature. The stigma is known to be in a fit state for fertilizing by a viscous exudation of its sutures; when this is the case, no time should be lost in completing the operation.

When the flowers have disappeared, and the seed-pods begin to swell and ripen, they should be protected from birds; they should remain on the trees till perfectly ripe, and turning black, when they should be gathered to be sown, and buried in the earth, either in pots or in the ground, taking care that proper numbers are attached to them, so that there may be no doubt as to the origin of the expected progeny.

The operation of cleaning the seeds is performed by rubbing them out between the hands, preparatory to sowing in February or March. As soon as the seed-pods are broken up, lay the seeds out in the sun to dry, so that the pulp and husk may be entirely removed, and the seeds sifted and winnowed ready for sowing. Some prefer sowing the seed immediately on its reaching maturity, and this M. Boitard considers the best plan. In this case they come up the following spring, with all the fine growing season before them.

The hardier kinds may be sown in the open borders, selecting a sunny spot, with an eastern aspect, and sheltered from the afternoon sun: the more tender varieties require a frame and glass.

The more delicate roses should be sown in pans thoroughly drained and filled with equal parts of leaf-mould and yellow loam well mixed together. Water, when sown, as directed above, and cover with half an inch of the same soil mixed with a little sand. In each case a sprinkling of soot or lime scattered over the bed will be a necessary precaution against insects and worms. About May some of the seeds will germinate, and others will come up from time to time till autumn; they now require constant care in shading, weeding, and watering when the soil is dry.

Roses may be increased by budding, by cuttings, by grafting, and by suckers. Special directions for budding will be found under that head, and need not be repeated.

By Cuttings.—Most roses may be increased by cuttings; but all are not alike calculated for being thus propagated, bottom-heat being indispensable for the more tender varieties. Where it is necessary to propagate a number of plants from one, however, the method is invaluable; it is especially useful in the case of Chinese and Indian roses, where the young branches are inclined to be woody. Summer and autumn are the best seasons for cuttings. The shoot made in spring is taken with a small portion of last year's wood attached, and cut into lengths of five or six inches, selecting such as have two lateral shoots, with five or six leaves to each. An inch of the old wood should be inserted in the soil, leaving at least two leaves above. From four to six of these cuttings may be placed round the inside of a 60-pot, in soil consisting of equal parts of leaf-mould, turfy loam chopped fine, and silver-sand, watering them well with a fine rose-pot, to settle the earth round the roots. When the water is drained off, and the leaves dry, remove to a cold frame, or place them under hand-glasses, shade them from the sun, excluding the dew, and sprinkling them daily for a fortnight. If threatened with damping off, give air and sun. In a fortnight *callus* will be formed. At this time they are greatly benefited by bottom-heat; they root more rapidly, and may soon be shifted singly into 60-pots, and removed back to the cold frame for planting out in August, without bottom-heat. They must be kept in the frames till the spring.

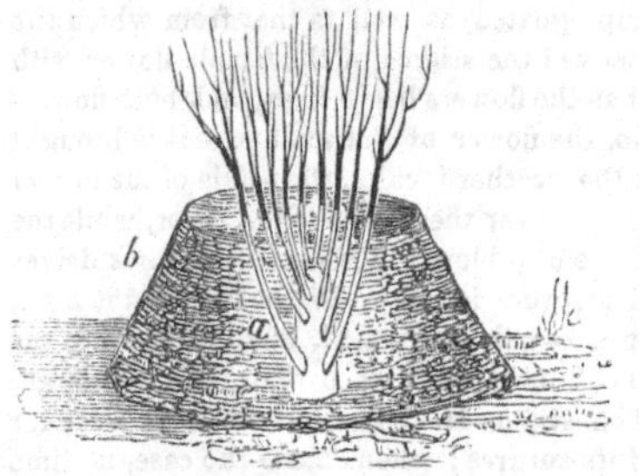

Grafting is performed by cutting the top of the stock to a proper height by a clean horizontal cut; then make a longitudinal V-shaped cut down the centre, one, two, or three inches long, according to the size of the stock. In this slit place the graft, after having cut the lower end of it to fit the cut in the stock. Having inserted it, bind the whole up with clay or grafting-paste, as directed in budding.

It has been asked whether roses *grafted* are equal to budded ones. Mr. Paul says, when the junction is perfect they are quite as good; but the scion and stock do not always coalesce, owing to which, more imperfect plants are raised in this way than by budding. The best time for grafting roses in pots is January; and July, in ordinary seasons, is the best month for budding; but that depends upon the season. Some operators prefer a moist gloomy day for the operation.

By Suckers.—Roses—some kinds much more than others—push their roots in a lateral direction under ground, and throw up young shoots or suckers from them; these suckers, separated from the parent root by the cut of a sharp spade, form flowering plants the same season, if separated in the spring and transplanted to suitable soil. When a rose-tree is shy with its suckers, it may be stimulated by heaping earth round the roots.

65.—ROSES IN POTS.

The best compost for roses in pots is a good stiff loam one bushel, rotten dung from an old hotbed one peck, and a half gallon of pigeon-dung or double the quantity of sheep-dung, well decomposed. Another good soil is a turfy loam, rather stiff, and well decomposed cow-dung in about equal parts. These, with good drainage in the pots, gentle forcing, careful examination for insects, and keeping the trees well balanced and neatly trained and tied down to their supports, with constant attention to watering with liquid manure, as well as pure water, will produce abundance of bloom. All the strong-growing varieties, such as Hybrid Provence, Hybrid China, Hybrid Bourbon, French Moss, Alba, Hybrid Perpetuals, and the strong-growing Bourbons and Noisettes, should be potted in autumn; the delicate Bourbons and Noisettes, together with the Tea-scented and Chinas, should be left till the spring. In the autumn, before potting, examine the stocks minutely, particularly among the roots, cutting out any decayed parts, all knots, and the remains of suckers; lop any wounded or bruised roots, and clean all well, as no such opportunity will again offer. When potted, cut back the head about a third, and at the end of February or the beginning of March cut back to from two to four eyes.

The pots should be plunged to their rims in coal ashes, or any garden soil, on an inverted flowerpot, leaving a cavity underneath each to prevent the ingress of worms. In this state they should remain till about the middle of May, when they will require a shift. The same soil as before should be used, but, if possible, more rough, and they should be again plunged to the rim. During summer they will require constant attention as to watering; they should have liquid manure at least once a week, using soft water at other times.

66.—AYRSHIRE ROSE.

The rapidity with which this rose covers a wall or pillar, added to its intrinsic beauty, renders it invaluable to the gardener. Where its growth is encouraged, it climbs to the summit of the tallest trees, from which its long graceful shoots hang in festoons. The Ayrshire seems to have been first grown in the garden of the Earl of Loudon, at Loudon Castle, Ayrshire. It soon found its way into the nurseries in Scotland, whence, in 1811, it was transplanted to London by Mr. Ronalds, of Brentford. It was at first considered a variety of *R. arvensis*, afterwards of the evergreen rose, *R. sempervirens*. It differs in many respects from both; its long slender branches grow so rapidly as to throw out shoots thirty feet long in one season. *R. arvensis* never grows with this rapidity, and its branches are stronger and thicker. The evergreen rose far less resembles the Ayrshire in these properties.

There is more than one tradition connected with the introduction of this rose: that which is most likely to be authentic is supported by Mr. Reilly, who states, in his monograph in the *Edinburgh Philosophical Magazine*, that the seeds from whence it was obtained were part of a packet received from Canada, from which several plants were produced. Another account traces it to a Mr. Power, from Yorkshire, who brought the plant from some gentleman's garden in that county. Whatever truth there may be in these statements, it is pretty certain that no rose having the slightest resemblance to the Ayrshire has since been discovered on the American continent. It also appears that at different periods, two very distinct varieties have been grown in nurseries under the same name; one being descended from *R. arvensis*, and the other from the real Ayrshire, with very long slender branches.

The Ayrshire is the hardiest of climbing roses, and its cultivation and management are very simple. Layers of its long pendulous shoots root readily, and it strikes easily from cuttings; it will grow rapidly where other roses will scarcely exist, and when placed in good rich soil, its growth is so rapid that a large space is covered by it in the second season of planting. It forms an admirable weeping rose when trained on wires. It is useful for trellis, verandah, or alcove, as well as in rough places of the park or shrubbery. Its luxuriant growth soon turns a rough and dreary waste into a flowery bank.

Like the other roses, the Ayrshire has yielded many hybrid varieties:—

Ayrshire Queen,—dark purple-crimson.
Bennett's Seedling,—pure white; forms a beautiful pendulous tree as a half-standard.
Dundee Rambler,—white, edged with pink; well adapted for a half-standard for the lawn.
Ruga,—pale flesh colour; very fragrant; a hybrid between Ayrshire and a tea-scented rose.
Splendens,—creamy white, approaching flesh-colour when full; crimson in the bud; large, double, and globular; one of the finest pendulous roses.
Queen of the Belgians,—creamy white; cupped, large and double; very sweet-scented.
Alice Grey,—creamy salmon-blush.
Countess Lieven,—creamy white; cupped and double; of medium size.

67.—BOURBON ROSE.

The *Bourbon Rose*, a variety of *Rosa indica*, was discovered in the Isle of Bourbon in 1817, by M. Breon, who flowered it and sent the seeds to M. Jacques, at Château de Neuilly, near Paris. It was growing among a lot of seedlings raised to form a hedge. This parent of a numerous and lovely progeny, graduating from pure white to the darkest tints, has been hybridized with the Chinese, with the Noisette group, and with other Bourbon roses; and in the latest edition of Mr. Paul's "Rose-garden" they number 251 varieties.

The distinguishing characteristics of Bourbon roses are brilliancy and clearness of colour, large and smooth petals, falling in numerous and graceful folds. They are perfectly hardy, and thrive under the ordinary culture, delighting in a rich soil, like most of the roses, and requiring close pruning, except the more vigorous kinds. They are of slow growth, however, in spring, and thus they are best adapted for autumn-flowering roses.

"The sorts of vigorous growth," Mr. Paul tells us, "form handsome umbrageous trees, with heads as large as summer roses; they also look chaste and elegant trained on pillars. The moderate growers are pretty as dwarf standards. The dwarfs form striking and beautiful objects when grown on their own roots. A great many are excellent for pot-culture, and are beautiful objects in the forcing-house."

Sir Joseph Paxton,—flowers bright-rose, shaded with crimson; large, full, and cupped, with fine foliage.

Comice de Seine-et-Marne,—flowers open, crimson-scarlet, changing to rosy purple; produced in clusters, very double and cupped. One of the most beautiful of the group for standards, but surpassed by another raised from it at Lyons in 1852.

Mrs. Bosanquet,—creamy white; a profuse bloomer from June to November; moderate grower. This rose appears to be intermediate between the Chinese and what are called Bourbons. Will form a small clump, if possible more unique and beautiful than any other variety. A truly splendid rose.

Bouquet de Flore,—brilliant carmine; opens freely, and blooms profusely from June to November; a strong, vigorous grower.

68.—EVERGREEN ROSES.

These are the progeny of *R. sempervirens*, which abounds throughout Europe in a wild state, and, like the Ayrshire, are employed as climbing and weeping roses. Their beautiful dark-green leaves grow on to the depth of winter, which has procured them the name, although, strictly speaking, they are deciduous. They are mostly trees of vigorous growth and abundant bloomers, adapted for pillar-roses; their small, but very double flowers, hanging in graceful cymes of fourteen to twenty on a branch. They require much thinning in the pruning season; the shoots left being merely cut at the points, the others cut close to the base.

Banksæflora,—creamy white, with yellow centre; cupped and double; produced in clusters.

Carnea grandiflora,—pale flesh-colour; large and double; cupped.

Donna Maria,—pure white, of medium size; full and cupped; a choice variety.

Félicité perpétuelle,—flesh-colour, changing to creamy white; produced in graceful trusses drooping from their own weight; a superb sort.

Fortune's five-coloured,—creamy white, striped with carmine.

Fortune's double yellow,—bronze and yellow; large and very distinct.

Myrianthes,—pale peach, the centre white; hanging in graceful trusses of full-cupped flowers; a fine pillar-rose.

Princess Louise,—creamy white, with black petals shaded with rose; large, double, and cupped; a fine climber.

Rosea major,—rosy flesh-colour, changing to white; large, double, and cupped, with fine glossy foliage.

69.—ROSA MULTIFLORA.

Of this there are many varieties, chiefly of Italian origin: it is a delicate rose, and often killed to the ground by the frost. Covered with mats, it shoots so early that it cannot endure the spring frost. Grevillea, or Seven Sisters rose, is a vigorous climber of this family, blooming in clusters, of shades varying from rose to purplish crimson; the flowers change from crimson at first coming out, to pale rose and purplish crimson. The protection recommended by Mr. Rivers, is to thatch over the pillar in November with green furze, which admits air and keeps off the severity of the frost; continuing this covering till March, and then removing it by degrees, so as to inure the plant to the cold before full exposure to it. In this way, Alba, or Double White, a pretty pale flesh-coloured rose; the Double Red, and Hybrida, or Laure Davoust, a most elegant and beautiful hybrid, with large flowers and beautiful foliage, will bloom in perfection.

70.—PILLAR ROSES.

There is no form in which the rose grows so gracefully as when rambling over rocks and climbing up trees or trellis-work, or over the alcove. In the garden, well-planted pillars may become objects of great beauty. To make roses grow with the necessary luxuriance, "each plant," says Mr. Rivers, "should have a station at least two feet in diameter to itself. In the centre of this space a stout stake of yellow pine, tarred at the bottom, should be driven two feet into the soil, and stand eight feet above the surface; the upper part painted green. If the soil be poor, it should be dug out three feet in depth, and filled up with rotten manure and loam, laying this compost about a foot above the surrounding surface to allow for settling. In wet soils they will grow the better for being on a permanent mound; but such soils should be well drained. In the centre of this mound plant a single rose, or, if a variegated pillar is desired, place three plants in the same hole,—a white, a pale-coloured, and dark variety. Having replaced the soil, cover the surface with manure, keeping the manure replenished, as it is drawn in by worms or washed in by rain. Water with liquid manure in dry weather, and there will be shoots, probably eight or ten feet in length, the first season. Three of the most vigorous should be fastened to the stake, and the spurs from them will, for many years, give abundance of flowers.

71.—ESPALIER ROSES.

Roses, as espaliers, may be made to assume a striking feature in the economy of an ornamental rose-garden. The espaliers should be formed of galvanized iron, five feet or so high, and of proportionate width. Hurdles of this material, if the bars be close enough, will answer very well; but the bars should not be more than six inches apart. This form of rose-training is especially suitable for varieties with weak footstalks, notably Maréchal Niel. The plants used should be of strong-growing habit, on their own roots of the Manetti, and planted upon the southern aspect, or as near to that as possible. Intertwine and mix the branches as thickly as may be, and tie them to the cross-bars with wire or tarred string. To construct a rose-terrace, let such espaliers as described above form the back rows of sloping beds; an ascent may thus be charmingly laid out in stages, with path and terrace alternately to the level ground.

The following climbing roses may be used in the espalier form:—

Laura Davant,—pink, double.
Russelliana,—dark crimson, double.
Madame D'Arblay,—white, blooming in clusters.
The Garland,—nankeen and pink, semi-double.
Frederick the Second,—rich crimson-purple.
Juno,—pale rose, very large.
Madame Pantier,—pure white.
Madeline,—pale flesh

But it is useless to multiply names, the varieties of climbing roses are almost endless. The rose-grounds of Messrs. Paul & Sons, of Cheshunt, present an abundant choice for the most fastidious; the only difficulty is, amidst so many beauties, which to choose.

72.—AUSTRIAN BRIARS.

Austrian Briars, and other yellow roses, are difficult to flower, and require careful management. The old double yellow is remarkably capricious, and the Cloth of Gold, or Chromatella rose, in which rose-growers expected to find an unrivalled yellow, blooms in perfection only in very few places. "The general direction given," says Mr. Saul, of the Durdham Downs Nursery, Bristol, "is to grow it in a poor soil, as it is a strong grower." Now, this I consider very questionable advice. It belongs to the group of roses called Noisettes, many varieties of which have been crossed and raised from very dissimilar groups. The large section of tea-scented Noisettes will be found to differ most materially from the original Noisettes, from their affinity to the tea-scented, from which they have been raised. To this section belong the Cloth of Gold, Solfaterre, Clara Wendel, Lamarque, Triomphe de la Duchère, and many other fine roses; Solfaterre being very little inferior to the Cloth of Gold, both being raised from the same parent. The whole section requires peculiar treatment, approaching in some degree to what we give to the strong-growing tea-scented kinds.

73.—BOURSAULT ROSE.

Boursault Rose owes its name to M. Boursault, an eminent French grower; it is a cultivated variety of the Alpine rose; the shoots are very long, flexible, and smooth, in many instances entirely without spines, and the eyes are further apart than in most other kinds. The flowers are produced in clusters suitable for pillars, and from their naturally pendulous habit they may be trained to form weeping roses. They should be well thinned out in pruning, but flowering-shoots should only have the points cut off. Mr. Paul describes the following as blooming from May to July:—

Amadis,—deep crimson-purple, shaded with lighter crimson; large, semi-double, and cupped; the young wood whitish-green.

Black Boursault,—flowers whitish-blush, with deep flesh centre; very double and globular; of pendulous habit; excellent as a climbing rose in a good aspect.

Drummond's Thornless,—opens a rosy carmine, changing to pink; flowers large, double, and cupped; habit, pendulous.

Elegans,—flowers in clusters of semi-double rosy crimson; sometimes purplish, often streaked with white; erect in habit, and suitable for a pillar.

Gracilis,—flowers early; cherry, shaded with lilac-blush; full-formed and cupped; of branching habit; spines long and large; foliage a rich dark green.

74.—BANK'SIA ROSE.

The flowers of this elegant rose are produced in small umbels, each of three, four, and up to twelve flowers, at the extremities of small lateral shoots, branches of the preceding year's growth. The peduncles or stalks are slender and smooth, and about an inch and a half long, so that the flowers have a drooping habit when fully expanded. The scent of the flower is agreeably fragrant, not unlike to that of the sweet violet. This delicate rose was imported from China in 1807, and was named in compliment to Lady Banks. It was long treated as a greenhouse plant; but in 1813 two plants were turned out into the open ground at Spring Grove. In two years, although one of them died, and was replaced, they had covered both walls against which they were planted. In 1818 and 1819 these plants had covered 40 feet of wall, each shoot producing at its extremity an umbel of flowers, which began to open in April, and the whole wall was covered with blossoms until the middle of June. This vigorous growth and bloom was produced by planting the rose in a rich sandy loam, and against a wall with a south or west aspect, nailing its shoots close to the wall: and when the wall was completely covered to the extent proposed, cutting away all the strong shoots as they appeared, leaving only those intended to produce flowers in the following spring. From August to February the only care required is to nail in all young shoots, only removing those that are superabundant.

75.—NOISETTE ROSES.

This beautiful group of roses was originally a hybrid between the Bengal rose (*R. indica*) and *R. moschata.* It was sent from North America in 1817, by M. Philippe Noisette, to his brother, Louis Noisette, c Paris; and the letter which accompanied it states that it was obtained by the artificial fecundation of the former with the pollen of the latter, applied by camel-hair pencil. M. Boitard thinks this statement may be doubtful, pointing out how easily the hybridizer may deceive himself in carrying out the details of his operation. "Our cultivators," he adds, "obtain every day new varieties resembling the Noisettes by merely sowing the seed of *Rosa indica,* without any attempt at artificial fecundation." It is not very obvious, however, why this doubt is raised: it is quite certain that the original type of the Noisette is becoming scarce, while the new sorts approximate to other groups.

76.—PERPETUAL ROSES.

These have been so crossed that their real origin is very indistinctly known. A French author, quoted by M. Boitard, traces their wood and spines to the old Damascus roses, their leaves and foliage to the Bengal, their clustering flowers to the Bourbons, and their odour to Centofolio ;.proving, as M. Boitard observes, if all these characteristics be true, the correctness of his opinion on the question of species. Mr. Paul divides them into groups; founded—1. on the Four Seasons rose, "a branch of the Damask now little known, and chiefly valued as the type of the lovely Damask Perpetuals." 2. The Trianons, a group formed by M. Vibert, of Angers, "obtained," as he tells us, "from several varieties which acknowledge the Rose de Trianon as their type." The flowers of this group are produced in small clusters, flowering in the summer; the leaves gathering in tufts near the end of the shoots. A third group—3. the Damask Perpetuals, "chiefly descended from the old monthly and Four Seasons rose; the varieties being more remarkable for the delicious fragrance of their flowers than for their size or symmetry of form; but which," he adds, "are more properly described as Damask roses blooming in autumn." The fourth group—Hybrid Perpetuals. These lovely additions to the rose-garden have been raised in great part by M. Laffay, between the Bourbons, the Chinese, and the Damask. "Princess Helena was the first introduced in 1837; Queen Victoria followed; and in 1840," says Mr. Paul, "there were above twenty varieties enumerated in the rose catalogue, one-fourth of which, however, were Bourbons. The others might be called," adds Mr. Paul, "hybrid Chinese blooming in autumn." They are, indeed, fine roses, quite hardy, and very sweet. They thrive under the common treatment, and are suited alike for standards and dwarfs, for pot-culture and forcing.

The following lists give a few of the best varieties of Hybrid Perpetual Roses:—

Earl of Derby,—scarlet, flowers deep rose, tinted with purple; large and semi-double.

Baronne Prevost,—bright pale rose-colour; glossy; a very large, full, and compact flower; free grower, blooming freely from June till November.

Dr. Marx,—rich glowing carmine colour; large full-cupped flower, and very fragrant. A very beautiful rose.

Géant des Batailles,—dazzling crimson, the nearest approach to scarlet in this class; very free grower, and one of the most abundant bloomers, flowering from June till December.

La Reine,—brilliant glossy rose-colour, shaded with lilac, and sometimes with crimson; cupped and very large; has the appearance of a true perpetual cabbage, but much larger.

Madame Laffay,—bright crimson, inclining to purple; large, cupped, and very double; very fragrant; free, vigorous grower; and an abundant bloomer from the end of May to the middle of December.

Standard of Marengo,—brilliant crimson.

William Jesse,—light crimson, tinged with purple; very large, double, and cupped; a large and fine flower; strong grower; forms a splendid mass.

77.—CHINA ROSES.

The first introduction of the Chinese roses dates from 1789, in which year both the monthly rose and the crimson Chinese were introduced. The varieties which have sprung from these are innumerable. They have also been hybridized extensively with the tea-scented. Unlike other roses, also, the Chinese roses get a deeper and more brilliant tint by exposure to the sun. "Archduke Charles, and some other Chinese roses," Mr. Paul tells us, "are distinct and beautiful as variegated roses; they expand rose and white, soon become mottled with, then change entirely to crimson, from the action of the sun's rays."

The common and Crimson China are very beautiful, either grown in beds or on walls. Among groups of bedding-plants mixed with Cérise unique, or Lady Middleton geranium, the common China rose, edged with the crimson, and surrounded with a white band of Alyssum or *Cerastium tomentosum*, is very effective, distinct, and striking. Towards the end of May or beginning of June, they will be in full beauty, and the mass of blush pink, with the setting here described, is peculiarly soft and beautiful. By cutting off the flowering-stems as soon as they begin to fade, a succession of flowers will be secured throughout the summer. If, however, a short hiatus should intervene, the geraniums will fill up the gap; and the varieties here named have the merit of harmonizing nicely with the roses. Several other China roses form beautiful groups for the flower-garden. The best for this purpose are, perhaps, the following:—

Cramoisie supérieure,—bright crimson.
Clara Sylvain,—pure white, large.
Prince Charles,—bright carmine.
Mrs. Bosanquet,—pale flesh, similar in habit to Bourbon Queen.
Aimée Plantier,—bright fawn and rose.
Eugène Hardy,—creamy pale blush.
Marjolin,—rich dark crimson.
Madame Bréon,—bright rose, large and good.

78.—BOX EDGING.

The box used for the purpose of edging is *Buxus sempervirens*. It is readily propagated by dividing the old plants, and it will grow in any soil not saturated with moisture. The best time for clipping box is about the end of June. To form edgings of box properly is an operation in gardening that requires considerable care. Mrs. Loudon gives the following excellent directions:—"First, the ground should be rendered firm and even: secondly, a narrow trench should be accurately cut out with the spade in the direction in which the edging is to be planted: thirdly, the box should be thinly and equally laid in along the trench, the tops being all about an inch above the surface of the soil: and fourthly, the soil should be supplied to the plants and firmly trodden in against them, so as to keep the edging exactly in the position required: the trench should always be made on the side next the walk, and after the soil is pressed down and the walk gravelled, the gravel is brought up over the soil, close to the stem of the box, so as to cover the soil at least an inch in thickness, and to prevent any soil being seen on the gravel-walk side of the box. This also prevents the box from growing too luxuriantly, as it would be apt to do if the trench were on the border side, when the plants would lean against the gravel, and the roots being entirely covered with soil, would grow with so much luxuriance that the plants would be with difficulty kept within bounds by clipping. A box edging once properly made and clipped every year, so as to form a miniature edge, about three inches wide at bottom, three inches high and two inches wide at top, will last ten or twelve years before it requires to be taken up and re-planted; but if the edging be allowed to attain a larger size—say, six inches wide at bottom, six inches high and three inches wide at top, it will last fifteen or twenty years, and probably a much longer period.''

79.—TURF.

The appearance of a garden depends greatly on the quality of the turf and the way in which it is kept. Close cutting and continual rolling is the secret of good turf. On good soil little else is requisite; but on poor sandy soil the verdure must be maintained by occasional waterings with liquid manure and a dressing with guano or soot, if the lawn be not so near the house as to render such applications objectionable.

80.—FERNS.

Ferns are by many considered to be a remnant of the vegetation of a past era in the history of the earth, and the treatment they require would seem to confirm this view. Most of them delight in a loose soil, abundant moisture, and a warm humidity in the air. Many kinds are hardy; some are natives of our own climate; yet it is observed by experienced cultivators that those species which are found in temperate climates grow much finer in tropical ones, &c.: the sorts that live in England in the open air grow luxuriantly in a stove. However, in the cultivation of them, it is not necessary to go beyond what is ordinarily done. Hardy sorts may be grown out of doors, and those that will stand a greenhouse temperature are great ornaments in the house. Some kinds are naturally large, and the larger the more valuable—as the Dicksonias or tree-ferns: these require a large house to grow them properly. Too much cannot be said in praise of this interesting class of plants. It is impossible here to give lists of the varieties and enter upon their merits. We can only advise our readers to consult the catalogues of Messrs. Williams, Messrs. Veitch, and others of our best growers, where they will find ample information on the subject, and to visit their nurseries, where they will see such specimens as may provoke envy without doing harm.

81.—FERN-CASES.

In planting fern-cases, it is best to choose sorts of very dwarf, compact habit. This will allow for a little drawing up of the fronds from confinement. A mixture of loam, leaf-mould, and sand will be best for them to grow in; but it should be liberally mixed up with broken pieces of sandstone, or broken flowerpots, and small pieces of charcoal. These will hold a supply of moisture without stagnating; but water should not remain unabsorbed, so that the cases must be provided with suitable drainage; and although the confinement of the case will agree with many ferns, yet some ventilation is necessary to prevent damp and mouldiness, otherwise every frond or dead piece of leaf that touches the glass will cause a general decay.

82.—FERNERIES.

Out-door ferneries are usually constructed in a shady nook, a damp situation on the margin of ponds, or in low-lying situations, where they would naturally receive that shade and moisture which agree with them: a shady bank, where water can be supplied artificially, makes a good fernery, and it requires a great deal of moisture to preserve the healthy greenness of them during a dry summer. If once allowed to flag, they get spotted, which diminishes the beauty of them, and they do not recover it the whole season. In building a fernery, make a liberal use of rough blocks of wood, gnarled roots, and blocks of stone, burrs, &c. Let there be no appearance of labouring to produce an effect, but let the whole be as natural as possible. Plant the ferns so that they show in niches and crevices, and not so thickly as to hide the burrs, &c. For out-door culture none can surpass hardy British ferns. Some of these grow much larger than others:—*Lastrea Filixmas*, *Filix fœmina*, *rigida*, *osmunda*, *regalis*, are among the largest growers. The common brake should not be admitted. There are ferns, however, that have a similar habit to that on a much smaller scale—various sorts of polypody; these are very useful for the front places. There are a great many ferns of dwarf habit, known as natives of this country, which require careful management to induce them to grow in a fernery; the Rue fern, for instance, seldom lives many years, except in the hands of very clever cultivators.

Greenhouse ferneries may be made in every way similar to those out-doors, only on a smaller scale, with the same view as to shade and moisture. If in an open spot, the top must be furnished with blinds, and the syringe be kept at work; but if close and warm, the ferns will grow most luxuriantly.

83.—ANNUALS.

Plants so termed, as their name implies, spring up and flower and die in one single year, and they naturally divide themselves under three heads,—hardy, half-hardy, and tender. Let us take them in order, merely affirming, on the outset, what every seed farm in the country proves, that a rich and varied floral display may be maintained by annuals from April to November. A judicious selection of varieties is, of course, necessary; but this may be left to any respectable seedsman. What the public require to know is, what to do with the seeds when they have got them. First, then, as to—

Hardy Annuals.—*Culture.*—*Time and Manner of Sowing.*—The proper time for sowing hardy annuals depends entirely upon the period at which it is desired that they should bloom. For a spring display, sow early in September (a fortnight earlier or later according to soil and locality), in beds 4 feet wide, with an alley 1 foot wide between the beds, in rows 9 inches apart, and if the weather be dry, well water the drills before sowing, and cover the seed very lightly with fine soil. Before winter, the plants, if too thick, must be slightly thinned out, and early in March carefully lifted in patches containing three or four plants, and planted where they are intended to bloom, regulating the distance between the patches by the known habit of the plants; but as this sowing is expected to produce an early and effective display, rather than a long continuance of beauty, they may be planted rather thickly. To furnish a succession, a piece of ground proportioned to the quantity of plants required should be set apart in an open situation, and laid out in beds 4 feet wide, with alleys of 18 inches between; the soil should be prepared with a dressing of decayed leaves, and if stiff, some sharp sand should be added; manure had better be avoided, except in light soil, where a moderate dressing of thoroughly decayed manure would be most suitable. On this ground successive sowings of the best kinds, should be made at short intervals, keeping the ground always occupied, and a good stock of plants ready for transplanting when required; thus a supply will always be at hand for removal into the beds or borders, to fill up vacancies occasioned by the removal of bulbs or autumn-sown annuals. Many of the spring-sown annuals, if properly attended to, will continue in bloom throughout the season, or as long as almost any other plants. Those intended for the principal summer display should be sown from March to May, either where intended to bloom, or in the reserved ground. By adopting the system of raising the plants in the reserve beds, and transplanting, the beds and borders of the flower-garden can be kept filled with plants in full beauty.

HALF-HARDY ANNUALS.—These are less liable to suffer from drought than the hardy varieties; they do not arrive at maturity so quickly; therefore, with these, successive sowings cannot be practised with any advantage. Sow early in May where intended to bloom, or early in April on a hotbed or on nicely-prepared soil in a frame, or under a hand-glass, and transplant them when sufficiently strong and the weather is favourable.

When sowing in a hotbed is necessary, place a layer of four inches of leaf-mould on the top, then a layer of the same depth of fine sifted soil, consisting of equal parts of loam, leaf-mould, peat, and sand. Sow the seeds in drills, formed about a quarter of an inch deep, with the point of a stick; for very small seeds, one-eighth of an inch will suffice. Carefully sow, label, and cover the seeds as you proceed. If the soil is in a proper medium state in reference to moisture, and it is shaded from the midday sun, no water will be necessary until the seeds appear. The frame must not be allowed to rise above 45° or 50°, and should never sink below 40°. The atmosphere should be changed daily by the admission of air, and the surface of the soil looked over frequently, to see if any mould or fungus is making its appearance on the surface; for if so, it must be at once removed, and the spots where it appeared dusted over with quick-lime. With proper treatment, most of the plants will appear in from a week to a fortnight of the time of sowing.

Transplanting and after-management of Annuals.—The plants should be carefully taken up, so as not to injure the roots, retaining as much soil about them as possible, and the roots should not be exposed to the air longer than necessary. If the ground is dry at the time of transplanting, the bed in which the plants are growing should have a good soaking of water at least twenty-four hours before taking them up; they must also be liberally watered as soon as planted, and in dry, hot weather this must be repeated as often as necessary to keep the ground moist; the plants, indeed, should be moistened overhead every evening until they get established. Those transplanted early in spring will seldom require water; but if it is necessary, it should be given in the morning. In transplanting for summer blooming, the plants must be allowed sufficient space to develop their natural size; if closely planted, they will not throw out the side branches, which furnish a succession of bloom. Some plants will require four or six inches, others three or four times that distance from plant to plant. This must be regulated by the habit of the plant, and to some extent by the nature of the soil and locality; but it is better in all cases to have the plants too far apart than too close. In dry weather, watering will be indispensable to keep the plants, whether of hardy or half-hardy annuals, in vigour, and secure a continuous bloom. When necessary to water in summer, the soil should be thoroughly saturated, and as soon as the surface is dry, it should be slightly stirred between the plants. Occasional waterings of *weak* guano-water will heighten the colour of the flowers and increase the growth.

TENDER ANNUALS.—These are plants which must be kept the whole period of their growth under glass; they may be sown in pots of light, rich earth, and plunged in a hotbed early in February or March. To bring the plants to perfection, they must be pricked out when they are an inch up, then as soon as possible be transplanted simply into the smallest-sized pots, and continually shifted as their roots increase. During the whole process of shifting they should be kept in a temperature of from 70° to 80°, and so abundantly supplied with water that the air should be constantly charged with moisture. When, however, they begin to flower, they may be removed to a dry and airy situation, and the temperature gradually lowered. This will greatly improve both the size and colour of the flowers.

84.—PERENNIALS.

Those plants are called perennials which do not require annual cultivation. They are not woody, and generally die down to the ground every year, and spring up again the year following. Some, however, do not die down, but retain their leaves, as pinks, carnations, saxifrages; and these, on this account, are called evergreen perennials.

85.—HERBACEOUS PLANTS.

The modern bedding system has well-nigh banished herbaceous plants from our gardens: nevertheless many of them are very beautiful, and a collection containing Phloxes, Asters, Campanula, Delphinium, Aconitum, Pentstemon, Helleborus, Aquilegia, Cheiranthus, Gentiana, Iberis, Lathyrus, Lupinus, Monardia, Pæonia, Potentilla, Primula, Salvia, Saxifrage, Sedum, Silene, Spiræa, Iris, Statice, Chelone, Lychnis, Alyssum, Acanthus, Fraxinella, Achillea, Orobus, Yucca, Scabiosa, Œnothera, Dianthus, Dielytra, Veronicas, Myosotis, Sachys, &c., &c., arranged according to their height, colour, and time of flowering, and intermixed with bulbus-rooted plants, such as the Crocus, Narcissus, Hyacinths, Cyclamen, Oxalis, Tulips, Ornithogalum, Scilla, Fritillaria, Lilium, Cypripedium, Gladiolus, Allium, Anemone, Ranunculus, Galanthus, Tritonia, Orchis, Colchicum, Pancratum, and Tigridia, make a very fine display.

Most herbaceous plants are easily increased, by dividing the roots, by offsets, by cuttings from the side-shoots or roots, and by seeds. Many of them, from their rapid growth, require taking up and dividing every second year. To insure good flowers, a few strong stems of such plants as Phloxes, Asters, &c., must be secured, in preference to a multiplicity of smaller ones. Consequently, they require frequent and severe subdividing; the early months of the year, up to April, being the best period for performing this operation. Others, again, such as Gentians, Iberis, Alyssums, Achilleas, and similar flowering plants, thrive best without being often disturbed, and must be increased by small-rooted offsets. These should be planted in the reserve garden for the summer, and transferred to their flowering quarters next year. Double rockets, scarlet lychnis, hollyhocks, and other double-flowering plants, are often increased by cuttings. These can generally be obtained either by thinning the young shoots in the spring, or by securing all that appear at the bottom of the flowering-stems in the autumn. Whenever taken, they should be inserted in sandy soil, covered with a hand- or bell-glass, and receive a gentle warmth until rooted. If in the spring, they will of course be transferred to the reserve garden as soon as the rooting and gradual hardening processes are completed. If in the autumn, they will be safer under shelter until the end of April. In either case, with liberal treatment, they may flower the following summer. To induce autumn-struck cuttings to do this, however, they must be potted off in rich soil as soon as rooted, shifted into separate pots, and receive the stimulus of a genial atmosphere and shelter from the weather. All the mints, galiums, and other plants with running or creeping roots, are so easily and obviously increased as to require no instructions.

All single-flowering herbaceous plants may be increased readily by seed. This may either be sown as soon as ripe, or after September in pots. If sown in early spring, on beds of light soil, and the plants carefully transplanted two or three times during the summer, they may be transferred to their blooming quarters in November: if they have been properly treated, they will flower profusely the following season.

86.—DECIDUOUS FLOWERING SHRUBS.

Almost all these, which are very ornamental, are easy of cultivation; but they require an amount of care and observation, in order that they may be made to produce abundant and handsome flowers. Lilacs and also laburnums may, with advantage, occasionally be thinned out, but they ought never to have their shoots shortened; while Guelder roses, syringas, ribes, spireas, and others, will be benefited by almost any degree of pruning, and can only be made to bear fine handsome flowers by having their spring shoots well cut in about the end of May and beginning of June. Appended is a list of—

Choice Deciduous Shrubs, &c.

Althæa frutex,—bears various-coloured flowers, like single hollyhock; blooms early.
Almond,—dwarf and tall; bears pink flowers; very early.
Azaleas,—These may be had of many colours,—scarlet, red, orange, yellow, pink, bronze-colour, and shaded; bloom early.
Buddlea globosa, — orange - coloured flowers.
Berberry,—flowers in bunches, yellow, and succeeded by scarlet berries.
Calycanthus floridus,—allspice-scented wood, with brown flowers.
Chimonanthus fragrans,—dull-coloured; flowers in the winter, highly scented, and before the leaves appear.
Cherry,—double-flowering.
Clematis (various). The common is very hardy and fragrant.
„ Siebaldii, — white, with blue centre.
„ Azurea grandiflora,—rich purple.
Cydonia (*Pyrus*) japonica, — scarlet flowers, blooming several months in the year.
Cytisus (Laburnum), — yellow, purple, and other varieties; flowering early in the spring.
Daphne Mazereum,—red and white.
Deutzia scabra,—flowers white.
Guelder rose,—bears white balls of bloom.
Thorne's Cratægus, — red, double and single; pink do.; white do.; and many varieties of fruit and foliage.
Jasminum nudiflorum,—yellow.
„ officinale,—white.
Lilac,—purple, pale, white, Persian, and other varieties.
Lonicera,—an immense number of species and varieties, and nearly all good, ranging from 3 to 15 ft.
Magnolia conspicua.
„ perpurea; and many other excellent species and varieties.
Peach,—double-flowering.
Philadelphus Syringa, — highly-scented white flower, leaves small, like the cucumber.
Ribes sanguineum,—crimson flowers.
" album,—white do.
„ aureum,—yellow do.
„ speciosum, — scarlet; a double variety, and many others.
Robinia (Rose-acacia), — pink flowers; several varieties.
Roses. — There are upwards of 1,500 varieties and species; but a list of these will be found elsewhere.
Spiræa,—many varieties, and all pretty.
Virginian creeper,—will grow to the top of a house, and the foliage turns crimson in autumn.

87.—BEDDING PLANTS.

Plants commonly known under the name of bedding plants consist of the common scarlet and other geraniums, verbenas, petunias, calceolarias, lobelias, and others; indeed all sorts of trailers and creepers used for covering beds in summer and autumn are so named. Most of these latter require pegging down; and in wet seasons, when the plants are apt to run too much to leaf, the lower extremities of the shoots may be slightly bruised, which will check their growth and promote flowering. Beds for all the ordinary bedding plants should be well drained, and the soil light and rich.

To grow bedding plants in perfection, the beds should have a dressing of manure annually, or a heavier application every second year. It would be almost as reasonable to attempt to grow two crops of cabbages in succession, without enriching the soil, as two crops of bedding plants. Many of them exhaust the soil more than any crop whatever; and to grow them rapidly, and in perfection, the beds must be liberally manured.

88.—AMERICAN PLANTS.

Under this general name are included rhododendrons, azaleas, kalmias, ledums, andromedas, and others, which are supposed to require what is called bog-earth. This, however, is not absolutely necessary to their successful cultivation.

The following materials, all of them within the reach of most persons, may be made to form a compost adapted to their culture. Rotted leaves, spent tan, sawdust, old thatch or straw, weeds, grass-mowings, and vegetable refuse of all kinds—old manure, even the bottoms of old wood-stacks. Any or all of these in a decomposed state, blended with a certain proportion of garden soil, may be rendered fit to grow American plants; but as it is the character of all decomposed vegetable matter, fit to enter rapidly into the composition of the vegetable fabric, to subside rapidly, this must be guarded against by employing also such organic matter as tree-leaves, lumps of peat, peaty turf, or other vegetable matter, which will take long before decay takes place. Where old tan or sawdust is liberally used, the leaves should be fresh; those which have been used as linings for hotbeds by preference, from their tendency to mass together. Old thatch, or litter, forms an excellent basis for the whole clump, and weeds and other vegetable refuse, when burnt or charred. This compost, with a subsoil sufficiently retentive of moisture, and situation not too much exposed to the direct influence of the sun, will grow these beautiful shrubs in great perfection.

Mr. Errington, a good authority in such matters, says:—"In consequence of some alterations at Oulton Park, an old rubbish-yard had to be turned into the ornamental portion of the grounds, and an elevated mound of American shrubs planted on the site. The first proceeding was to burn, or rather char, the decayed vegetable matter in the yard, and throw the ground into shape; and a coating of coarse clayey matter happening to be the nearest commodity, was spread over the surface a foot thick. Next we applied several loads of leaf-mould, and, finally, 3 inches of fine heath soil from Delamere Forest. All these materials were thoroughly blended together to a considerable depth, the raw and undecomposed rubbish, as it came up again, being buried in the bottom of the trench. On the mound thus raised the shrubs were planted, and nothing can exceed their robustness. The ericas, especially, exceed all I have ever seen, from the admixture of the clay with heath and leaf-mould."

These materials, with their proportions, stand nearly as follows:—Sand one part, clayey subsoil matter two parts, decomposed vegetable matter two parts, undecomposed vegetable matter five parts, and at the bottom of all a quantity of rubbish and raw vegetable matter, which the roots of the more vigorous plants will reach, and find food in abundance in due time.

89.—AQUATIC PLANTS, for ornamenting Ponds, the Sides of Running Streams, and other Artificial Pieces of Water.

Vegetation of some sort, and generally of an interesting character, is sure to be found in the neighbourhood of water. There are, however, many plants which are to be preferred to others when we undertake to cultivate a piece of water on the margin of a stream. Flowers for water, as Mr. Loudon reminds us, are of two kinds: aquatics, plants for the water, and marsh plants, to place round the banks. Aquatic plants are propagated by seed, and frequently by division of the roots. The seeds when sown must be placed under water; in other respects they require the same general treatment as other herbaceous plants. We extract from Mr. Loudon's large work on the practice of gardening the following useful list of plants, with the colour of their different flowers and times of flowering.

Aquatic plants with showy flowers.

MAY. *Red.*—Equisetum fluviatile, Hydrocolite vulgaris. *White.*—Nasturtium officinale, Ranunculus aquatilis. *Yellow.*—Ranunculus aquatilis. *Blue.*—Veronica Beccabunga.

JUNE. *Red.*—Equisetum palustre, Butomus umbellatus. *White.*—Hydrocharis morsusranæ, Phellandrium aquaticum. *Blue.*—Myriophyllum spicatum, M. verticillatum, Pontederia cordata, Veronica anagallis. *Green.*—Potamogeton densum. *Brown.*—Potamogeton lucens, P. pectinatum.

JULY. *Red.*—Hippuris vulgaris, Polygonum amphibium. *White.* — Alisma damasonium, A. natans, A. plantago, Calla palustris, Nymphæa alba, N. odorata, Poa fluitans, Stratiotes aloides. *Yellow.* — Iris Pseud-Acorus, Villarsia nymphæoides, Nuphar advena, N. lutea, Utricularia minor. *Purple.*—Utricularia vulgaris, Trapa natans, Sagittaria sagittifolia. *Green.* — Ceratophyllum demersum, Cicuta virosa.

AUGUST. *Red.*—Hydropeltis purpurea, Polygonum hydropiper. *White.*—Cerastium aquaticum, Poa aquatica. *Yellow.*—Potamogeton natans. *Blue.*—Alisma ranunculoides, Lobelia Dortmanna. *Brown.*—Potamogeton perfoliatum, Scirpus fluitans, S. lacustris, S. triqueter.

Marsh plants with showy flowers, for the borders of ponds, the margins of streams, and ornamental pieces of water. Time for flowering, May and June.

HEIGHT, FROM 0 TO ¾ OF A FOOT. *White.* — Pinquicula lusitanica. *Blue.*—Pinquicula vulgaris. *Brown.* — Carex dioica.

FROM ¾ TO 1½ FOOT. *White.*—Œnanthe peucedanifolia. *Yellow.* — Carex flava, Ranunculus flamula, R. repens. *Green.*—Carex disticha, C. pulicaris, C. precox, C. stricta, C. muricata, C. elongata. *Brown.* — Carex cæspitosa, C. digitata, Schœnus nigricans.

HEIGHT FROM 1½ FOOT TO 2½ FEET. *White.*—Œnanthe fistulosa. *Purple.*—Comarum palustre. *Brown.* — Carex paludosa, C. riparia. Juncus conglomeratus.

HEIGHT FROM 2½ FEET TO 3½ FEET. *Green.*—Carex pseudo-cyperus, C. vulpina, Cyperus longus, Juncus compressus.

HEIGHT FROM 3½ FEET UPWARDS. *Red.*—Scrophularia aquatica. *White.*—Œnanthe crocata. *Yellow.*—Senecio paludosus.

Marsh plants, &c., which flower in July and August.

HEIGHT FROM 0 TO ¾ OF A FOOT. *Red.*—Teucrium scordium. *White.*—Littorella lacustris, Samolus Valerandi, Schœnus albus. *Yellow.* — Hyperocum elodes. *Blue.* — Schœnus marisous. *Variegated.*—Scutellaria minor. *Brown.*—Schœnus compressus, Scirpus acicularis, S. cæspitosus.

HEIGHT FROM ¾ TO 1½ FOOT. *Red.*—Menyanthes trifoliata. *White.*—Galium palustre, G. uliginosum, Pedicularis palustris. *Yellow.*—Hottonca palustris, Rumex maritimus. *Purple.* — Pedicularis sylvatica, Triglochin maritimum, T. palustre. *Brown.* — Juncus squarrosus, Schœnus nigricans, Scirpus palustris, S. sylvaticus.

HEIGHT FROM 1½ FOOT TO 2½ FEET. *Yellow.*—Acorus calamus, Myosotis palustris, Rumex palustris. *Blue.*—Phormium tenax. *Purple.*—Aster tripolium. *Brown.*—Scirpus maritimus, Rumex crispus, Juncus sylvaticus.

HEIGHT FROM 2½ FEET TO 3½ FEET. *White.*—Rumex obtusifolius. *Yellow.*—Cineraria palustris, Senecio aquaticus. *Green.*—Juncus effusus. *Brown.*—Scirpus holoschœnus.

HEIGHT FROM 3½ FEET UPWARDS. *Red.*—Malva sylvestris. *White.* — Dipsacus pilosus, Selinum palustre. *Yellow.*—Sonchus palustris. *Variegated.* — Angelica sylvestris. *Brown.* — Rumex aquaticus, Cyperus longus, Juncus acutus, J. maritimus.

To those who desire to ornament any piece of artificial water, the above lists are invaluable. The situation best adapted for hardy aquatics will be found to be in accordance with their height. Many that are not hardy may be introduced for summer decoration in pots sunk either wholly or half deep in water. These can be removed in winter to warm-water tanks under cover of glass. Every year something new in aquatics is being introduced, for the water-plants of the tropics are inexhaustible, and very many of them supremely beautiful. Most of the plants which have been mentioned in the above lists prefer a shady and sheltered situation, and will be found to flourish best when protected by overhanging trees. Rock-work and root-work form admirable receptacles for plants on the margins of streams, and afford all the protection they require, when properly arranged. We shall have more to say on this subject when we give lists of those plants which are suitable for rock- and root-work.

90.—FLOWERING GRASSES.

That splendid importation from the river Plate, *Gynerium argenteum*, or Pampas-grass, resists the cold of our ordinary seasons, but was in many instances killed during the severity of the winter of 1860. Old plants seem hardier than young ones, arising probably from the larger top affording more efficient protection to the roots. The old leaves should not be removed until the end of April, as they afford the best possible protection as far as it goes; experience, however, shows that of themselves they will not preserve life: hence the necessity of a little extra litter. This is decidedly the king of all the grasses, and deserves a place in every garden. As the centre group of a *grassery*, or placed in a rich shady dell, contiguous to rocks or water, it finds a congenial home, and imparts a charming effect. A rich alluvial soil, at least a yard deep, abundance of space to unfold its large, graceful leaves, and throw up its flower-stems, and an unlimited supply of water, are all the conditions its successful culture demands. Packets of seed can be bought at one shilling each, and plants that will flower the second year, from nine to twelve shillings a dozen. With liberal treatment, seedlings will flower the third or fourth year. By sowing thinly in February or March in pots, and planting out in prepared beds in May, a season may almost be gained in the growth of the plants. Like all the grasses, the seed should be barely covered with soil, and the surface kept moist, until germination is insured. There seem, however, to be several varieties of this grass, in addition to its sexual distinctions. When practicable, it is therefore best to purchase divided plants from those who have grown the finest flower-stems: it can be rapidly increased by division. Plants divided are more tender than others, and will require more protection, until they are thoroughly established. Few plants, however, are more worthy of attention, as they have a fine appearance when growing; and if the flower-stem is cut before it begins to fade, it looks almost as noble when dry. In addition to the localities here pointed out for them, they also contrast admirably with large masses of yews or other dark-foliaged shrubs. A companion grass to this, with broad-striped foliage and large feathery flowers, is the *Eranthis Ravenna*. The Tussack-grass, and some of the common reeds and rushes, also form beautiful features in connection with these. There are also eight or ten varieties of the *Holcus saccharatus*, or sugar-cane, whose elegant leaves, stately stems, and various-coloured heads of corn, are highly ornamental. These are half-hardy annuals, and should be sown in a gentle heat, and pushed rapidly forward to secure strong plants for planting out in May. The variegated, white-striped, and beautifully marbled *Zea*, or maize, requires similar treatment, and has a charming effect. The pretty millet-grass (*Milium effusum*), charming love-grasses (*Eragrostis elegans*, *E. Memaquensis*, and *E. Pursthi*), and the several varieties of *Briza*, or quaking grass, should be sown either in pots or on a rather sheltered bed out of doors. Two more beautiful annual grasses are the *Brizopyrum siculum*, whose branches rival in beauty the deciduous cypress, and *Bromus brizoporoides*, so useful for bouquets. The two feather-grasses, *Stipapinnata* and *S. gigantea*, hardy perennials, and the hardy biennial *Hordeum jubatum*, are also most useful for mixing with other flowers, and very elegant in themselves. The smallest feather-grass almost rivals the *Festuca glauca* for edgings. The handsome silver foliage of the *Festuca* contrasts beautifully with red gravel walks, and is said to harbour fewer vermin than box or any other living edging whatever.

91.—CLIAN'THUS (nat. ord. Legumino'sæ).

A genus of magnificent free-flowering shrubs, with elegant foliage and brilliantly coloured and singularly-shaped flowers, which are produced in splendid clusters. *C. magnificus* and *C. puniceus* blossom freely out of doors in summer, against a trellis or south wall; *C. Dampierii* succeeds best planted in the border of a greenhouse, and is one of the finest plants of recent introduction; seeds sown early in spring flower the first year; succeed best in sandy peat and loam. They may also be raised easily by cuttings.

Clianthus Dampierii,—brilliant scarlet, with intense black spot in the centre of the flower; exceedingly handsome and showy, from New Holland.

Clianthus magnificus,—scarlet; beautiful, 4 ft., from New Zealand.

Clianthus puniceus,—scarlet, 4 ft.

92.—ESCALLO'NIA (nat. ord. Escallonia'ceæ). Half-hardy.

Handsome evergreen shrubs, with rich glaucous leaves and bunches of pretty tubular flowers; they succeed against a south wall, if protected in the winter, and thrive best in sandy peat and loam.

Escallonia floribunda,—white, 3 ft., from Granada.

Escallonia macrantha, — purple-scarlet, 3 ft., from Granada.

93.—DAPH'NE (nat. ord. Thymela'ceæ).

Beautiful shrubs, remarkable for the elegance of their flowers and for their bright red poisonous berries. *D. Mezereum* is the most common variety. The dwarf daphnes are somewhat tender: they bear pink flowers very fragrant. There is a Chinese daphne, *D. odorata*, which is a great ornament in the greenhouse, but too tender for the open air.

94.—CANDYTUFT (nat. ord. Crucif'eræ).

The Candytuft, or Iberis, as it is also called, derives its name from the island of Candia, whence it first came. It springs up readily from seed sown in any light rich soil. Autumn is the best time for sowing.

The improved varieties of this favourite flower are exceedingly beautiful, and it may be questioned if there be two more effective annuals than Dunnettii and the sweet-scented; the one a rich crimson purple, the other pure white: they succeed in any soil.

Candytuft, Dunnettii, new dark,—crimson, exceedingly beautiful hardy annual, 1 ft.

Candytuft, Normandy,—lilac, 1 ft., from Normandy.

95.—EP'ACRIS (nat. ord. Epacrida ceæ).

These are heath-like shrubs, natives of New Holland. They all require a fine gritty peat soil, and flourish best in double pots, with moist moss between them. The pots should be well drained; but the roots of the plants must never suffer for want of moisture. They must have plenty of air and light, and just sufficient heat to keep them from frost. Cuttings of the young wood strike easily in sand with a little bottom heat.

These plants should be freely cut back as soon as they have done flowering; and after the shoots have grown afresh, 2 or 3 inches long, is the best time for potting them. Place them in a close pit, but by no means warm, for a few weeks; gradually inure them to the air, plunge in a sunny situation: see that the wood is brown and hard by the end of September. Remove to conservatory-shelf in October, and you will have a charming profusion and succession of tiny tubes of colour.

As no one ever saw, or ever will see, an ugly epacris, it seems almost a loss of time to specify varieties. However, all are not equally beautiful; therefore select the following, which can be bought from 9*s*. to 18*s*. per dozen:—

Epacris alba odorata, Albertus campanulata, Impressa, Alba carnea, Atleana delicata, Impressa alba, Impressa floribunda, Alba coccinea, Alba floribunda, Kinghornii magnifica miniata, Kinghornii splendens, Kinghornii compacta, Kinghornii major, Viscountess Hill. The following generally flower in the autumn: —Racemosa, Picturata, Mont Blanc, Exquisite, and Fireball.

96.—DELPHIN'IUM (nat. ord. Ranuncula'ceæ).

A genus of profuse-flowering plants of a highly decorative character. Planted in large beds or groups, their gorgeous spikes of flowers, of almost endless shades, from pearl-white to the very richest and deepest blue, render them conspicuous objects in the flower-garden and pleasure-ground: they delight in deep, highly enriched soil. With the exception of *D. cardiopetalum* they are all hardy perennials.

97.—BEGO'NIA (nat. ord. Begonia'ceæ).

There are no plants more worthy of admission into a conservatory than begonias, and the facility with which they may be cultivated is equal to their beauty. All they require is a good rich loamy soil, mixed with a little sand, and a little heat to start them. Either hotbed or stove answers every purpose, provided there is a conservatory or greenhouse in which they can be flowered; the chief requirements being heat, moisture, and shade.

There is a delicious fragrance about some of the species, which particularly recommends them for cultivation; others are recommended by their richly-variegated foliage and graceful habit, and they all hybridize with great facility.

98.—CINERA'RIAS (nat. ord. Compos'itæ).

Unless for exhibition, it is best to grow them annually from seed. The first sowing should be made in March, in pans filled with equal parts of peat and loam and one-sixth part sand. They should be well drained, made firm, and the seeds lightly covered, and placed on a slight bottom-heat. Keep the pans and young plants, when they appear, partially shaded from the bright sun; put them into 3-inch pots as soon as they will bear handling, return them to the same place, and renew the same treatment until they are thoroughly established in their pots. Then gradually harden them by giving plenty of air, and place them in a sheltered situation out of doors towards the end of May. As the roots reach the sides of the pots, shift them into larger, giving them their final shift in September. The first flower-stems should be cut out close to the bottom when large plants are desired. This will induce them to throw out from six to twelve side shoots;—these may be reduced, or all left, at the option of the grower. Towards the end of September, they should be returned to a coal-pit, and they will begin to flower in October. No soil is better for growing them than equal parts rich loam, leaf-mould, and thoroughly rotted sheep- or horse-dung, liberally mixed with sharp sand or charcoal-dust, and used in a roughish state. They also luxuriate under the stimulating regimen of *rich* manure-water. Another sowing may be made in April, and a third in May, for very late plants. The treatment of old plants may be similar to this. Cut them down as soon as they have done flowering; shake them out, and pot each sucker separately in March; then proceed as above in every respect.

List of the best Varieties.

Acme, Adam Bede, Beauty, Brilliant, Duke of Cambridge, Bridesmaid, the Colleen Bawn, Handel, Incomparable, Mademoiselle Parepa, Magenta, Maid of Honour, Miss Marnock, Mrs. Livingstone, Perfection, Queen Victoria, Queen of Lilies, Regalia, Royal Marine, Solferino, Slough's Rival, the Wizard, and Wonderful.

99.—CACTUS (nat. ord. Cacta'ceæ). Greenhouse Perennials.

An extremely curious and interesting genus, many of the varieties producing magnificent flowers of the most brilliant and striking colours. They succeed best in sandy loam, mixed with brick and lime rubbish and a little peat or rotten dung.

100.—HYDRANGEA (nat. ord. Saxifa'ceæ).

The common Hydrangea is a Chinese shrub, half-hardy, imported into England about the year 1790, by Sir Joseph Banks. It thrives best in a rich soil and requires plenty of water. When the plant has done flowering, its branches should be cut in. Blue hydrangeas are much admired. It is some peculiarity in soil and situation which produces this variety. Blue flowers may in general be procured by planting in a strong loam and watering freely with soapsuds, or, what is better with a solution of alum or nitre.

101.—ERICAS (nat. ord. Erica'ceæ).

This important genus of greenhouse plants, which includes five or six hundred described species, and as many varieties produced by cultivation, is the great ornament of the greenhouse at a time when other flowering plants are scarce.

The soil best adapted for heaths is that obtained from a locality where the wild heath grows luxuriantly, taking care it is not dug too deep: the turf must not exceed four inches; if deeper, it is more than probable that the good and nutritious upper soil will become deteriorated by an admixture of inert and mischievous subsoil. The summer is the proper season to procure and store up a heap which may safely be used after having a summer and winter's seasoning.

The next matter of importance is the selection of healthy dwarf-growing, robust plants, taking care to avoid anything like meagre, stunted plants, which might live for years, but cause nothing but disappointment to the cultivator.

To prepare the soil for potting or shifting, it should be cut down from the heap so as to disarrange it as little as possible, breaking the lumps well with the back of the spade, and afterwards rubbing the soil through the hands, which is far better than sifting, as it leaves more of the fibrous decomposing vegetable matter in it; add to this one-fifth good white sand, and well incorporate the two.

To convert plants into handsome well-grown specimens in a moderately short space of time, they must have a liberal shift. Young plants in a 60- or 64-sized pot may be shifted into a 24- or 8-inch pot, taking care that plenty of potsherds be used for drainage. Care must be taken, also, that the soil is thoroughly mixed, by pressing with the fingers in the fresh pots all round the ball of the plants, so as to make it quite firm and close. After being set away in a cool frame or pit, let them be well watered. It is impossible to give a list of named varieties, they are so numerous.

102.—BER'BERIS (nat. ord. Berberida'ceæ). Hardy Shrubs.

Highly ornamental free-flowering shrubs; covered in spring with a profusion of rich yellow flowers, and in autumn with fruit of a very attractive character. The common Berberis is a most elegant plant when trained to a single stem and then allowed to expand its head freely on all sides. So treated, the branches become drooping, and have a fine effect every spring, when they are covered with their rich yellow blossoms, and in autumn from their long red bunches of fruit, which rival the fuchsias.

103.—ACHIME'NES (nat. ord. Gesnera'ceæ).

A genus of truly splendid plants, suitable either for the sitting-room, greenhouse, or stove, and especially adapted for hanging-baskets.

Culture.—Use a compost of peat, loam, and leaf-soil; or leaf-mould, loam, and silver-sand, and secure good drainage. Plant five to seven tubers in a five- or six-inch pot, with their growing ends inclining towards the centre, and their root ends towards the circumference of the pot, and cover them with about an inch of the compost. While growing, they should be well supplied with liquid manure; start them when convenient in heat, and when an inch and a half high they may be removed to the greenhouse.

104.—CEANO'THUS (nat. ord. Rhamna'ceæ). Half-hardy Shrubs.

An extremely handsome, free-flowering genus of highly ornamental shrubs, suitable either for conservatory decoration, or for covering fronts of villas, walls, or trellis-work in warm situations: they succeed best in peat and loam.

Ceanothus Americanus,—white, 4 ft., from New Jersey.
Ceanothus azureus grandiflorus,—sky-blue, flowers in bunches, 4 ft., from Mexico.
Ceanothus Baumannii, 3 ft.
,, Californicus, — blue, tinged with lilac, 4 ft., from California.
Ceanothus cæruleus microphyllus,—blue, small leaves, very pretty, 5 ft., from North America.
Ceanothus Delilianus, — white, shaded pale blue, beautiful, 5 ft., from North America.
Ceanothus floribundus,—dark blue, remarkably pretty, 3 ft.
Ceanothus Fontanesianus,—blue, shaded red, 4 ft., from North America.
Ceanothus grandiflorus roseus,—reddish blue, handsome, 4 ft., from North America.
Ceanothus Hartwegii,—blue, 6 ft., from Mexico.
Ceanothus ovatus,—white, elegant, 6 ft., from North America.

105.—CY'CLAMEN (nat. ord. Primula'ceæ).

A genus of charming winter and spring blooming bulbous roots, with very pretty foliage, and flowers the most beautiful, graceful, and ladylike, so easily cultivated withal, that any one may enjoy these floral bijous either in the sitting-room window, conservatory, or greenhouse, from October to May, by a little management in the period of starting them into growth.

Culture.—Plant one bulb in a 5- or 6-inch pot, using a rich soil composed of loam and leaf-mould, rotted dung, and a little silver sand, and, to secure good drainage, place at the bottom of the pot an oyster-shell or hollow potsherd, and over that some pieces of charcoal: the bulb should not be covered more than half its depth.

When the blooming season is over and the bulbs are at rest, plunge the pots in a shady well-drained border, and there let them remain till the leaves begin to grow, when they should be taken up, turned out of the pots, and as much

Pl. III.

OUR COLOURED PLATE.—No. 3.

19. *Gladiolus.*—These are noble flowers, especially such as belong to the Gandavensis family; with a little management in successional planting a fine display may be kept up from August to November. The bulbs, however, should not be suffered to remain in the ground later than the end of that month, and the earliest planting should be made in February.

20. *The Hollyhock* is not so much prized as it used to be—perhaps cultivators think that they have done enough for it, for they have certainly produced some magnificent and stately plants. It is a useful flower to come in as a screen between the flower and kitchen gardens.

21. *Iris.*—What an immense variety there is included under the genus Iris! The wild sorts, of which there are several, are very beautiful, and the hardy sorts in cultivation are among our gayest flowers of spring or early summer. They all prefer a loamy soil and delight in moisture.

22. *Lily of the Valley.*—This well-known plant will never cease to be a favourite; the flower in growth and arrangement is so elegant, and the scent so delicious. To have fine blooms, and an abundance of them, the plants must be treated liberally, and not allowed to grow too thick.

23. *Mimulus.*—The Mimulus, or the Monkey Flower, as it is sometimes called, has much improved of late under the florist's care. Many of the seedlings are highly ornamented, and may be grown either in the open ground or in pots for greenhouse decoration.

24. *Narcissus.*—Few spring flowers can equal the Narcissus, either in beauty of growth or in sweetness of scent. The varieties are either double or single; our illustration belongs to the latter class, and is usually known as the Poet's Narcissus.

25. *Enothera.*—The Evening Primrose is showy while it lasts, but far too short-lived. The plants so called now are mostly perennials or biennials; a large number of the old-fashioned annuals known as Enothera, are in the present catalogues called Godetia.

26. *Pæonia.*—The old red Pæony here illustrated is to be found in almost every garden; there are other choicer varieties, but none more showy. The shrubby kinds, generally known as Tree Pæonies, are very effective when grown as specimen plants on lawns.

27. *Petunia.*—This valuable bedding plant has been brought to great perfection under cultivation. Many of us who can remember the first flower so called may well wonder how they could be tolerated. Still in their day they were accounted beautiful. Beauty, however, in all things is a matter of comparison; and who can say whether, after a few years, our present beautiful Petunias may not be surpassed by others which shall be as much more beautiful than they are, as they at present exceed in beauty the Petunias of the past?

soil removed as can be done without injury to the roots, and replaced with the compost already mentioned.

106.—FUCHSIAS (nat. ord. Onagra'riæ). Under Glass.

Whoever has a glasshouse 2 yards square, or a window free from dust, may grow one or more fuchsias. In fact they have become quite window-plants, and no plant is better adapted for the purpose. To grow them, however, in perfection, requires some judgment and skill. Cuttings should be inserted in pots filled either with loam and leaf-mould, or peat and silver-sand, in equal parts, to within an inch and a half of the top. Place over this three-quarters of an inch of silver-sand, and level the surface to make it firm; then insert the cuttings,— about 1 inch long is the proper length, and plunge the pots in a bottom-heat of 60°, either in a pit or propagating-house: if the latter, cover them with a bell-glass. In three weeks they may be potted in 3-inch pots, and replunged in the same bed, keeping them at a temperature of from 50° to 60°. As soon as the roots reach the sides of the pots, the plants should be shifted into fresh pots, until they receive their final shifts into 6-, 9-, or 12-inch pots, towards the end of June. The size of the pot must be regulated by the period when they are wanted to bloom. If in July, a 6- or 9-inch pot will suffice; if in September or October, a 12 will not be too large. During the period of growth, the plants will require stopping at least six times, care being taken never to stop the shoots immediately preceding or directly after the operation of shifting into larger pots. If the pyramidal form of growth, which is the best of all forms for the fuchsia, is adopted, the plants, from the first, must be trained to a single stem, and all the side-shoots stopped, to make the pyramid thick and perfect. If the bush form is wanted, the whole of the shoots should then be stopped at every third joint, until branches enough are secured to form the bush, and then be trained into the desired shape. A regular moist, genial temperature must be maintained during the entire period of growth, never exceeding 60°. During bright sunshine, the glass should be slightly shaded with tiffany or other material: the delicate leaves are easily injured, and the plant should never receive the slightest check by being allowed to flag. Fuchsias will grow in almost any soil. Garden-loam and leaf-mould in equal proportions, with some broken charcoal and sand, do very well. It is better to feed them with manure-water than to mix dung with the soil. After they are well rooted, they should never be watered with clear water.

107.—FUCHSIAS. Hardy.

These make the best show when planted together in beds upon a lawn, the colours being judiciously blended. Those fuchsias which trail upon the ground should be grown with a wire hoop supported by three legs underneath them, so that their branches may be made to bend over the hoop. Several of the more hardy sorts may be trained on one stem, so as to appear as standards in the bed. Many large and beautiful varieties of fuchsias are hardy, and will stand our winters in the open ground, especially in a well-drained light soil having a large portion of peat in it; and a great many that are looked upon as tender varieties may be preserved if covered 3 or 4 inches with dry cinder ashes at the first approach of frost. The best plan is to cover the whole fuchsia-bed at that time with a good coating. The dead branches should not be cut off, nor should the ashes be removed until the fuchsias begin to shoot in the spring.

108.—PÆO'NIA (nat. ord. Ranuncula'ceæ).

For late spring or early summer flowering few plants are more useful than Pæonias. Every flower-garden should have some of them. They are mostly very hardy, and in colour vary from pure white, blush, salmon, and rose, to the most intense and brilliant scarlet. The Chinese tree varieties (*Pæonia Moutan*) are also hardy and early flowering. Bedded upon lawns they have a beautiful effect. In a shrub-like form they rise from 3 to 5 feet in height, and branch out in a good rich soil to 10 or 18 feet in circumference. There are many varieties, and the colour is extremely rich. They are most of them profuse flowerers.

109.—NASTUR'TIUM (nat. ord. Tropæola'ceæ).

The dwarf improved varieties of Nasturtium are amongst the most useful of garden flowers for bedding, massing, or ribboning; and rank with the Geranium, Verbena, and Calceolaria; their close, compact growth, rich-coloured flowers, and the freedom with which they bloom, all combine to place them among first-class bedding plants. The scarlet, yellow, and spotted Tom Thumb are distinguished favourites, as are also the old crimson, and the new Crystal Palace Gem.

Seed, 3*d.* per. packet.

110.—MIM'ULUS (nat. ord. Scrophularia'ceæ).

A genus of extremely handsome profuse-flowering plants, with singularly-shaped and brilliantly-coloured flowers, which are distinguished by their rich and strikingly beautiful markings. Seed sown in spring makes fine bedding plants for summer blooming, and seed sown in autumn produces very effective early-flowering plants for greenhouse decoration, &c.

Mimulus cardinalis,—scarlet, 1 ft., from California.
Mimulus cardinalis atrosanguineus maculatus,—dark red, 1 ft.
Mimulus cardinalis Lehmannii,—bright rose, 1 ft.
Mimulus cardinalis Napoleon III.,—rich purple, 1 ft.

111.—VIN'CA (nat. ord. Apocyna'ceæ).

This is the classical name of the common Periwinkle, and of its class there are many beautiful varieties. *V. major elegantissima*, *V. major reticulata*, *V. minor argentea*, and *V. minor aurea*, are all variegated and very showy. They grow in any soil, and look well on rock-work. Under the name Vinca also are included many choice greenhouse evergreens, as remarkable for their shining green foliage as for their handsome circular flowers. Plants raised from seed which has been sown early in spring will be found useful for the ornamentation of flower-beds and borders in warm situations.

Vinca rosea,—rose, 2 ft., from East Indies.
" alba,—white, with crimson eye, 2 ft., from East Indies.
Vinca alba pura nova,—pure white, 2 ft.
" lutea nova —pure white, with yellow eye, 2 ft.

112.—BUD'DLEA (nat. ord. Scrophularia'ceæ).

Deciduous shrubs, natives of India and South America. They are not quite hardy enough to endure very severe winters with us out of doors; but in greenhouses they flower profusely. A loamy soil, mixed with peat, suits them best.

113.—AGERA'TUM (nat. ord. Compos'itæ).

Useful half-hardy annuals. The shades are blue, white, and red. The seed should be sown in a warm border in a light soil in April or May. Price *3d.* per packet.

114.—MIS'TLETOE.

To many persons the cultivation of the mistletoe is looked upon with as much doubt as we are told the ancient Romans looked upon the cultivation of mushrooms. It may, however, be very readily cultivated by attending to the following directions:—Make an incision in the bark of an apple-tree (many other trees, as the pear, oak, whitethorn, and even laurels, will answer equally well), and into this incision, in the spring of the year, insert some well-ripened berries of the mistletoe, carefully tying the bark over with a piece of bass, mat, or woollen yarn. This experiment often fails, from the birds running away with the berries from the place where they have been inserted, for they are very fond of them. To prevent this, the incision in the bark should be made on the underside of a hanging branch, where birds are not likely to rest.

115.—HEATHS.

Under the term Erica will be found an account of the best varieties of greenhouse heaths. Our remarks now will be confined to the hardy sorts. As a rule, however, all thrive best in a hard, sandy, gritty peat. Bog-peat is hardly fit for their growth, unless it is liberally mixed with sharp sand and the *débris* of freestone rocks. For their culture in the garden, from 18 inches to 2 feet of such soil, resting on a dry bottom, would be desirable. Beds or groups of hardy heaths make a charming display. Such groups harmonize well with the different fir-trees. Nothing can exceed their beauty, congruity, and adaptability, as furnishings for rock-work. Peat-earth could easily be placed among the crevices, and the heaths introduced there. They thrive admirably in such situations, and contrast well with ferns and other rock plants.

The following are the varieties most generally cultivated:—

Erica australis.
" cinerea alba.
" " rosea.
" " rubra.
" herbacea.
" " carnea.
" lanceolata.
" mediterranea.
" " hibernica.
" Mackaiana.
" stricta.
" Tetralix rubra.

Erica Tetralix alba.
" vagrans carnea.
" vulgaris alba.
" " minor.
" " coccinea.
" " decumbens.
" " Hammondii.
" " pumila.
" " pygmæa.
" " variegata.
" " rigida.
" " variabilis.

116.—ORCHIDS (nat. ord. Orchida'ceæ).

The plants of this extensive genus are generally divided into four classes: orchids from the tropics, which require a stove; orchids from the Cape of Good Hope, which are suited to a greenhouse; those from the south of Europe, which need only a slight protection during winter; and our own native varieties. We need hardly add that the more tender sorts are the most curious and beautiful. Full directions as to treatment will be found in the calendar of operations for each month.

Those who take an interest in the cultivation of orchids, and do not grudge expense, will do well to add to their collection some of the latest varieties, which have been introduced by Mr. B. Williams, of Holloway.

117.—PASSIFLORA—Passion-flower (nat. ord. Passiflora'ceæ).

A genus of magnificent ornamental twining shrubs, with flowers at once interesting, beautiful, and curious, which are produced in the greatest profusion and in succession during the greater part of the year under glass, and out of doors during summer and autumn. They are among the most important and effective of plants for training in conservatories or covering the fronts of cottages and villas in town or country. For indoor culture, P. Countess Guiglini is one of the most important, and for outdoor decoration *P. cærulea.*

118.—SNOWDROP.

This is one of the most elegant and interesting of spring flowers: it may be had in bloom in-doors at a very early period; its swan-white blossoms contrasting beautifully with the rich hue of the crocus, &c. A row of snowdrops is often very effective in juxtaposition with a row of blue crocuses. October is the best month for procuring and planting them, although they may be inserted much later. The following are among the best varieties of this charming bulb.

Aletta, Wilhelmine, Argent, Bride of Albion, Blucher, Bride of Lammermoor, Charles Dickens, Caroline Chisholm, Cloth of Silver, David Rizzio, General Todtleben, La Plus Belle, Lord Raglan, Mrs. B. Stowe, Prince of Wales, Passelonto, Sir J. Franklin (very large, but loose), Sir W. Scott, and Victoria Regina.

119.—PRI'MULA (nat. ord. Primula'ceæ).

This genus is a very large one, including, as it does, some of the most popular florists' flowers, viz., the Auricula, the Polyanthus, and the Primrose. We cannot here do more than call the reader's attention to the greenhouse varieties of primula so useful for winter decorations. Mr. Williams, to whom the public are so much indebted for the care he has bestowed on these beautiful plants, says, "Taking them all in all, these are the most valuable winter-flowering plants in existence." His directions for sowing and culture are as follows:—"Sow in March, April, May, June, and July (with great care, for although so easily raised in the hands of some, it is nevertheless a great difficulty to others, who, in many instances, too hastily condemn the quality of the seed), in pots filled to within half an inch of the top with sifted leaf-mould, or what is better, with thoroughly-rotted manure, which has been exposed to all weathers for a year or two. Leave the surface rather rough, and sprinkle the seed thinly upon it, not covering with soil. Tie a piece of paper over the top of the pot, and leave it in a warm house or hotbed. When the seed becomes dry, water the paper only: the seed will then germinate in two or three weeks. After this remove the paper and place in a shady place, potting off when sufficiently strong into small pots: place the pots near the glass in a frame or greenhouse . . . . One caution is necessary—never use peat mould or any soil liable to cake on the surface or turn green, as the loss of the seed is a certain consequence."

120.—PETU'NIA (nat. ord. Solana'ceæ).

Highly ornamental and profuse-flowering half-hardy perennials, easily cultivated both from seed and cuttings. The brilliancy and variety of the colours, combined with the duration of the blooming period and the capability of the flowers to bear the atmosphere of large towns, render petunias invaluable. Seed sown in March or April makes fine bedding plants for a summer and autumn display: they succeed in any rich soil.

In growing petunias, it is best to adhere to a few distinct, well-defined varieties, such as the Bride Rose, with white eyes; the Countess of Ellesmere, and Lady Emily Peel, for violet rose. The following are also good:—

Brilliant,—fine velvety rich crimson.
Eclipse,—distinct and novel.
Rosy Circle,—good rose-coloured bedding variety; better than the Queen, which is also good, and of similar character.
Prince Albert,—crimson.
Prince of Wales, — crimson, often blotched, and edged with white.
Crimson King,—a good bedder.
Fascination,—white circle; bright rose; good habit for bedding.

121.—RHODODENDRONS.

Since the rise of the great Bagshot and Knaphill nurseries, rhododendrons have not only increased and multiplied, but improved so rapidly by crossing, that they are now, without exception, the most magnificent of all our hardy shrubs. They are also so cheap as to be brought within the reach of all, and yet many fine specimens are so valuable as to continue the luxuries of the rich. They vary in price from *15s.* a hundred to fifteen guineas a plant. Nothing equals the common Ponticum for underwood in plantations, or furnishing cover for game. There are about eighteen or twenty varieties of this class alone, including almost every shade of colour. The splendid Catawbiense variety has been almost equally fruitful in hybrids, and presents its formidable list of albums, roseums, purpureums, splendens, &c.

Amidst hosts of other hardy hybrids, from the scarlet arboreum, perhaps the following dozen are as good as any:—Blandyanum, Aclandianum, Atrosanguineum, John Waterer, Nobleanum, *Maculatum grandiflorum*, *Roseum picturatum*, Mrs. John Waterer, Victoria, Towardiana, Duchess of Sutherland, and Parryllianum. Doubtless there are a hundred more almost as good. Unless you have considerable experience, it is best to leave the selection to the nurseryman, stating the price per dozen or hundred. Fine plants of the above will range from *5s.* to two guineas each. All rhododendrons require bog. They bear frequent removal, but care must always be taken not to break the ball of earth or loosen the soil from the stem.

122.—VIOLETS.

The common violet, *V. odorata*, is a native of our own island. It is found wild, both purple and white. "White violets," says Mrs. Loudon, "are generally found in calcareous soils, and the sweetest I ever smelt were, I think, those I have gathered growing among the limestone rocks in the woods of Dudley castle."

Violets may also be grown in pots, by placing two or three runners or offsets in a pot in May, and keeping them in the frame slightly shaded from the hot sun in summer. Loam and leaf-mould suit them admirably. Russian violets, and sometimes the Neapolitan, will flower all the winter. True violets flower in March and April.

123.—AZA'LEA (nat. ord. Rhodora'ceæ).

Beautiful flowering plants, natives of North America, Turkey, and China. The azaleas common in our gardens are deciduous shrubs, varying in height from 2 feet to 6 feet. The loftiest of them is the *arborescens*, which will grow from 10 feet to 15 feet in height. With azaleas, as with rhododendrons, the best garden varieties are hybrids. The following are among the choicest named sorts:—

Ambrosia,—red, orange blotch.
Aurantia major,—orange-scarlet.
Beauty of Flanders,—fawn and pink, orange blotch.
Buckii,—fine scarlet.
Coccinea major,—fine scarlet.
Emperor of Russia,—pink, fawn blotch.
Honneur de la Belgique,—bright orange.
Julius Cæsar,—dark crimson.
Louis Bonaparte,—fawn and yellow
Minerva,—salmon.
Nudiflora colorata,—bright pink.
Pontica,—large, yellow, very showy.
Prince of Orange,—fine orange.
Rosea rotundifolia,—rosy buff, yellow blotch.

For conservatory decoration the Chinese and Indian azaleas are most important. These plants require a similar growing season, after flowering, to the camellia, and until the shoots are sufficiently numerous, or the plants as large as desired, they can be grown on throughout the entire year, and stopped four or five times during that period. This *pushing* treatment will, however, sacrifice the blossom; but if the plants are started early, they can be stopped twice, and yet the terminal buds be sufficiently matured in the autumn to develop flower-buds. After the growth is made, the plants should be gradually hardened off, and be placed during September full in the sun's rays out of doors, to thoroughly ripen their wood. Two parts of peat, two of loam, a sprinkling of sand, and one-sixth part of charcoal that has been steeped in manure-water suits them well. While growing, they will also bear watering with clear weak manure-water every time that they become dry. Before housing them for the winter, examine the plants, and dip them into a tubful of equal parts soot-water, made by throwing half a bushel of soot in soap-suds, and tobacco-water. Repeat this dose three times, and every thrip will either take itself off or die. Indian azaleas bear forcing well, and by inducing an early habit by the aid of the forcing-pit, the luxury of their beauty may be enjoyed in conservatory or sitting-room for six or eight months of the year.

There is an immense variety. The following azaleas, nice plants of which can be bought from 18*s.* to 42*s.* a dozen, according to size, are perhaps as good as any:—

Admiration, Alba, Alba magna, Alba striata, Albertii, Amœna lateritii, Ardens, Beauty of Europe, Beauty of Reigate, Blondin, Broughtonii, Chelsonii, Coronata, Criterion, Danielsiana, Distinction, Duc de Nassau, Duke of Devonshire, Empress Eugénie, Exquisita, Exquisita pallida, Fulgens, Gem, Gladstonesii, Gladstonesii excelsa, Gladstonesii formosa, Glory of Sunning Hill, Grande Duchesse Hélène, Formosa variegata, ditto alba cincta, ditto alba suprema, ditto elegans, Magnet, Mars, Optima, Petuniæflora, Prince Albert, Queen of Whites, Queen Victoria, Leopold, Rosea punctata, Rosy Circle, Smithii coccinea, Striata formosa, the Bride, and Triumphans.

124.—RUSSIAN VIOLETS.

To have an abundance of fine flowers in the autumn and early spring, these should be planted in beds under a wall, in a warm aspect. The soil should be light, but very highly manured, with a large quantity of sand about four inches underneath the top soil. The roots should be planted in rows about three or four inches apart, and well watered. Every year, about Valentine's day, immediately after they have done flowering, the beds should be broken up, the soil renewed, and fresh plants set for another year. Even if the old beds have not done flowering, fresh beds should not be delayed beyond the last week in February.

125.—NEAPOLITAN VIOLETS.

These are the best for forcing. Place a frame over a small, well-prepared hotbed, and cover the bed a foot thick with prepared soil, consisting of the remains of an old cucumber-bed with a little leaf-mould added. The violet plants should then be carefully removed, with as large a ball of earth round the roots as possible, and planted in rows close together, but not touching each other, and so arranged that the foliage may be close to the glass without touching it, as the bed will settle an inch or two after the lights are put on. When planted, give them a copious watering, and in warm showers take the lights off: this will give them a clean, healthy appearance. The lights may be kept off all night with advantage when there is no appearance of frost, and all dead, decayed, or turning leaves, should be removed as soon as they appear. Plants thus treated will yield a supply of violets from November to April.

Where it is desired to have violets in summer and autumn, runners should be laid either in pots or on a hotbed where they are growing, in February, selecting the strongest runners, and pegging them down, with a little soil over the runner, and keeping them moist. These will be ready to plant out early in April, each with its bundle of roots, and will come in a month or six weeks earlier than the others. But they must be placed in their winter quarters early in September.

126.—NEMO'PHILA (nat. ord. Hydrophylla'ceæ).

This is perhaps the most charming and generally useful genus of dwarf-growing hardy annuals. All the varieties have a neat, compact, and uniform habit of growth, with shades and colours the most strikingly beautiful, so that ribboned, sown in circles, or arranged in any style which the fancy may suggest, the effect is pleasing and very striking. *N. maculata* and *N. phacelioides* are distinct; the latter is a beautiful hardy perennial; the former is more robust in growth, and has larger flowers than the other varieties. They all grow well from cuttings and seeds,—the latter from 6*d.* to 3*d.* per packet.

127.—LOBE'LIA (nat. ord. Lobelia'ceæ).

Exceedingly pretty profuse-blooming plants: the low-growing kinds make the most beautiful edgings. *L. speciosa* forms a delightful contrast to *Cerastium tomentosum* and the variegated alyssum; *L. gracilis erecta compacta*, from its bush-like habit and profusion of celestial-blue flowers, is equally beautiful in pots, beds, or used as an edging: all the varieties of *L. Erinus* are valuable for hanging-baskets, rustic-work, or vases, over the edges of which they droop in the most graceful and elegant manner. The perennial species, with their handsome spikes of flowers, are exceedingly ornamental, and are valuable from their blooming in autumn, along with gladioli, *Cilium lancifolium*, tritomas, &c. All the varieties grow freely from seed, and most of them from cuttings.

128.—MARIGOLD (nat. ord. Compos'itæ).

Well-known, free-flowering plants, with handsome double flowers, of rich and beautiful colours, producing a splendid effect, whether planted in beds, borders, or ribbons. The African, the tallest, are also the most striking in large beds, mixed flower and shrubbery borders. The dwarf French, in beds, or used as a foreground to taller plants, are invaluable, while the new brown and new orange miniature French varieties make splendid compact edgings to beds or borders.

129.—GENTIA'NA.

All the gentians are beautiful. *G. acaulis*, with its large deep mazarine-blue blossoms, looks well as an edging plant. It requires a pure air and rich light soil.

130.—LEPTO'SOPHON (nat. ord. Polemonia ceæ).

A charming tribe of hardy annuals. *L. densiflorus*, with its pretty rose-lilac flowers, and *L. densiflorus albus*, with its pure white blossoms, are exceedingly attractive in beds or ribbons. *L. hybridus*, *L. aureus*, and *L. luteus* make pretty low edgings, and are very suitable for rockwork; they all make nice pot plants, and succeed in any light rich soil. Seed, 3*d.* to 4*d.* per packet.

131.—SWEET-WILLIAM—Dianthus barbatus (nat. ord. Caryophylla'ceæ). Hardy Perennials.

A splendid free-flowering garden favourite, producing an unusually fine effect in flower-beds, borders, and shrubberies. Amongst these, Hunt's auricula-flowered varieties are remarkable for their rich, varied, and beautiful colours and immense heads of bloom, which surpass in effect even the handsomest of the perennial phloxes.

Sweet William, Hunt's magnificent auricula-flowered, choice mixed, 1 ft., 1*s.* per packet.

Ditto, in twelve splendid colours, 2*s.* 6*d.*

132.—TACSO'NIA (nat. ord. Passiflora'ceæ).

A grand genus of the Passiflora family, yielding to no twining shrub in cultivation for the dazzling brilliancy, size, and beauty of its flowers, which are produced in great profusion for months in succession. As some difficulty has been experienced in this country in blooming the varieties of this magnificent genus, the Sardinian correspondent of Messrs. Barr & Sugden has supplied the following information:—"To flower the Tacsonia successfully, it should be frequently stopped, as the flowers are produced upon the lateral shoots; it should be grown in rich soil, and frequently syringed during warm weather, to induce a vigorous growth: thus treated it will cover a large space in an incredibly short period, and bloom most profusely." He adds, "I have had *T. ignea* with from 150 to 200 gorgeous scarlet flowers open at the same time."

133.—PHLOXES (nat. ord. Polemona'ceæ).

This magnificent genus of plants, both annual and perennial, is unrivalled for richness and brilliancy of colours, and profusion and duration of blooming. The *P. Drummondii* varieties make splendid bedding and pot plants; the *P.*

decussata, French perennial varieties, produce a fine effect in mixed borders; no garden should be without them. They succeed best in light rich soil.

Seed from 4*d.* to 6*d.* per packet.

134.—MIGNONETTE-TREE.

The tree mignonette is formed by training a vigorous plant of common mignonette for about three years. Sow the seed very thin in April, draw out to a single plant. Next autumn remove all the lower shoots and shape the plant into a tree. Somewhat later shift into a larger pot in good loam. Keep it in a warm greenhouse and in a growing state, carefully removing all flowers. In the spring, it will appear woody. Treat it in the same manner the next year, removing all branches except those that are to form the head of the tree. By the third year it will have bark on its trunk, and form a handsome shrub; and by stopping the flowers as they appear during summer and autumn, it may be made to blossom freely during winter and spring for many years in succession.

135.—GLOXIN'IA (nat. ord. Gesnera'ceæ).

A superb genus of stove plants, producing, in great profusion, flowers of the richest and most beautiful colours; thrive best in sandy peat and loam. The varieties are

Gloxinia hybrida.
" " erecta.
" tubiflora, —white flowers in trusses, resembling the single tuberose; fragrance powerful and delicious, a single plant being sufficient to perfume a conservatory, 1 ft.

136.—HU'MEA (nat. ord. Compos'itæ).

A remarkably handsome plant, invaluable for decorative purposes, whether in the hall, the conservatory, or dispersed in pots about the lawn, pleasure-grounds, or terraces. Planted in the centres of beds or mixed borders, its majestic and graceful appearance renders it a most effective and striking object; in long mixed borders, if placed at intervals in irregular positions, it breaks that monotonous appearance which most persons so much dislike. Indeed, in any position, this plant stands unrivalled as a garden ornament. The leaves, when slightly rubbed, yield a powerful odour. When well grown, we have seen it 8 feet high and 4 feet in diameter. Succeeds best in light rich soil.

Humea elegans,—red, 8 ft. half-hardy biennial, from New South Wales.

137.—GAILLAR'DIA (nat. ord. Compos'itæ).

These strong flowers, natives of America, are both annuals and perennials. The former are splendid bedding-plants, remarkable for the profusion, size, and brilliancy of their blossoms, continuing in beauty during the summer and autumn: they thrive in any light rich soil.

Gaillardia alba marginata,—white-edged, half-hardy annual, 1½ ft.
Gaillardia grandiflora hybrida, — rich crimson and yellow, flowers remarkably large and attractive, half-hardy annual, 1½ ft.
Gaillardia grandiflora hybrida, — half-hardy annual, 1½ ft.

138.—HELIOTRO'PIUM (nat. ord. Boragina'ceæ). Half-hardy Perennials.

Profuse-flowering and deliciously fragant plants, valuable for bedding, ribboning, rustic baskets, and pot-culture; seeds sown in spring make fine plants for summer and autumn decoration; succeed in light rich soil. The best plants are obtained from cuttings in the same way as verbenas and bedding calceolarias. All the heliotropes are very sensitive of frost.

139.—MESEMBRYAN'THEMUM (nat. ord. Mesembrya'ceæ). Half-hardy Annuals.

A brilliant and profuse-flowering tribe of extremely pretty dwarf-growing plants, from the Cape of Good Hope, strikingly effective in beds, edgings, rock-work, rustic baskets, or vases in warm sunny situations; also for in-door decoration, grown in pots, pans, or boxes.

140.—LANTA'NA (nat. ord. Verbena'ceæ). Half-hardy Perennials.

A genus of dwarf bushy shrubs, from 12 in. to 18 in. in height, thickly studded with pretty miniature verbena-like blossoms of varied colours and changing hues,—from *snow-white* with *primrose* centres to delicate *pink* and *rose* with *white* discs, and from bright *rose-lilac* to *orange* and *scarlet* with *creamy* centres. For the conservatory and flower-garden they are alike valuable. Seeds sown in March make fine summer and autumn blooming plants. They succeed best in dry, warm situations, and in light rich soil.

141.—LARKSPUR (Delphi'num, nat. ord. Ranuncula'ceæ). Hardy Annuals.

For the guidance of amateurs, we may simply notice, that the stock-flowered Larkspur is of the same habit as the old Dwarf Rocket, but has longer spikes and much larger and more double flowers. The tall stock-flowered variety is of the same style as the branching, but with more compact spikes and larger and more double flowers. The tall-growing varieties scattered in shrubbery borders produce a charming effect when backed by the green foliage of the shrubs.

142.—LAVATE'RA (nat. ord. Malva'ceæ).

Very profuse-blooming showy plants, which are exceedingly attractive as a background to other plants, or for woodland walks and wilderness decoration, growing freely in any soil.

Lavatera trimestris,—rose, pink-striped, 3 ft., hardy annual, from Armenia.
Lavatera trimestris alba,—white, 3 ft., hardy annual.
Lavatera arborea (Tree Mallow),—violet, 5 ft., hardy biennial, from Italy.
Lavatera Armeniaca,—lilac, 4 ft., hardy biennial.

143.—WALLFLOWER, Cheiranthus (nat. ord. Crucif eræ).

For spring gardening these hardy perennials are as indispensable as the Crocus or the Tulip and from the delicious fragrance of their beautiful flowers they are especial favourites, producing a splendid effect in beds or mixed borders. On account of their variety, much interest is excited in raising them from seed. Packets are from 2*d*. to 1*s*. each.

144.—VIRGINIA STOCK, the Malco'mia.

A pretty little annual, the seeds of which may be sown at almost any season. It is sure to grow and bloom abundantly.

145.—VIRGINIAN CREEPER (nat. ord. Vita'ceæ).

A favourite plant for covering an ugly wall or shed. Its flowers are very insignificant; but this defect is amply compensated for by its beautiful leaves, which assume a most brilliant scarlet colour in autumn. Its growth also is very rapid: by some persons it is known as the Five-leaved Ivy.

146.—DIAN'THUS (nat. ord. Caryophylla'ceæ).

A beautiful genus, which embraces some of the most popular flowers in cultivation. The Carnation, Picotee, Pink, and Sweet William, all "household words," belong to this genus. *D. Sinensis* and its varieties may be considered the most beautiful and effective of our hardy annuals; the double and single varieties, with their rich and varied colours in beds or masses, are remarkably attractive; while the recently introduced species, *D. Heddewigii*, with its large rich-coloured flowers, 3 to 4 inches in diameter, close compact habit, and profusion of bloom, is unsurpassed for effectiveness in beds or mixed borders.

147.—GESNE'RIAS (nat. ord. Gesneria'ceæ).

Showy hothouse plants with scarlet flowers. They are propagated by cuttings, which require a little bottom-heat.

148.—TU'BEROSE (nat. ord. Hemerocallid'eæ).

Bulbous-rooted plants from the East Indies; flowers white, very odoriferous. They require to be started in a pit. In January plant the bulbs singly in very small pots in sandy loam; plunge them in a pit of moderate heat;

give little or no water till they have made a start, then water sparingly. When they have filled their pots with roots, shift them and re-plunge them until they show bloom, when they may be removed to the greenhouse, where they will last in flower about two months.

149.—TE'COMA (nat. ord. Bignonia'ceæ).

Magnificent ornamental greenhouse evergreen twining shrubs, with large and beautiful flowers; require the same cultural treatment as Bignonia.

Tecoma radicans, from North America.

Tecoma jasminoides grandiflora,—white flowers, with beautiful crimson throat.

150.—STOCKS—Mathi'ola (nat. ord. Cucif'eræ).

The 10-week Stock (*Mathiola annua*) is the most universally cultivated: it usually blooms ten or twelve weeks after being sown, grows from 6 to 15 inches high, and when cultivated in rich soil and occasionally watered with very weak guano-water, throws out an immense quantity of lateral spikes of bloom, so that a plant forms a perfect bouquet: it would indeed be very difficult to surpass the grand effect produced by these exquisite floral gems.

The Imperial or Emperor, sometimes called Perpetual Stocks.—They are half-hardy biennials, hybrids of the Brompton, growing 18 inches high, and of a robust branching habit. Sown in March or April, they make splendid "autumn-flowering stocks," and form a valuable succession to the summer-blooming varieties. Should the winter prove mild, they will continue flowering to Christmas. Sown in June or July, they flower the following June, and continue blooming through the summer and autumn months.

The Brompton and Giant Cape are generally called winter stocks, on account of their not flowering the first year: the former is robust and branching, the latter possesses the characteristic so much esteemed by some, viz., an immense pyramidal spike of bloom. These are half-hardy biennials.

The Intermediate Stock is extensively cultivated for Covent-garden market; it is dwarf and branching, and in the early summer months constitutes the principal feature in furnishing jardinets, &c., for "the London season." It is also of great value in filling flower-beds for an early summer display. Half-hardy biennials.

151.—STOCKS, TO SELECT DOUBLE FLOWERS.

Reject from the seed-bed all those plants which have a long tap-root (these will almost invariably prove single), and reserve for bedding only those which have the largest quantity of delicate fibres at the roots: experience shows that these, in general, prove double.

152.—IPOMŒ'A (nat. ord. Convolvula'ceæ).

Beautiful climbing plants of various colours. The seed should be raised under glass in April and the young plants set out in May. They look very well over the handles of baskets and in contrast with canariensis. They require a rich light soil. In some situations the plants will shed their seed and come up from year to year in the open ground.

153.—LILIUM.

The Lily in all its varieties is equally adapted for ornamenting the conservatory and the sitting-room as it is for the flower-borders.

Culture in-doors.—Use a good mellow soil, composed of equal parts of leaf-mould and loam, with a little peat, and one-sixth of silver sand. A 12-inch pot, with six bulbs planted in it, will furnish a group of no ordinary beauty: smaller-size pots will require fewer bulbs. Place at the bottom of the pot a piece of potsherd, and over it some pieces of wood charcoal and rough fibry soil to secure good drainage, then fill up with the compost. When planted, the bulbs should be covered one inch, and the soil made close by pressure: they should be treated in their first stage of growth precisely as hyacinths grown in pots, except that they should remain buried in ashes or cocoa-fibre till they begin to indicate a top-growth. Those intended to flower early should be placed under glass, while such as are for late blooming should remain out of doors in a sheltered situation, the pots plunged to the rim in ashes or cocoa-fibre.

Culture out of Doors.—If the land be of an adhesive nature, it should be removed to the depth of two feet, and replaced with a rich free soil, or else the bulbs should be planted in 5-inch pots, and early in May turned out where intended to bloom. Light or medium soils will only require deep digging and well working, with the addition of some thoroughly rotted manure. Plant the bulbs five inches deep, and for the first winter place on the surface a few dry leaves. The bulbs should not be disturbed oftener than once in three years, as established patches bloom much more profusely than those taken up and divided annually.

The varieties best adapted for in-door culture are *Atrosanguinium maculatum, Lancifolium album, L. punctatum* and *roseum, L. rubrum* and *longiflorum.* These vary in price from 20*s.* to 5*s.* per dozen. The Martigon varieties are very effective in borders; so also are the common White Lily and Chalcedonicum, or Scarlet Turk's Cap. These can be purchased for 5*s.* to 3*s.* 6*d.* per dozen.

154.—LILY OF THE VALLEY.

To grow these spring favourites to perfection, the roots should be set in bunches one foot apart and covered with a dressing of well-rotted manure before the winter sets in. They can hardly be treated too liberally. If grown in pots for the greenhouse, by a little management a succession may be kept in bloom till June. Keep the pots perfectly dry and in a cool shady place until their natural season is past, and by watering they soon come into foliage and flower.

155.—CAMEL'LIA (nat. ord. Ternstromia'ceæ). In house.

Favourite winter and spring-flowering shrubs of great beauty. Almost any soil will grow camellias. Some persons grow them entirely in peat, some in strong loam, approaching to clay. The best soil, however, appears to be two parts fibry peat, one fibry loam, one-sixth part sharp silver-sand, and one-sixth part *rotten wood*, or clean leaf-mould. Keep them in a temperature of 55° to 60° until their growth is made and flower-buds formed. During this period they should be frequently syringed, and a humid atmosphere maintained. Towards the end of April gradually remove them by easy transitions, to a cool house or cold pit, and the last week in May to a sheltered situation out of doors, or they may continue in the same house or pit throughout the season. The pots must be placed on a hard bottom to prevent the ingress of worms, watered alternately with clean water and weak liquid manure, and finally removed under glass in October. With such treatment their blossoms will expand in November or December. As soon as the last flower has fallen, shift the plants into larger pots if the state of their roots require it.

Fine plants, from 18 inches to 2 feet high, of the following or equally good varieties, can be purchased at from 30*s*. to 42*s*. a dozen; larger plants, from 60*s*. to 120*s*. a dozen:—

Alba plena, Albertus, Amabilis, Archduchesse Marie, Beata, Chandlerii, Elegans, Countess of Ellesmere, Countess of Derby, Cup of Beauty, De la Reine, Duchesse d'Orléans, Eximia, Fimbriata, Lady Hume's blush, Grand Frederick, Grandis, Imbricata, Imbricata alba, Jenny Lind, Mathotiana, M. alba, Marchioness of Exeter, Princess Royal, Prince Frederick William, Reticulata, R. flore pleno (this is still expensive), the Bride, Tricolor, Tricolor imbricata plena, and Victoria magnosa.

156.—CAMELLIAS in the open Air.

Among the evergreen plants which are suitable for the shrubbery or border, none can excel the Camellia; and there are a few varieties of this beautiful shrub which do well in the open ground. It may be useful to enumerate some of the sorts which are known to flourish and blossom freely against a north wall, or upon a north border, and to make one or two observations upon their culture and the soil best adapted for them. The sorts best suited to open-air cultivation are Carolina (double white), Pæoniflora, Prince Leopold, Perfection, Eclipse, Dahlia-flora, *Imbricata alba*, Duchess of Orleans, and Bealii. The soil in which they are planted should be a mixture of peat, leaf-mould, and cow-dung, about two feet deep. Great care should be taken that the plants never suffer from drought. After flowering they should be freely watered with liquid manure, especially if the season be dry. The surface of the ground just round the stems of the plants may frequently, with very good effect, be paved with small stones, which assist in keeping the roots cool and moist. As a general rule, the borders on which camellias are planted should not be disturbed more than is necessary to remove the surface weeds. A top-dressing of fresh soil may, with advantage, be given to them every winter. So treated, the sorts of camellias mentioned above will be found as hardy as most of our common evergreens, and require no protection, except, perhaps, in an unusually severe winter, when a few fir boughs may be placed before or around them. The snow should never be allowed to rest upon their branches. Some growers of camellias in the open ground bind straw round the stems of their plants, about five or six inches from the ground, when winter sets in: this is found a very efficient protection against frost.

157.—ALYS'SUM (nat. ord. Crucif'eræ).

Free-flowering, useful, pretty little plants for beds, edgings, or rockwork. The annual species bloom nearly the whole summer; the perennials are amongst our earliest and most attractive spring flowers. The varieties are:—

Alyssum argenteum.—yellow, with silvery foliage, hardy perennial, 1 ft., from Switzerland.
Alyssum atlanticum.—fine light yellow, very ornamental, hardy perennial, ¾ ft.
Alyssum Benthamii.—white, very fine, hardy annual, 1 ft.
Alyssum sexatile,—yellow, extremely showy, hardy perennial, 1 ft., from Candia.
Alyssum sexatile compactum.—golden-yellow, very compact, free-flowering and beautiful, hardy perennial, ½ ft.
Alyssum, sweet (Kœnigia maritima).—white, very sweet, hardy annual, ½ ft., British.

Price of seed, *3d.* to *6d.* per packet.

158.—ANDROM'EDA (nat. ord. Eri'ceæ).

Deciduous heath-like shrubs, natives of North America; many of them are also evergreen. They delight in bog, like all other so-called American plants, and their roots should never be suffered to become quite dry; for if this occurs, the plant has seldom vigour enough to send out sufficient quantity of new roots, and in general dies. All the kinds may be propagated by layers.

159.—BELLADON'NA LILY.

An amaryllis, the flower white, flushed with rose-purple, very handsome. The bulbs may be purchased at *5s.* per dozen.

160.—CHOROZE'MAS (nat. ord. Legumino'sæ).

A most interesting genus of plants from Australia, which bloom almost the whole year, more especially in the winter and early spring, and are consequently most acceptable additions to our greenhouses and conservatories. They are not very difficult to manage, and are alike useful for decoration and as cut flowers for bouquets, at a time when such flowers are valuable. They delight in a rich turfy peat, mixed with fibrous loam and leaf-mould and gritty sand. When recently potted, they require a close pit or the warm part of a greenhouse, and cautious watering, until they get into free growth. When thoroughly established, water with clear liquid manure twice a week.

Chorozemas are propagated by cuttings of the half-ripened young wood, taken off in July or August, taking the short, stiff, and weak, or medium growth, but avoiding twigs of a robust habit. These, after being trimmed, should be about 1 inch long, and must be inserted in sand, under protection of a bulb-glass. In preparing the pots for the cuttings, take care to drain thoroughly, by half filling them with potsherd; then place fibrous peat about an inch deep over the drainage, and fill up with clean sand. After the cuttings are in, place the pots in a close cold-frame, water when necessary, and wipe the condensed moisture from the inside of the glass twice or thrice a week. Here the cuttings must remain until they are cicatrized, when they may be removed to a warmer situation, and the pots plunged in a very slight bottom-heat, and in a few weeks they will be ready to pot off. If it is late in the season before the cuttings are ready to pot off they should remain in the cutting-pots through the winter, and be potted off in February; but if they are ready for single pots in September, they will be much benefited by being potted off early.

Attention must be paid to stopping the rude shoots, so as to induce close, compact, and healthy growth. If the plants progress as they ought, they will require a second shift during the season. They should be kept growing until the winter fairly sets in, at which time they should be brought to a state of rest. In the second year some of the plants will have a nice head of bloom; but in order to produce rapid growth, remove the bloom-buds when quite young, and keep the plants vigorously growing through the second season. If they are in good health and the pot full of roots, a shift any time between Christmas and October will not hurt them: never shift a plant until the pot is full of vigorous roots, and also take especial care that the roots do not become matted before you shift.

Manure-water in a weak state may be used with advantage; but use it with caution, and not more than twice a week. That prepared from sheep's dung and soot is best, and it *must* be used in a perfectly clear state.

161.—CROCUS.

For in-door decoration the Crocus properly managed is very useful; for the flower-garden, indispensable. When used as an edging to beds, borders, &c., the colours, being nicely blended, they are remarkably attractive; while in the arrangement of any fancy designs, or grouped in distinct colours of from six to fifty, or even a hundred in a group, the effect is all that could be desired. We have seen large beds of Crocus planted in circles, each circle one colour: with the sun shining on them, nothing could possibly be more brilliant, more especially when there chanced to be a good proportion of them yellow.

Culture in-doors, same as the Hyacinth, and, like it, they may be grown in any ornamental contrivance; but it is absolutely necessary, to insure success, that they be kept well supplied with water, close to the glass, and have abundance of *fresh air*, or they will produce leaves only.

Culture in the open ground.—They can be successfully grown in almost any soil and situation. Plant 2 to 3 inches deep, and not more than 2 inches apart.

Crocuses, however, are very accommodating in reference to the depth at which they are planted. When planted in beds devoted to bedding-plants, they will reach the surface and flower, if inserted four times that depth. As the young bulbs are formed on the top of the old ones, they thus possess a self-elevating power. Crocuses will flower freely for many years without being disturbed. The best growers, however, recommend dividing and re-planting every third or fifth year. To secure perfect blooms, the foliage must be left to die down of its own accord. If planted in ribbon-beds, the following arrangement would look well:—Back row, David Rizzio, large purple; 2nd row, largest yellow; 3rd row, Flos niger, fine large blue; 4th row, Mont Blanc, or any other pure white; or the order may be reversed, or confined to three colours only,—yellow, blue, and white, or *vice versâ*. On wide borders, the same order may be repeated as often as necessary. The great point in this style of planting is to choose distinct colours only, and not varieties of colour.

162.—TRITO'MA.

An exceedingly showy free-flowering plant, with long graceful leaves and majestic flower-spikes, three to seven feet in height, crowned with densely-flowered spikes of bloom, which are produced during the autumn months, 18 to 27 inches long.

Culture.—Dig and well work the soil to the depth of two or three feet, adding plenty of rotted manure. The crown of the plant should not be more than an inch and a half in the soil; for winter protection surround the plant with two inches of sawdust, firmly trodden. Remove this early in May: from then till the plant is in bloom, weak liquid manure must be applied in large quantities, especially during dry weather.

Tritoma glaucescens,—rich scarlet.

Tritoma J. grandis—bright scarlet, tall and late.

163.—TROPÆ'OLUM (nat. ord. Tropæola'ceæ). Half-hardy Annuals.

A tribe of elegant-growing, profuse-flowering, and easily cultivated climbers, combining with these important qualities great richness and brilliancy of colour, with finely-formed and beautifully-marked flowers. For pillars and rafters, in the greenhouse or conservatory, they are invaluable; for covering trellises, verandahs, and bowers out of doors, they are of equal importance; while for bedding purposes we have only to remind our readers of the important part they annually play on the terraces of the Crystal Palace. When used for bedding, they should be regularly and carefully pegged down, interlacing the shoots, and occasionally removing the large leaves. In pleasure-grounds, where the beds are sometimes protected with fancy wire-work against the depredations of rabbits and hares, the tropæolums are invaluable for covering it: they grow rapidly, are easily trained, and continue flowering the whole summer and autumn.

We may remark that all the "Lobbianum" varieties bloom beautifully through the winter months in the greenhouse or conservatory, so that where cut flowers are in demand they will be found an invaluable acquisition. Grow freely in light rich soil.

164.—AMARAN'THUS (nat. ord. Amarantha'ceæ).

Half-hardy annuals, very graceful, with highly ornamental foliage. *A. ruber*, with dark carmine foliage, is a most strikingly beautiful plant for bedding, ribboning, or massing. Other varieties are bicolor, leaves crimson and green, and tricolor, red, yellow, and green.

Culture.—Sow in heat in early spring; plant out in May and June in very rich soil. Price of seed, 3*d.* to 6*d.* per packet.

165.—BALSAM (nat. ord. Balsamina'ceæ). Half-hardy Annuals.

Magnificent plants either for conservatory or out-door decoration, producing gorgeous masses of brilliant flowers. When grown in pots, and large specimens are desired, they should be shifted into 10 or 12-inch pots, using the richest compost at command, and the pots plunged in spent hops or tan, and liberally supplied with manure-water; when used for out-door decoration, the soil should be rich, the plants supported with neat stakes, and liberally supplied with manure-water.

Balsam, mixed, saved from all the most choice double varieties, including Webb's, Smith's, and Glenny's.

Balsam,—aurora-coloured, striking and very beautiful, 2 ft.

Balsam,—camellia-flowered, ten magnificent double varieties:—pink spotted, lilac spotted, purple spotted, scarlet spotted, rose spotted, rose spotted white, fine bronze spotted, crimson spotted, bright pink spotted white, beautiful crimson spotted white. Each colour separate, or the collection from Messrs. Barr and Sugden.

166.—FOLIAGE PLANTS.

These have become very fashionable of late. For illustrations and for culture we refer the reader to Lowe's "Beautiful-leaved Plants," and a companion volume by Mr. Shirley Hibberd.

167.—AGAPAN'THUS (nat. ord. Homerocallida'ceæ).

An African lily blooming in August, combining graceful foliage with large handsome heads of blossom. In flower beds or masses, the blue variety is lovely; planted in a strong rich soil, it produces a splendid effect, and when mixed with Gladioli, either of the Ramosus or Gandavensis sections, the effect is unique. Protect the bed or patch during winter with a thatched frame.

Culture.—A nine-inch pot will be ample for a strong plant, but a large pot or tub is required for several plants; and this is the most effective and more usual way of growing the Agapanthus. Use a strong rich loam, and during the summer months give abundance of water, and liquid manure twice a week. In winter protect from severe frost, and water sparingly.

168.—AUTUMN FLOWERS.

In addition to the ordinary bedding plants—geraniums, verbenas, &c.,—a good display of flowers in autumn requires a free use of autumnal roses, hollyhocks, dahlias, *Lilium lancifolium*, delphiniums, phloxes, foxgloves, the hardy bamboo, the holy thistle, pampas grass, *Arundo Donax*, tritoma, yucca and for foliage, sundry kinds of ferns.

169.—CALANDRIN IA (nat. ord. Portulaca'ceæ).

Very beautiful free-flowering plants. *C. discolor* and *grandiflora* have large handsome flowers, and are fine for edgings; *C. umbellata* is of a trailing habit, and produces profusely its glowing *rosy violet* flowers in bunches; is invaluable for rock-work and dry hot banks, or similar situations, where it will stand for many years. They succeed best in a light rich soil. They may be raised from seed.

Calandrinia discolor,—rose-lilac, hardy annual, 1 ft., from Chili.

170.—ANTIRRHI'NUM (nat. ord. Scrophularia'ceæ).

The Antirrhinum, popularly called Snapdragon, is a hardy perennial, and one of our most showy and useful border plants. Amongst the more recently improved varieties of this valuable genus are large finely-shaped flowers of the most brilliant colours, with beautifully marked throats; they succeed in any good garden soil, and are very effective in beds. *A. nanum* and varieties are valuable for rockwork and old walls. Choice mixed seed, 3*d.* per packet.

171.—CAMPAN'ULA (nat. ord. Campanula'ceæ).

A genus of exceedingly beautiful plants, characterized by the variety of their colours, profusion and duration of their bloom. Some of them are remarkable for their stately growth, others for their close, compact habit; of the former, *C. pyramidalis* grown in pots, placed about terraces, gravel walks, or the margins of lawns, produces a most striking effect. Visitors to Paris may have been struck with the free use made of this plant in the public gardens. Of the dwarf varieties, *C. Carpatica* is a most valuable bedding-plant. The whole genus deserves a prominent place in every garden.

172.—AUBER'GINE—EGG-PLANT (nat. ord. Solana'ceæ). Half-hardy Annuals.

Several varieties are eatable, and extensively cultivated in the South of Europe. As pot plants they are curious and interesting, being covered in autumn with beautiful egg-shaped fruit: the scarlet variety is a great novelty. In warm localities they succeed out of doors on a south border. Seed from 3*d.* to 6*d.* per pack.

173.—BRACHY'COME (nat. ord. Compos'itæ). Half-hardy Annuals.

Beautiful free-flowering dwarf-growing plants, covered during the greater portion of the summer with a profusion of pretty cineraria-like flowers. very effective for edgings, small beds, rustic baskets, or for pot-culture; succeeding in any light rich soil.

Brachycome iberidifolia,—blue, ½ ft., from Swan River.

Brachycome iberidifolia albiflora,—white, ½ ft., from Swan River.

174.—CLARK'IA (nat. ord. Onagra'ceæ). Hardy Annuals.

Cheerful-looking flowers, growing freely from seed and blooming profusely under almost all circumstances. When planted in rich soil and properly attended to, they rank amongst the most effective of bedding plants, more especially *C. integripetala, alba,* and *marginata, C. Tom Thumb,* and *C. T. T. integripetala:* the large handsome flowers of the former render them strikingly attractive, and the latter, with their fine bushy habit, make useful plants for bedding or conservatory decoration. The new double variety, *C. pulchella,* is of a rich magenta colour and very handsome. These new varieties are decided acquisitions.

175.—ACAN'THUS (nat. ord. Acantha'ceæ).

Perennials attractive for the beauty of their foliage: natives of Southern Europe. The most common varieties are *A. mollis* and *A. spinosa.* From the former of these the original idea of the capital of the Corinthian order of architecture is said to have been derived.

Culture.—All the sorts grow readily from seed, or they may be increased by dividing the roots. They require a sandy soil and free space.

176.—BUGLOSS.

The Bugloss (*Anchusa*) is a fine showy plant, mostly with large blue flowers They may be propagated by slips, and by dividing the roots into as many plants as there are heads, when they have done flowering, as well as by seed saved in the autumn, and sown on a warm border in the spring.

177.—GERA'NIUM (nat. ord. Gerania'ceæ).

These well-known floral favourites are not less indispensable for outdoor than for indoor decoration. No plants are more universally cultivated, and of none are there greater varieties. Under the heads of Fancy and French and Spotted Geraniums will be found useful lists, and descriptions for the guidance of purchasers of these different sorts. With geraniums for the green-house, in order to secure profusion of bloom, early growth and under-potting are of the first importance. No matter how robust a plant is grown, one eighteen months old cannot be made to flower as freely as one four or five years old. Plants to flower in May should be cut down by the end of the previous June;

they should have broken, been reduced, repotted, and encouraged to grow two or three inches in a close cold frame, for a fortnight, and have received their final stopping by the end of July, and be placed in their blooming-pot by the 1st of November. Success depends upon their chief growth being completed before Christmas. No after-management can compensate for the neglect of early growth. Any size of plant or leaf may be obtained at any period; but flowers will be scarce unless early growth is secured. Under-potting is the next great point. Plants in general, and geraniums in particular, flower best when they are pot-bound, and some varieties will scarcely flower at all unless their roots are in this condition. The reason seems to be, that whatever tends to check the extension of other parts, favours the development of flowers.

178.—GERANIUM, BEDDING VARIETIES.

Of Bedding varieties, the following list will be found to embrace the most useful. Strong healthy plants may be had at from 4*s.* to 6*s.* the dozen.

Auber Henderson,—white.
Baronne Souget,—fine scarlet.
Boule de Feu,—fine scarlet.
Boule de Neige,—white.
Brown's compactum.
Carmine,—mossy.
Cerise Unique,—cerise.
Christine,—rosy pink.
Comte de Morny,—rosy scarlet; blush edges, suffused with pink; fine.
Countess of Bectivo,—deep salmon, with dark brown zone.
Dawn of Day,—brilliant scarlet.
Defiance,—large.
Emperor of the French,—scarlet.
Frogmore scarlet.
Garibaldi,—bright scarlet, with white eye.
Harkaway,—deep habit; perfect bloomer; very distinct.
Henri de Beaudot,—very large; salmon-coloured blossoms, belted with white.
Imperial crimson,—mossy.
,, Stella,—ditto.
Lady Middleton,—fine rose.
Madame Chardine,—salmon-rose, white centre; large.
Madame Vaucher,—white.
Minnie,—pink; white spot on upper petals.
Paul l'Abbé,—clear rosy salmon; fine.
Purple,—mossy.
Red,—ditto.
Richmond Gem,—light orange-scarlet.
Rose superbe,—fine rose.
Rubens,—fine salmon-colour.
Scarlet shaded and Spottishall,—varieties.
Shrubland,—scarlet; fine large.
Silver variegated,—mossy.
,, Stella,—ditto.
Tom Thumb,—scarlet.
Tom Thumb's Bride,—good scarlet.
Trentham rose,—fine rose.

Silver and gold varieties, one thousand:

Alma,—scarlet flower; silver lines.
Annie,—scarlet; sulphur-tinged.
Bijou,—strong-growing scarlet flower; silver leaves.
Brilliant,—slightly silvered; good scarlet; fine bedder; not a strong habit.
Cloth of Gold,—unique.
Countess of Warwick.—scarlet; white margin; leaves brown-white.
Culford Beauty,—scarlet flower; sulphur variegated.
Dandy,—a neat silver-edged gem.
Gold Ivy-leaved.
Golden Cerise,—unique.
Golden Tom Thumb.
Hendersonii,—like Flower of the Day, but scarlet flowers.
Lady Plymouth,—a neat silver-edged gem.
Miss Emily Domville,—salmon flowers, silver-zoned; good.
Perfection,—scarlet flower; white-margined leaves.
Pink-flowered,—ivy-leaved.
Rainbow,—scarlet flowers; silver, with red zone.
Scarlet-flowered,—ivy-leaved.
Silver Queen,—pink flowers; smooth foliage.
Silver,—ivy-leaved.
St. Clair,—pink flowers; silver habit.
White-flowered,—ivy-leaved, &c.

179.—GERANIUM CUTTINGS.

Cuttings of all sorts of geraniums for bedding the following year should be struck early: from the last week in July to the end of the first week in August is very good time. They should be taken in dry weather, when the parent plant has had no water for some days, and they should be kept to dry twenty-four hours after they have been prepared for potting. The more succulent sorts, and any that appear difficult to strike, may with advantage be touched at the end with a small paint-brush dipped in collodion, which will serve to hasten the callus which the cutting must form before it will throw out roots. They may be potted four or six in a pot, according to size. It is essential that the pots be well fitted with drainers, that the soil be light and sandy, and that it be pressed tight round the joint of the cuttings, which should be buried in it as flat as possible. When potted, they may be sunk in the ground on a south border, and well watered in the evening, when the sun is off. They will require no shading, except the sun be very scorching; and in this case, they must not be kept from the light, but merely screened from the scorching rays of the sun. They may flag a little, but this is of no importance; in two or three days they will recover, and put forth roots. If they grow too freely before it is time to take them in for the winter, the top shoots should be broken off, and in this way they will make strong bushy plants.

180.—GERANIUMS, TO PRESERVE OLD ONES.

Take them out of the borders in autumn, before they have received any injury from frost; and let this be done on a dry day. Shake off all the earth from their roots, and suspend them, with their heads downwards, in a cellar or dark room, where they will be free from frost. The leaves and shoots will become yellow and sickly; but when potted about the end of May, and exposed to a gentle heat, they will recover and vegetate luxuriantly. The old plants, stripped of their leaves, may also be packed closely in sand; and in this way, if kept free from frost, they will shoot out from the roots, and may be re-potted in the spring.

181.—FRENCH OR SPOTTED GERANIUMS.

The following list contains a most excellent selection, those marked * being the most expensive.—Average cost about 18*s.* per dozen.

French or Spotted Varieties.

Admiration,—rose, edged with white; dark spots on upper petals.

Adam Bede,—deep pink, under petals crimson, margined with the bottom colour.

*Arthur Henderson,—very dark maroon; black spots; light centre.

*Archimede Rose, — crimson; maroon spots.

*Bertie,—deep rose; rich spots on lower petals; upper ditto carmine; fine good show variety.

Bracelet,—large rose-maroon spots; dark blotch on upper petals; fine.

Beadsman,—pale pink; maroon spots on centre petal.

Black Diamond,—dark crimson, margined with rosy scarlet.

*Constellation,—quite new; beautiful.
*Edmond Bossier,—carmine-rose, lower petals, upper, black maroon; black spots; purple veins.
Endymion,—bright rose; all the petals spotted with black.
Eugène Duval,—fine purple.
Eole,—mulberry lower petals, upper ditto with dark blotches.
Fairy Queen,—rose-maroon blotch; white centre; fine.
*Fair Rosamond,—white, with purple spots; good.
Guillaume Peveryns,—lilac-maroon spots.
Gustave Odier,—crimson-lake; crimson blotch; good habit.
Géant des Batailles,—white, shaded with crimson-scarlet spots; fine large flower.
Hortense,—salmon-rose; crimson blotch.
Impératrice Eugénie,—pure white; dark violet spot.
James Odier,—bright carmine, tipped with rose; violet spot.
*Leo,—orange-rose spot on lower, and blotch on top petals.
Madame Furtado,—blush spot; rosy crimson.
Madame Pescatore—rose-crimson spot; dark blotch on upper petals.
*Mara,—rose, with white edges; dark spots, shaded with violet.
Osiris,—rich crimson; dark spots on upper petals.
Oscar Lisible,—carmine; light maroon blotch; white centre.
Pandora,—scarlet, with light margin.
Perrugino,—superb pink, with large spots.
*Phiton,—deep crimson; dark maroon spots; white centre.
Painted Lady,—pink; white edges; maroon spots.
*Rupee,—purple lower petals; black upper ditto; fine purple margin.
Riflemen,—crimson scarlet; black spots; fine free-grower.
Rubens,—flower crimson; scarlet blotch; maroon spots.
Sanspareil,—bright rose; crimson spot; good form.
Salvator Rosa,—rose, shaded violet; brown spots; white centre.
Scaramouch,—fine exhibition variety.
*Senior Wrangler,—peach; maroon spots; shaded margin.
Spotted Pet,—lilac; maroon spots; large blotch on upper petals.
*Telegraph,—crimson lower plates; upper, bright maroon; deep habit; free bloomer.

182.—AS'TERS (nat. ord. Compos'itæ). Half-hardy Annuals.

This splendid class of plants is not only one of the most popular, but also one of the most effective of our garden favourites, producing in profusion flowers in which richness and variety of colour are combined with the most perfect and beautiful form. The Aster is indispensable in every garden or pleasure-ground where an autumnal display is desired. In our flower-beds and mixed borders it occupies a deservedly prominent position, whilst for grouping or ribboning it stands unrivalled.

The Aster may be divided into two sections—French and German. The French, as improved by Truffaut, has flat petals either reflexed or incurved; the former resembling the Chrysanthemum, whilst the latter, turning its petals towards the centre of the flower, forms, when well grown, a perfect ball, and is best described by its resemblance to the Pæony. The German varieties are quilled, and the most perfect flowers are surrounded by a circle of flat or guard petals, as in the Hollyhock. The flowers of these are particularly admired for the exquisite symmetry of their form. The dwarf bouquet varieties of this beautiful plant grow from six to nine inches high, and are particularly adapted for small beds, edgings, or for pot-culture; they often flower so profusely as entirely to hide their foliage. All the varieties delight in a deep, rich, light soil, and in hot, dry weather should be mulched with well-rotted manure, and frequently supplied with manure-water: this labour will be amply repaid by the increased size, beauty, and duration of the flowers. Seed from 4*d.* to 1*s.* per packet.

183.—OX'ALIS (nat. ord. Oxalida'ceæ).

A genus of exceedingly pretty bulbous plants, all of which have beautiful green foliage, which forms a fine contrast to their richly-coloured blossoms. They are admirably adapted for pots, borders, and rock-work, succeeding in any light soil.

Messrs. Barr & Sugden offer the following varieties from 1*s.* 6*d.* to 3*s.* 6*d.* per dozen :—

Oxalis bipunctata,—lilac.
„ Bowieii,—crimson, large trusses.
„ cernua,—yellow.
„ „ fl. pl.,—double yellow.
„ floribunda rosea,—rose.
„ grandiflora alba,—white.
„ hirta rosea,—rose.
Oxalis hirta rubra,—red.
„ rubella,—red.
„ speciosa,—rosy-purple, very showy.
„ tetraphylla,—purple.
„ versicolor,—scarlet and white.
„ fine mixed, per 100, 10*s.* 6*d.*

184.—ACA'CIA (nat. ord. Legumino'sæ).

Elegant-growing plants, natives of Australia, the East Indies, and Mexico: nearly all are evergreen. During winter and early spring they flower freely in greenhouses, but they are not hardy enough to endure our climate unprotected except in the summer, when they may be plunged with their pots in borders with good effect. Of Acacias there is a great variety; all may be grown readily from seed, which is best imported, and from cuttings in pots of very fine mould, set in a hot-bed. Among the most beautiful varieties are *A. albicans*, with fine silver foliage; *A. balsamea*, yellow, about 6 ft. high; *A. Drummondii*, yellow; *A. Ixiophylla*, golden yellow; *A. odoratissima pendula*, yellow, with fine long pendent blossoms, very fragrant; *A. spinosa*, rose-coloured and white The roots of the Acacia have generally a very poisonous smell, which renders. them unpleasant and unhealthy in confined houses.

Culture.—All kinds require a sandy loam, well drained.

185.—DEUT'ZIA (nat. ord. Philadelpha'ceæ).

A beautiful hardy shrub, when in bloom covered with pretty snowdrop-like flowers, exceedingly valuable for the spring decoration of the conservatory.

Deutzia gracilis,—pure white, very graceful, 3 ft., from Japan.

186.—BROMPTON STOCKS. (Mathiola incana).

These are biennials. The seed should be sown early in May in a light sandy border with an eastern aspect. It succeeds best sown thinly in drills about six inches apart. As soon as the plants show their second leaves they should be watered every evening with a fine rose pot. When about three inches high they should be thinned out to at least six inches apart and the other plants removed to another bed. In about a month's time they should be thinned again and the alternate rows taken up, so as to leave the remaining plants about a foot apart every way. These may be suffered to flower where they stand, or they may be transplanted to the flower-borders in March or April. Great care is necessary in transplanting not to expose the roots, and the new soil should be of the richest description possible. The plants will require shading till they are established, and watering with liquid manure till they begin to flower. Thus treated the flowers will be splendid.

187.—AMARYL'LIS (nat. ord. Amaryllida'ceæ).

Flowers of rare beauty, whose large, drooping, bell-shaped, lily-like blossoms range in colours from the richest crimson to pure white, and striped with crimson or scarlet. They are easily cultivated, and with a little management a succession of bloom may be secured throughout the year. Some varieties do not require heat. A moderate supply of bulbs will serve the purpose.

Varieties which do not require heat are,—

Bella-Donna purpurea (Bella-Donna Lily),—white flushed with rosy purple.
Bella-Donna blanda.
Formosissima (Sprekelia formosissima),—rich crimson.
Longifolia alba (Crinum capense alba),—white, sweet-scented.
Longifolia rosea (Crinum capense alba),—rose, sweet-scented.
Lutea (Sternbergia lutea),—yellow, flowers in autumn.

Others must always be started in a hotbed; as,—

Aulica,—reddish-brown, with greenish stripes, flowers very large.
Cleopatra,—dark red, margined with white, of great substance and very showy.
Crocea grandiflora,—vermilion.
Crocea superba,—bright orange, large flowers.
Johnsonii,—scarlet, with pure white stripes, very showy.
Johnsonii, striata,—striped.
Prince of Orange,—bright orange, large and handsome.

188.—CALLIOP'SIS (nat. ord. Compos'itæ).

The Calliopsis, or, as it is sometimes called, Coreopsis, is one of the most showy, free-flowering, and beautiful of hardy annuals. The tall varieties are very effective in mixed borders and fronts of shrubberies; the dwarf kinds, from their close compact habit of growth, make fine bedding-plants, and are valuable for edgings. The different varieties make very pretty ribbons. Amongst the tall varieties, *C. filifolia Burridgii* is the most graceful and beautiful, and *C. bicolor grandiflora* the most showy and effective in mixed borders. All are hardy annuals except *C. Ackermannii*, which has a yellow, crimson centre, and is a hardy perennial; 3 ft., from North America.

189.—BENTHA'MIA (nat. ord. Corna ceæ).

An ornamental profuse-flowering half-hardy shrub; the flowers succeeded by reddish-yellow, strawberry-like fruit, which is eatable: succeeds against a south wall in any good soil.

Benthamia fragifera,—large cream-coloured flowers, 10 ft., from the East Indies.

190.—CU'PHEA (nat. ord. Lythra'ceæ).

Profuse-blooming plants, equally valuable for the ornamentation of the conservatory, drawing-room, and flower-garden. *C. eminens* is of a graceful branching habit, covered with splendid long scarlet and yellow tubular flowers; *C. Zinampinii* is covered with red-violet, and *C. ocymoides* with rich purple-violet flowers. The perennial species, if sown early, can be used for bedding-plants the first year; the annual kinds may be treated like ordinary half-hardy annuals.

191.—VARIEGATED PLANTS.

Much attention has been paid of late years to plants of variegated foliage, and certainly they are a great success. Few things are more attractive than the different sorts of caladiums, begonias, *Coleus alocasea*, &c. For a plant to be truly variegated, that is, able to retain its different colours under propagation, the edges of the leaves must be well defined. We frequently meet with party-coloured leaves which are mere sports, and which revert to their original hue if any attempt is made to increase the plants by cuttings, &c.

192.—CALCEOLA'RIA (nat. ord. Scrophularia'ceæ). Half-hardy Perennials.

Plants of a highly decorative character, indispensable for the ornamentation of the conservatory and flower-garden. The herbaceous varieties are remarkable for their large, finely-shaped and beautifully spotted flowers; these are cultivated exclusively for in-door decoration. The half-shrubby kinds grow more compact, have smaller flowers, bloom more profusely, and by many are more highly prized than the herbaceous sorts; they are alike useful for in and out-door decoration, and succeed in any light rich soil. The calceolarias mostly used for bedding are the different shades of yellow and brown. These admit of a very easy cultivation.

Take the cuttings early in October, and having prepared a piece of ground in a north border, the soil of which must be well drained, and made light with a large admixture of sand, place the cuttings in, and press the earth well round them, water them well, and cover with a hand-glass, or place the cuttings in pots, and having sunk them in a north border, under a wall, place a hand-glass over them. In this way they may be kept without further attention till the following spring, unless the weather should be very frosty, in which case it may be well to throw some covering over the hand-glass. In the spring the cuttings should be repotted, and will soon become fine plants. It is to be observed that the state of atmospheric influence most favourable to all cuttings is when a change to moist growing weather succeeds, within two or three days, the warm dry weather during which the cuttings have been taken.

THE KITCHEN GARDEN.

No better form can be devised for a kitchen-garden than a square, subdivided by two centre walks, as in fig. 1, or a long parallelogram, as fig. 2. Something like fig. 3 has been recommended by Mr. Loudon and others, the rounded part being a fruit-garden. The same figure might also be rounded at both ends. The centre walk should pass through close at each end. *a* re presents the wall; *b*, fruit-tree border, ten feet wide; *c*, walk, six feet wide;

and *d*, border for dwarf trees or bushes, or the culture of strawberries, &c., six feet wide. Whatever shape be adopted, borders should always be introduced on each side of the main walks. Nothing tends more to relieve the heavy appearance of large masses of vegetables than such borders. They are separated from the main vegetable compartments by small walks, from 18 inches to two feet wide. These walks can be edged with pebbles, and have a sprinkling of gravel, or simply cut off as alleys, and left solid earth. If formed of some hard substance, all the wheeling can be performed on them instead of on the main walks.

Perhaps the nearer to a level a kitchen-garden can be formed, the better. A slight inclination to the south-east, south, or west, might be an advantage; on no account should it incline to the north. Some gardens, however, are formed on the side of a hill, and prove very productive. Where the garden is nearly level, it may often be desirable to give fruit-tree borders a considerable inclination, to get the benefit of the sun's rays and insure thorough drainage. Borders against the wall may be sloped in directions opposite to those which line the inner side of the walk. These borders have also a good effect laid on in round ridges. It is also in kitchen-gardens a good plan to throw up sloping banks or zig-zag ridges for early and late crops. The south front of such banks, especially if a thatched hurdle or some other check to the wind is placed on the top, is equal to a south border; and the north side is equally useful for late strawberries, salading in hot weather, &c. Such banks are also most useful for training peas, &c., on table-trestles, within one foot or 18 inches of the surface. Some of the borders at the side of the walk might also be occupied by iron wire for training trees. One should be devoted to raspberries, planted three feet from the walks, and trained to a hand-rail at the side of the walk, from three to four feet high.

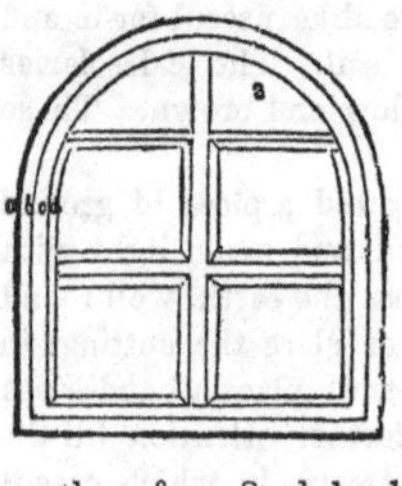

The size of the kitchen-garden must depend upon the demands upon it, and the mode of culture adopted. It is bad policy to have it too large. It should be kept in the highest state of cultivation, and its productive powers stimulated to the utmost by liberal dressings of manure. The soil should be trenched at least four feet deep, and drained a foot deeper. All the coarse vegetables, such as Jerusalem and globe artichokes, horseradish, rhubarb, &c., should be grown outside the walls, if possible, in a slip by themselves. Herbs should have a border devoted to them, and be grown in beds three feet wide. Thus cultivated, the back-garden becomes a source of interest and an object of beauty.

If the soil in the kitchen-garden is naturally good loam, no more is required than to mix a quantity of well-rotted dung with it before throwing it back into the trench, making the borders slope gradually towards the paths. If the soil requires improving, get a quantity of friable loam, mix rotten dung with it in the proportion of one part dung to three parts loam, and mix this again with the soil of the border where the trees are to stand. Plant healthy young trees of peach, nectarine, and apricot, and, if desirable, some grape-vines and figs: these ought to be placed 12 or 15 feet apart. We have seen a very convenient plan of growing grapes on a wall between the peaches. The latter were placed 15 feet apart, and a vine planted in each space halfway between; the vine was carried in a single stem to the top of the wall, where it divided into two stems, which were trained right and left under the coping; and as they were pruned on the spur system, they took up little room, and did not interfere with the other trees. On the east and west walls plant trained trees of plums, cherries, pears, and mulberries, after the same rule, but without the same precaution as to soil, as these are not so particular.

In draining the kitchen-garden, one of the drains ought to run the whole length of the south border; for where peaches, nectarines, and especially apricots are to be cultivated, the ground should be thoroughly drained.

193.—KITCHEN-GARDEN SEEDS.

In most large kitchen-gardens seeds of all the different vegetables in use are saved from year to year. This practice is recommended not only by economy, but by every consideration of good management; for, in this way, sorts that have been found to suit the soil and situation are effectually preserved. To save seeds, however, is a work of some trouble. It causes a great waste of ground, exhaustion of the soil, and also involves much labour. Moreover, in a thickly-wooded country, the birds are generally so troublesome, that, if they do not prevent the saving of seeds, they add much to the expense of it. Wherever, therefore, ground is limited, birds troublesome, and there are no spare hands, we should be inclined to give up the practice, or, at least, to limit the saving of seeds to a few favourite sorts. Good seeds can now be purchased at a very reasonable rate, and novelties in every kind of vegetable are continually being introduced.

But whether seeds are saved or bought, great care must be taken in storing them. They should be kept until wanted for use in some dry, airy situation. It is not well to commit seeds to boxes and drawers; the safest plan is to hang them up in small paper or muslin bags. Peas and beans, which are subject to maggots, should be looked over occasionally and kept clean. Seeds of some sorts of vegetables will keep good for years; but, of course, it is best to use new seeds; and there is always a saving of time in so doing, for old seed, when good, does not germinate as quickly as new. Most seedsmen publish lists of seeds with prices, suitable in quantity to gardens of all sizes, and as the demand for seeds is an annual one, no man of character will venture to hazard his reputation and his interest by sending out bad seed. It is very easy to test the growing qualities of seeds, and this should be done before they are packed up and offered for sale. We are assured that it is done with all seed supplied by Messrs. Barr and Sugden, of 12, King-street, Covent-garden, London. Their collections of kitchen-garden seeds are so varied in sorts, and so moderate in price, that the generality of gardeners may well be spared the trouble of saving seeds, and use their land for some useful crops. The following is a year's supply, with quantities and prices to suit gardens of six different sizes:—

Collections of Seeds for One Year's Supply.

	No. 1.	No. 2.	No. 3.	No. 4.	No. 5.	No. 6.
LEGUMINOUS SECTION.	12/6	15/6	21/0	31/6	42/0	63/0
Peas, including those best suited for succession, the earliest, most productive, and the finest flavoured	3 pt.	3 qt.	5 qt.	7 qt.	10 qt.	18 qt.
Broad beans, best sorts, for succession ...	1 pt.	1½ qt.	2 qt.	3 qt.	4 qt.	7 qt.
French Beans, including Dwarf and Runners	1 pt.	1½ pt.	1½ pt.	3 pt.	3½ pt.	6 pt.
EDIBLE-LEAVED AND EDIBLE-FLOWERED SECTION.						
Borecole, or Kale, including Cottager's Kale	2 pkt.	2 pkt.	3 pkt.	3 pkt.	4 pkt.	5 pkt.
Brocoli, the best successional varieties ...	2 pkt.	4 pkt.	5 pkt.	6 pkt.	7 pkt.	9 pkt.
Brussels Sprouts, imported Seed	1 pkt.	1 pkt.	1 pkt.	1 pkt.	1 pkt.	1 pkt.
Cabbage, the best varieties...	1 pkt.	3 pkt.	5 pkt.	6 pkt.	7 pkt.	9 pkt.
Cabbage, Savoy, best sorts	1 pkt.	1 pkt.	2 pkt.	3 pkt.	3 pkt.	4 pkt.
Couve Tronchuda, or Seakale Cabbage ...	1 pkt.	1 pkt.	1 pkt.	1 pkt.	1 pkt.	1 pkt.
Cauliflower, including Covent Garden early	...	1 pkt.	1 pkt.	1 pkt.	2 pkt.	2 pkt.
Spinach, summer and winter	2 oz.	3 oz.	4 oz.	1 pt.	2 pt.	3 pt.
EDIBLE-ROOTED SECTION.						
Beet, the best varieties	1 pkt.	1 pkt.	1 pkt.	2 pkt.	2 oz.	3 oz.
Carrot, best for forcing, and general crop	1 oz.	2 oz.	3 oz.	7 oz.	10 oz.	16 oz.
Leek	1 pkt.	1 pkt.	1 pkt.	1 pkt.	1 pkt.	1 oz.
Onion, including White Spanish, syn. Reading	1 oz.	2 oz.	3 oz.	6 oz.	8 oz.	13 oz.
Parsnip, including selected Hollow Crown	1 oz.	1 oz.	2 oz.	3 oz.	4 oz.	6 oz.
Salsafy, syn. Vegetable Oyster, an excellent vegetable	...	...	...	1 pkt.	1 pkt.	1 pkt.
Scorzonera,—a most valuable and fine-flavoured vegetable	...	...	...	1 pkt.	1 pkt.	1 pkt.
Turnip, including best varieties for succession crop	2 oz.	2 oz.	3 oz.	6 oz.	8 oz.	16 oz.
EDIBLE-FRUITED SECTION.						
Capsicum and Chili	...	...	...	1 pkt.	2 pkt.	2 pkt.
Cucumber, including Barr's Pearl Gem ...	1 pkt.	2 pkt.	2 pkt.	3 pkt.	3 pkt.	4 pkt.
Melon	...	...	1 pkt.	2 pkt.	2 pkt.	3 pkt.
Tomato	...	...	1 pkt.	1 pkt.	2 pkt.	2 pkt.
Vegetable-Marrow	1 pkt.	1 pkt.	1 pkt.	2 pkt.	2 pkt.	2 pkt.
SALAD SECTION.						
Celery, including Covent Garden variety	1 pkt.	1 pkt.	2 pkt.	2 pkt.	2 pkt.	2 pkt.
Corn Salad, valuable for winter use ...	1 pkt.	1 pkt.	1 pkt.	1 pkt.	1 pkt.	1 pkt.
Cress, including Curled, Plain, and Australian	3 oz.	3 oz.	4 oz.	6 oz.	8 oz.	12 oz.
Endive, best kinds	...	1 pkt.	1 pkt.	1 pkt.	2 pkt.	3 pkt.
Lettuce, including Covent Garden Giant White Cos	2 pkt.	2 pkt.	3 pkt.	3 pkt.	4 pkt.	6 pkt.
Mustard	2 oz.	3 oz.	4 oz.	½ pt.	1 pt.	1 qt.
Radish, suitable sorts for succession ...	3 oz.	4 oz.	6 oz.	9 oz.	14 oz.	16 oz.
POT, SWEET, AND GARNISHING HERB SECTION.						
Herbs, Pot and Sweet	2 pkt.	2 pkt.	2 pkt.	3 pkt.	4 pkt.	6 pkt.
Herbs for Garnishing	...	...	1 pkt.	2 pkt.	2 pkt.	2 pkt.
Parsley	1 pkt.	1 pkt.	1 oz.	1 oz.	2 oz.	3 oz.

			£	s.	d.
No. 7.	Extra large Collection of Vegetable Seeds for One Year's Supply	...	4	4	0
" 8.	" " " "	...	5	5	0
" 9.	" " " "	...	6	6	0
" 10.	" " " "	...	8	8	0
" 11.	" " " "	...	10	10	0

194.—TOOL-HOUSE.

The tool-house may be attached to the gardener's cottage or placed at the back of the hothouses, if such a situation be found more convenient. In every well-planned tool-house there should be contrivances of different sorts for hanging up the tools,—rakes, hoes, spades, &c.,—which should all be well cleaned before they are put away. If many men are employed in the garden, each one should have a proper place for his own tools. Watering-pots, syringes, garden-engines, should have their movable parts separated, and be reversed, in order that they may drain and dry. The mowing-machine should be kept thoroughly clean and oiled, and so should all clippers and pruning instruments. A bench with a vice attached to it will be found very useful in a tool-house; also a grindstone and hones for the sharpening of different tools.

195.—TOOLS.

The tools required in a garden may be divided into implements, instruments, and tools; the former comprising the pick, spade, fork, hoe and rake, all of which require the use of both arms, and some of them the whole muscular force of the frame. They generally combine the principle of the lever and the wedge, the blade of all of them being employed to separate particles of matter by the application of lever power, which lies in the shaft or handle. Where the implement is intended to be grasped and held firmly at one spot, as in the spade, a handle is provided adapted for that end; where the hand is to slide along, the handle should be smooth and round: such is the form of the handles of the rake, the hoe, the pick, and all similar tools. Ash is the best material for handles requiring strength, willow being lighter, and strong enough for others, such as the rake, the Dutch hoe, &c., where the handle is required to be of some length.

From the catalogue of Messrs. Barnard & Bishop, the Walk, Norwich, we select the following tools, with their prices.

The Pick is a compound lever, the blade of which ought to be of the best wrought iron edged with steel, the handle of well-seasoned sound ash. The pick is made in various shapes; for garden use, one end is usually pointed, the other wedge-shaped, to adapt it for cutting through roots when they are met with in the soil. That called the mattock, having the edge axe-fashion, is used to chop up hard grassy surfaces, and to stub up whins, heather and other wild shrubs.

The spade is a broad blade of plate-iron, attached to a handle of tough root-cut ash. Two-thirds of the blade ought to be of steel, and the other of the best scrap-iron, well welded together. In some, the blade is perforated for the purpose of cleaning itself when employed on adhesive soils. The semicircular spade, used by "navvies," is useful in hard close soils and new ground. The shovel, having a broader blade, is useful in throwing up the loose soil at the bottom of a trench. Cast-steel spades, 3*s.* 6*d.* to 5*s.* each; shovels, 3*s.* 6*d.* to 4*s.* 6*d.*

The Fork, of which there are several kinds, is used for filling in dung litter and haum, and for levelling the surface, or stirring it round the roots of plants which might be injured by using the spade. The form of the fork is generally three-pronged,—those used for litter, and for forking up the earth round roots of trees, have only two prongs; but Parker's fork has five prongs, and in many cases supersedes the use of the spade. The three-pronged fork with broad tines is used for digging potatoes, and is otherwise a very useful implement in the garden. Three-pronged forks, 4*s.* to 4*s.* 6*d.*; five-pronged, 5*s.* 6*d.*

The Hoe, also, is of many forms. Its use is to stir the surface of the soil and destroy weeds in their first leaves. The draw-hoe is set at the end of a long handle at an angle, so that when drawn to the operator, it sinks also into the soil. In the other, generally termed the Dutch hoe, he thrusts it from the angle formed by the height of the hands from the ground, carrying it gently into the surface of the soil. This is used where the weeds are merely cut up and left in the soil. Hoes, 6*d.* to 1*s.* 6*d.* each; Dutch hoe, 9*d.* to 2*s.* 6*d.*; Prussian hoes, 1*s.* 6*d.* to 2*s.*

The Rake is a row of small tines inserted in a bar of iron from six to eighteen inches long, and fixed by a socket placed at right angles to the bar, to a handle five or six feet long. This is a most important implement to the gardener, being used to make his beds smooth and trim, and to remove loose weeds. It is also used in raking the lawn before rolling. Garden rakes, from 1*d.* to 2*d.* per tooth; daisy rakes for lawns, 5*s.* to 7*s.* each.

The common Garden-knife consists of a blade of prepared steel, fixed in a handle of horn or bone, without a joint, and carried in a sheath. It generally has a hooked blade, and is used for cutting cabbages and other vegetables trimming the roots of turnips, and similar purposes.

The Pruning-knife has the blade quite straight on the cutting side: it may be either with or without a joint. 1*s.* 6*d.* to 4*s.* each.

The Grafting-knife has a thinner and narrower blade, and is generally made, with a folding joint. 1*s.* 6*d.* to 4*s.* each.

The Budding-knife has the sharp edge of the blade rounded off, backwards; while the handle, of bone or ivory, is brought to a fine point at the end opposite the blade, so as to form a spatula for lifting up the bark of the stock for the insertion of the bud. Goodsall's budding-knife has a heart-shaped termination to the handle, which is found more convenient for lifting the bark. 1*s.* 6*d.* to 2*s.* 6*d.* each.

The Grafting-saw.—This has a blade of steel with double teeth, so that it may pass through green wood without choking up. For larger branches the forest-saw, having a handle six or eight feet long, is used to enable the operator to reach the loftier branches. When this instrument is used, the end left on the tree must be smoothed with the chisel or pruning-knife, and the wounds covered with some composition which will exclude the air. Saws from 3*s.* to 5*s.* each.

The Chisel.—There are two sorts in use,—the garden and the forest chisel: the first is chiefly used in grafting, and differs from the carpenter's chisel in being wedge-shaped, or bevelled on the edge on both sides; thus tapering to a central point. It is used with a mallet, to slit stocks for grafting when they are too large for the grafting-knife, and to smooth branches cut with the saw. 1*s.* to 1*s.* 6*d.* each.

The Bill-hook: a hooked blade sharpened on both sides, with a handle about a foot long: a useful instrument for cutting away branches too thick for the knife. With a long handle and sharpened on the hooked side, it is used to cut hedges, in place of clipping them. 2*s.* 3*d.* to 4*s.* each.

The Axe is a steel wedge, attached at right angles to a strong handle of ash timber, about two feet and a half long. The axe varies much in form: the long and narrow one, known as the American axe, is the most useful. It is used for cutting roots of trees which cannot be conveniently reached with the saw. 6*d.* to 6*s.* 6*d.* each.

The Scythe: a long blade of steel, attached to the end of a crooked wooden handle. Boyd's scythe is much used in gardens where the grass is kept under by mowing. Lawn scythes complete, 6*s.* 6*d.* to 7*s.* 6*d.* each.

Shears of various kinds are in use in the garden, both for clipping hedges and dressing the edgings of the beds. The turf edges are kept smooth by using those with long handles, moving upon a single wheel: shorter shears are used for cutting box-edgings and quickset hedges. Shears, 2*s.* 9*d.* to 7*s.* 6*d.* per pair; with wheels, 7*s.* 6*d.* to 8*s.* 6*d.*

The Garden-engine is of essential service where water cannot be distributed from a tank to all the garden by means of a hose. These vary in price according to contents—from 12 gallons to 30 gallons, price from £3 10*s.* to £5 15*s.*

With the Syringe, which is essential in the greenhouse, the garden-engine may be made to produce the nearest possible approach to a natural shower of rain, preceded by a misty dew. Price of syringes, 6*s.* to 13*s.* 6*d.* each.

The Garden-line consists of an iron reel turning on a spindle, terminating in a peg and a cord. The cord is wound round the reel, and, having a peg at the other end, the beds are trimmed and the rows directed by the line. Price 2*s.* 6*d.* to 3*s.* 6*d.*

The Level, as its name implies, is to guide the workman in all operations where either a perfect level, or, as in draining, a regulated fall, is required. The most useful is half a square, with an iron index, marked out with ninety degrees or divisions. Where a slope is to be laid out at a perfect level, the plummet will hang in the centre, or 45°, and on a slope it may hang at any lesser number, in ascending at any higher number, according to the steepness of the slope.

The Measuring-rod is made of wood, generally 10 feet long and an inch square, on which the feet are marked, the last foot having the inches also.

The Wheelbarrow, though last is not least; indeed, it is the most indispensable implement in the garden next to the spade, hoe, and rake, and is too well known to require description.

The Ladder, of various sizes, will be required in the garden.

196.—HERB-GARDEN.

The *olitory*, or herb-garden, is a part of horticulture somewhat neglected, and yet the culture and curing of simples was formerly a part of a lady's education. All the sweet herbs are pretty, and a strip of ground halfway between the kitchen and the flower-garden would keep them more immediately under the eye of the mistress. This would probably recover, for our soups and salads, some of the neglected tarragons, French sorrel, purslain, chervil, dill, and clary, which are only found now in the pages of the old herbals. Laid out after a simple geometrical design, the herb-garden might be rather ornamental than otherwise. Most of the herbs are propagated by slips in the autumn. *Basil burnet*, and other herbs, require to be sown early in spring, on slight hotbeds of about two feet in depth; but many cultivators leave them later, and sow in the open ground. Thyme, marjoram, savory and hyssop, chervil, and coriander, may be sown in dry mild weather, to be transplanted afterwards. Sow in shallow drills about half an inch deep and eight or nine inches apart, and cover in evenly with the soil. *Mint* may be propagated by separating the roots, and planting them in drills drawn with a hoe six inches asunder, covering them with an inch of earth, and raking smooth. They will quickly take root, and grow freely for use in the summer. This method may be applied to the several sorts of spearmint, peppermint, and orange-mint.

The whole family of borage, burnet, clary, marigolds, orach-root, carduus, dill, fennel, buglos, sorrel and angelica, may be sown about the middle of March, when the weather is open. Sow them moderately thin in drills or beds (each sort separate), in good light soil; if in drills, six inches apart; some of the plants may remain where planted, after a thinning for early use; others may be planted out in the summer.

197.—STORING SEEDS.

In collecting seeds, the greatest care is required to have them ripe, and that the bags into which they are put are correctly marked. All that is known of the parent plant should be added, if it is other than a common kind, including the soil in which it is found. When collected, before packing away, the seeds should be carefully dried. When they belong to pulpy fruit, separate the grains from the pulp as soon as decomposition begins, and dry before placing them in bags.

198.—STORING VEGETABLES.

There are several sorts of vegetables which require storing for winter:—potatoes, carrots, beet, and onions are the chief of them. Potatoes do best when harvested in clumps in the open ground, care being taken to protect them from rain and frost. A long ridge is the best form. The ground should be dry and thoroughly drained. The potatoes should be heaped on a ridge, tapering from a base of three feet to a foot and a half, or less, at the top, separating the different sorts by divisions in the ridge. It is usual to cover this ridge with a thatch of wheat-straw, and then with six or eight inches of mould; but some authorities highly disapprove of this. McIntosh recommends the tubers being covered with turf, and afterwards with soil; and in the absence of these, laying on the soil at once without any litter. After having laid on nine or ten inches of soil, thatch the whole over an inch and a half thick, with straw, fern leaves, or any similar non-conducting material; "the object being," he says, "first to exclude frost and wet, and, secondly, to exclude heat; for which purpose earth is not sufficiently a non-conductor of heat and cold."

If the weather is fine when the tubers are taken up, and the potatoes are required for early use, much of this labour may be dispensed with; but if for spring and early summer use, the precautions will be found necessary.

Carrots, beet, and other similar root-crops should be taken up before the frosts set in: they may either be stored in a dry cellar, covered with dry sand, or after the manner of the potato. The London market-gardeners winter their beet and carrots in large sheds, in moderately damp mould, and banked up with straw; "for," says Mr. Cuthill, "it is a mistake to pack them all in dry sand or earth for the winter; and the same may be said in regard to carrots, parsnips, salsafy, scorzonera, and other similar roots; and by this means," he goes on to say, 'the roots retain their natural sap, and the colour is preserved."

It is probably unnecessary to add that in roots and tubers, as with fruit, all cut or bruised ones should be thrown aside: when the skin is cut, or a bruise exists, the elements of decay are soon introduced, and all others within reach contaminated. A dry day should be chosen for lifting them, and they should be exposed a few hours before collecting into heaps, that the soil adhering to them may dry.

Onions should be lifted a little before they have altogether ceased to grow: the leaf turning yellow and beginning to fade will be the sign. As they are taken up, they should be placed in a dry airy place, but without being exposed to the sun. If they are thinly spread out on a dry floor or shelf covered with sand, or on a gravel walk partially shaded in fine weather, they will do very well. As they dry, the roughest leaves should be removed; when dry, they should be removed to a warm dry loft, where they can ripen more thoroughly. When in a proper state for storing, they should be gone carefully over and separated, the smallest ones for pickling, the ripest picked out, as likely to keep longest: those with portions of leaves to them are best stored by stringing and suspending them from the ceiling of the room, which promotes ripening. The stringing is done by twisting a strong piece of matting or twine round the tails of each in succession, so that they may hang as close together as possible without forming a cluster, until the string is about a yard long; when they are hung up, they occupy very little room, and have a good opportunity of ripening.

199.—PEAS.

Selection of Soil and Situation.—For heavy crops of this prime esculent a deep loamy soil must be secured; but ordinary garden soil, if properly prepared and well manured, will yield abundantly, For an early crop, plant in the warmest and most sheltered situation; but for the main crops choose an open airy situation; and instead of devoting a portion of the garden to peas alone, as is usually done, plant them in single lines amongst other crops: the plants will thus get more sun and air, and bear much longer and more abundantly.

Preparation of the Land.—Trench to the depth of two feet, and ridge up roughly, exposing as large a surface as possible to the action of the weather; and this should be done as long before sowing as convenient. The summer and autumn crops will require abundance of well-rotted manure; but the early crop will come sooner into bearing if planted in poorer soil, which should be deep and well pulverized.

Time and Manner of Sowing.—Sow the first crop about the middle of November, the second early in January, putting in a small breadth of a second early variety at the same time; and to secure a constant succession, sow once a fortnight from this time till the end of June, or yet later. After the beginning of March sow the best kinds of Wrinkled Marrows; but for the last two sowings use a free-cropping early, or second early variety, and when the ground is sufficiently dry to work kindly, sow in drills two inches deep and four inches wide, covering the seed with friable soil. If sown in successive lines, let the intervening space exceed the reputed height to which the variety grows by six or twelve inches. As the seed for the earlier crops will be some time in the ground exposed to the depredations of mice, &c., it should be sown thickly. The strong-growing branching kinds, which are used for the main crops, succeed better if sown thinly, but it is prudent to guard against loss from various causes by sowing all rather thickly. If the plants are found to be too close when fairly started, they can easily be thinned out.

After-management.—When the plants are about two inches high, draw the soil neatly towards them, and apply stakes of about the height to which the variety grows. Spruce fir or other evergreen branches will afford a useful shelter to early crops. Keep the ground between the rows well stirred and free from weeds. In dry weather mulch with manure for eighteen inches on each side the rows, giving a liberal supply of water when necessary, to keep the plants vigorous, and to prevent mildew: the growing crops should never be allowed to feel the want of water.

Peas, earliest Sorts. — Sutton's Ring-leader; Dixon's First-and-best; for later sowing, Laxton's Alpha; Laxton's Superlative; Veitch's Perfection; Hair's Dwarf Mammoth, Ne-plus-ultra, and Oxford Tom.

200.—PEA-PROTECTORS.

These are made on an arched frame, covered with galvanized iron wire netting, to protect peas when sown from birds. They are a yard long, and may be bought of Messrs. Barnard and Bishop, Norwich, at 15*s.* per dozen.

201.—POTATO.

Selection and Preparation of the Soil.—A deep, thoroughly drained light sandy loam, or peaty soil, is most suitable for the potato. The application of manure is now generally held to increase the liability to disease, to bring it on at an earlier period, and to produce large crops of imperfectly matured tubers, which if they escape the disease while in the ground, are more liable to be attacked after they are lifted. The ground selected, therefore, should be in fair condition, from having been moderately manured for some exhausting green crop in the previous season. But if the only land to be had is so poor as to render it necessary to apply manure in order to insure a fair crop, then use charred vegetable refuse, or a very light dressing of well-decayed farm or stable-yard manure. The ground should be trenched two spades deep, and ridged up early in autumn; if manure is applied, this should be well mixed with the soil. Charred vegetable refuse, however, may be applied about the sets when they are planted. We believe that a slight sprinkling scattered along the trench before planting, and then used in covering the sets, has proved a partial preventive of disease.

Time and Manner of Planting.—We have no hesitation in saying, that early planting has hitherto proved the best preventive against the attacks of disease; for, as the crops sooner arrive at maturity, they frequently escape altogether, or suffer comparatively little. Plant, therefore, as early in January as the ground can be found in fair working condition. A small breadth of the ash-leaved kidney should be planted on a south border, or in the warmest and most sheltered situation at command, to furnish an early supply. In planting, let the ground be nearly levelled, then, beginning at one side, dig it over about six inches deep, and put in the sets in the openings at proper distances, which must be regulated by the growth of the variety. The lines for the early kinds, as ash-leaved, &c., which form but small tops, may be about twenty inches apart, leaving about nine inches between the sets, but for the second early varieties two feet should be allowed between the lines, and ten inches between the sets. The late kinds will require an additional six inches between the lines. The sets should be covered about six inches, leaving the soil over them as open and loose as possible. On strong heavy land the ash-leaved and other weakly growers should not be covered more than four inches. Planting in autumn has been strongly recommended, and on light, well-drained land, it may safely be practised.

After-management.—When the tops are four to six inches above the ground, ridge the soil up neatly about them. In the case of the early varieties, which may be in danger of suffering from the frost, the soil should be kept ridged up round the shoots as soon as they appear above the ground, keeping them covered until they are four to six inches high, and all danger of frost is past. Before earthing up, fork the ground lightly between the lines, so as to pulverize the soil, then draw it to the plants with a hoe or spade. Keep the ground clear of weeds. When the crop attains maturity, lift and store; or if the disease is troublesome before the tubers are ripe, lay the shoots down along the top of the ridge and cover them with soil. This seems to be the most successful method known at present of checking the ravages of the disease. The crop should be taken up, however, as soon as possible after disease makes its appearance, and the tubers stacked and temporarily covered in some place where they can be examined occasionally. When the disease appears to have done its worst, pit them in the usual manner, covering them sufficiently to protect against severe frost.

The following varieties are most worthy of cultivation :—

Round Potatoes.

Early Covent Garden Prolific,—a first-class variety, very early, an extremely heavy cropper, fine-flavoured, with a dwarf compact top. We highly recommend this variety.
Early Oxford.
Early Shaw.
Regents.
„ Pink-eyed.
Sutton's Flour-ball.
Napoleon.
Headley's Seedling,—a very long keeper.

Kidney Potatoes.

Early Ash-leaf,—for forcing or very early use. A very fine stock.
Early Myatt's Covent Garden Prolific Ash-leaf,—a handsome and most desirable potato; an abundant cropper, but not quite so early as the above, nor so well adapted for forcing.
Webb's Imperial,—a great cropper, resists the disease better than most others, is very handsome, and may be cooked from July to June.
Daw's Covent Garden Matchless,—a very handsome and desirable variety, fine-flavoured and very prolific, ready in August, and much in demand in Covent Garden.
Flukes.
Lapstone.

202.—POTATOES IN FRAMES.

When potatoes are grown in a frame, the treatment is much the same as before; but they are also grown very successfully in this manner :—The frame being placed on a level piece of ground, the soil within is dug out to the depth of two feet, and banked round the outside of the frame. The pit thus formed is then filled with prepared dung; and on this three inches of soil is placed; then the potatoes; then six inches more soil. The potatoes, when planted, should be just starting into growth; but the shoots should never be more than half an inch from the tuber, or they do not grow so strong. It is advisable to pick off some of the shoots: three on each tuber are sufficient. They may also be forced under the stand in a greenhouse or hothouse, the potatoes being planted singly in large pots of very rich light soil. Each pot ought to yield a good dish.

203.—CABBAGES.

Cabbages, or *Brassicæ*, are the most important product of the garden, whether we look at them as a necessary or a luxury of life. They are also, except under a well-considered system of rotation cropping, the most exhaustive class of vegetables under the gardener's care.

This important family of vegetables is biennial, triennial, and nearly perennial in some of the varieties. It may be divided into—

1. The Cabbages proper, which have heads formed of the inner leaves growing close and compactly round the stem, which are thus blanched into a whitish yellow by the outer leaves.

2. Red, or Milan cabbage, which grows in the same form, but differs in colour.

3. Savoys, distinguished by their curly wrinkled leaves, but retaining the tendency to form a head.

4. Brussels sprouts, producing the sprouts, or edible part, from the stem in small heads, like very young cabbages.

5. Borecole, of which there are many varieties, having a large open head with large curling leaves.

6. Cauliflower and brocoli, in which the flower-buds form a close fleshy head of a delicate yellowish-white, for which both are cultivated.

Of the first of these there are many varieties, some of them valuable for their precocity, which adapts them for early spring cultivation; others for more enduring qualities. They are all propagated by seed sown for main crops twice a year—namely, in April, for planting out in June and July, for autumn and winter use; and in August and September, for spring use; but it is usual to make sowings of smaller quantities every month for succession.

The Romans propagated the Brassicæ by seeds and cuttings, by which choice varieties may be perpetuated with greater certainty than from seed. This is done by slipping off the sprouts, which all the tribes produce on the stem, when about four inches long, exposing them to the air for a day or two to cauterize the wound, then dipping them in caustic lime, and planting them where they are to grow. Pliny tells us they are fittest for planting or for eating when the sprout has six leaves.

The Cabbage.—The seed is sown on beds four feet wide, and long in proportion to the sowing. A bed 4 feet by 20 will take 2 oz. of seed. Cover the seed to an eighth or a quarter of an inch with rich light soil, and rake it in: the after-cultivation will be gathered from the monthly calendars. The cabbage requires a rich retentive soil, and is improved by early transplanting. When about two inches in height, the young plants should be removed into nursery-beds thoroughly prepared by digging and manuring, and, if dry, by watering, where they are planted four or five inches apart. Here they must remain till well rooted. Their next remove is usually to the place where they are permanently to grow; but they will be rather improved than otherwise by an intermediate shift to a second nursery-bed.

In final planting out, the ground being trenched and well manured, a drill is drawn, three inches deep, at a distance proportioned to the size and habit of growth of the variety; the small or early dwarfs at 12 or 15 inches apart in the rows, the larger sorts, as Vanack, at eighteen inches. The subsequent culture is confined to weeding and occasionally stirring the earth during summer, and drawing it up round the stem when about eight or nine inches high.

The best varieties of the white cabbage are the Early York, Early Battersea, Early Dwarf Sugar-loaf, the Late Sugar-loaf, Vanack, the Portugal, or *Couve Trunchuda*—of all of which there are many varieties; as Atkin's Matchless, Sutton's Dwarf Combe, Sutton's Imperial, Enfield Market, Shilling's Queen. The conical Pomeranian is singularly hardy and very compact. The Vanack, was subjected to experiment by the Horticultural Society, and Mr. George Lindley reported it as "always in season by timely sowings, making excellent spring coleworts; becomes white-hearted cabbage very early, and furnishes fine sprouts after the cabbage is cut." The red cabbage is chiefly used for pickling, and the varieties are confined to the Dutch, Aberdeen, and Dwarf Red. Their cultivation is in all respects the same as the white cabbage, and the vegetable is only gathered when the head is thoroughly formed, and when so gathered the stem is thrown away as of no further value.

204.—BORECOLE.

Borecole and Curlies are a numerous tribe of Brassicæ, culivated for their leaves in winter, and for their sprouts in spring. The first week in April or May, and again about the second week in August, is the time to sow. The borecoles are less exhausting to the soil than cabbages, and will follow peas without fresh manuring, if the ground is in tolerably good heart; or they may be planted between rows of peas or potatoes, to occupy the ground when these crops are removed.

The Horticultural Society have experimented on this tribe of Brassicæ, and issued a report; but the varieties are so numerous and so mixed, that the distinction between them is very indefinite. *Dwarf Curled Greens*, under half a dozen names, are the old Scotch curly, very dwarf in habit, and closely curled,—an excellent variety. The *Tall Green Curled*, also under a host of names, grow two or three feet high, stand severe frost, and afford the most delicate greens when frosted. *Purple Borecole* differs little from the preceding except in colour. Variegated borecole is a mere variety, very useful, and even ornamental, in the mixed gardens.

205.—ARTICHOKES.

Of Artichokes there are two sorts, the Jerusalem and the Globe. The former are most nutricious vegetables, and far too little known. They will grow in any light sandy soil, where few other things will grow. The tubers should be planted in March, and the roots raised and stored away in November.

206.—ARTICHOKES (GLOBE).

These are best propagated by offsets taken in March. The plants bear best the second or third year after planting; so that it is advisable to plant one or more rows every year, and remove the same quantity of old roots. The ground should be deeply worked and well manured: let the manure be incorporated with the soil, not laid in a mass at the bottom of each trench. It is better to trench the ground first, and fork the manure well into the surface-pit, which gives the plants a better chance of immediately profiting by it. The offsets may be dissevered with a knife, or slipped off and cut smooth afterwards, and planted with a dibber. Some plant in threes, a yard apart, and four feet from row to row; or they may be planted singly, two feet apart in the row, and four feet from row to row. They should be well watered, and the ground kept loose between.

207.—CELERY.

The celery-plant, *Apium graveolens*, is a biennial in its wild state, although the mode of cultivation adopted makes it an annual, except when grown for seed: it grows naturally in our marshy grounds, but we are indebted to Italy for this vegetable as well as the name, *celeri; ache* being the popular English name given to it by Ray and the older writers. It is propagated by seed, which is best obtained from the seed-shops. It may be sown in any month from Christmas to April. To get plants for the table in September, seeds should be sown in February in pans, which should be placed on a moderate hotbed: in about three weeks they will germinate, and, when about two inches high, the plants should be pricked out under glass, either in a frame or in pots, in a compost of loam, and three parts well-rotted dung. If in pots, shift them in April, and at the end of May plant them in shallow trenches in a warm part of the garden. If the trenches are dug out to the depth of two feet, six inches of hot dung placed in the bottom to stimulate the plants, the soil replaced, and the plants put in and covered with hand-glasses, an early crop will be the result. A second sowing should be made in March, still on a hotbed or on pans, or protected by sashes and mats until the plants are up; when fit to handle, they should be pricked out on a slight hotbed, or on a warm border. After a few weeks they should be again transplanted into a similar bed, and placed four or five inches apart each way. In July the plants will be fit to plant out in trenches for autumn use; a third sowing in April, treated in a similar manner, will be ready for winter use, pricking them out in fresh loam and decomposed leaf-mould when large enough to handle. When ready to plant for good in the trenches, mark out the ground into 4 feet clear spaces between the trenches, allowing 15 inches for the trench, if single rows, and 20 inches if double rows, are to be planted. Have not less than four feet clear space between the rows; dig out the trenches one spade deep, throwing the soil on the spaces between; then dig in the trenches a good dressing of rotten dung. If the weather is hot and dry, it is as well to wait for a shower of rain, unless the trenches can be watered both before and after planting. If in single rows, it is merely necessary to plant along the centre of the trench, a foot apart; if double, stretch a line along the centre, and plant a row on each side of it. Let the young plants be taken up and planted with a trowel: they should be well settled in with water. Before earthing-up, as it is called, it is always best to gather up each plant with the hand and press the soil about it, to keep the leaves together, which prevents the soil falling into the heart of the plant. Never earth-up till about three weeks or a month before it is wanted,—before the month of October, when a little more time must be allowed for blanching. Early in November the final earthing-up should be done, unless in wet soils, when the plants might be protected with litter and earthed-up for blanching at discretion, otherwise sharp frosts might injure them.

208.—CARROTS.

Sow broadcast on beds, and thin to three or four inches for the smaller sorts larger sorts are better sown in drills. If it is preferred to drill the seed, let the drills be one foot or 15 inches apart, as shallow as possible, and sow the seed continuously along the drill, or three or four seeds at intervals of six or eight inches: this economizes the seed, and admits of going amongst the plants without treading on them. Light ground should be trodden before it is drilled; the seed hangs together and should be separated by rubbing it up with soil, if sown broadcast; but this is unnecessary if sown in drills. The seed is very light, so that a calm day should be chosen for sowing: a little wind is apt to blow it anywhere but into the right place: it takes from one to three weeks to germinate. As soon as the plants are well above ground, use the small hoe unsparingly, and thin out to not less than six inches apart; as they advance, continue using the hoe both to destroy and prevent the growth of weeds, and also for the benefit derived from loosening the ground. Carrots may be drawn for table as soon as large enough; but the main crop for storing should not be taken up till quite the end of October, or even later, unless severe frosts set in. There are many different sorts; but the Dutch Horn is generally used for forcing and early crops; Intermediate for second or late crop; the improved Altringham is good for main crop; but much depends on soil and locality. To produce carrots and parsnips of an extraordinary size, make a very deep hole with a long dibble; ram the earth well round it while the dibble is in, and when it is removed, fill up the hole with fine rich earth. Sow a few seeds on the top, either parsnips or carrots, as may be required, and when up, draw out all except the one plant nearest to the centre of the hole. Prodigious carrots and parsnips may be produced by this means.

209.—CAULIFLOWERS (Brassica oleracea Botrytis).

The origin of the cauliflower, like all the family, is ascribed to the common wild cabbage, *Brassica oleracea.* It is of eastern origin, having been brought from Cyprus into France, and introduced from that country into England early in the 17th century.

With us the plant is treated as an annual, although it may, like all the race, be propagated from cuttings. In order to keep up a succession, three or four sowings should be made in the season, the first sowing being made on a slight hotbed in February, or very early in March.

Early in April a second and larger sowing should be made in the open ground, and a third and last sowing about the middle of August to stand through the winter. These sowings are made on beds of rich light soil, thoroughly pulverized by digging, and neither too dry nor too moist, four and a half feet wide, and long in proportion to the requirements of the garden, half an ounce of seed being sufficient for a ten-foot bed. In very dry weather, the seed-beds should receive a copious watering the night before sowing. In June the April sowings will be fit to plant out where they are to grow; in September they will be heading, and will continue to improve up to the frosts of early winter.

The autumn-sown plants are usually pricked out under frames for protection during winter, keeping them clear of weeds and decaying leaves, stirring the soil occasionally, and giving plenty of air in fine weather, protecting them from frost and rain. As they advance, and begin to head under hand- or bell-glasses, every opportunity should be taken of giving air; in severe weather, protect the frames and hand-glasses by packing litter round them.

When the heads begin to appear, shade them from sun and rain by breaking down some of the larger leaves, so as to cover them. Water in dry weather, previously forming the earth into a basin round the stem, and pour the water into the roots, choosing the evening in mild weather for so doing, and the morning when the air is frosty.

210.—BEET.

There are two or three sorts of beet for garden cultivation—one known as the Perpetual Spinach beet, the leaves of which may be gathered throughout the summer and used as spinach: another called Seakale beet, the mid-rib of the leaf of which is from two to three inches broad, very white, and delicate in flavour, and used as a substitute for seakale. The edible-rooted beet is, however, the most useful variety. For the first supply sow a small quantity early in April, and the main crop the first week in May; but where small roots are desired, sow as late as the middle of June. Sow in drills about one inch deep, and from 15 to 18 inches apart, covering with friable soil. If possible, select a dry day when the ground is in good working order for putting in the seed. Thin out the plants so that they may be from six to nine inches apart in the rows. Keep the ground free from weeds, and open by frequently stirring the surface. By the end of October the roots will have attained their full size, and they should then be taken up and stored in soil not over dry. In an airy cellar or shed they will keep perfectly well until spring; but when drying winds occur, they should be transferred from the latter into a damp, cool cellar. In pulling and cleaning, be careful not to wound the roots, or cut off any large fibres, as this would cause bleeding, which greatly injures the quality and tends to induce decay; neither should the leaves be cut off too close to the crown. In stacking place the crowns outwards.

Covent Garden,—extra fine, medium-sized, beautifully-shaped roots, of a rich deep blood-red colour; boils tender, and is of superior flavour.

Carter's St. Osyth's,—medium size, good shape, short top, rich deep blood-red colour; fine flavour.

Cattell's,—a useful sort.

Henderson's Pineapple,—compact short-topped variety, roots medium-sized, and of a fine deep crimson; boils tender, and is of very superior flavour.

White's Black,—large root, almost black.

211.—CAPSICUMS.

Pretty ornamental plants, especially in autumn, when covered with their light scarlet fruit. From the Capsicum cayenne pepper is made. The seed should be sown early in March, in well-drained pots filled with light sandy soil, and placed in a cucumber-frame, or wherever a temperature of about 65° is maintained. Cover the seed to the depth of about half an inch, and keep the surface constantly moist until the plants appear. When the plants are strong enough to handle, pot them off, placing two or three plants in a 5-inch pot, and replacing them in the warmth. Keep them rather close until they become established, then shift into 7-inch pots; and when they are fairly established in these, remove them, if intended for the open ground, to a cold frame, and gradually prepare them for planting out by a freer exposure to the air. Those intended to grow in pots under glass should be shifted into 10-inch pots as soon as they require more space for their roots, and be stopped, so as to cause them to form bushy plants; they must be liberally watered and syringed over head during droughty weather. Those intended for the open garden may be planted in properly prepared situations towards the end of May, protecting them by hand-glasses or any more convenient contrivance till they are fairly established. They must be liberally watered during hot, dry weather. In favoured localities most of the varieties do better planted out than when grown in pots under glass; but they will not succeed in the open air except in warm, dry situations.

212.—CARDOON.

A perennial in its native country,—the shores of the Mediterranean,—it becomes an annual in this country, the first sowing taking place in the beginning of March, on a very slight hotbed; in April, on the natural ground; and again in June, for next spring's crop. The trenches are dug as for celery, and moderately manured with well-decomposed dung. In sowing, two or three seeds are sown together in a clump 12 inches apart. Should each vegetate, remove all but one, when six inches high. When the plant is 18 inches high, put a stake to it, and tie the leaves lightly to it, earthing-up the stem at the same time, like celery. Throughout the summer, water copiously and frequently with soft water and a little guano, to prevent flowering. In September, the early crop will be fit for use.

213.—ASPARAGUS.

Preparation of the Land.—A deep, mellow, light loam, or sandy soil, is most suitable, but ordinary garden soil, if properly prepared, will yield fair crops. The ground should be well drained to a depth of four feet, and heavily manured, the surface being covered to a depth of at least three inches, with rich, well-decayed farm or stable-yard, manure. If the soil is of a clayey or strong, tenacious nature, sharp sand, or finely-sifted ashes, may be added with advantage. Trench to a depth of two feet six inches, well intermixing the manure as the work proceeds. The bottom of the trench should be loosened a spade's depth, still lower if the subsoil is such as will retain moisture, or otherwise benefit the plants; but if it is gravel it had better be left undisturbed. If the ground can be trenched in the autumn before planting, and ridged up roughly for the winter, levelling the ridges as early in spring as the ground may be fit to work, and then forking over the surface two or three times when in a dry state, the exposure to the weather which will thus be effected will be of great service to strong soils, and beneficial to any.

Time and Manner of Sowing.—As early in April as the ground can be found in fair working condition, sow in drills about an inch deep, scattering the seed very thinly and covering it evenly with the finest of the soil. The seed may be sown in drills fifteen inches apart, thinning out the plants so that they may stand four inches apart in the rows, to furnish plants for transplanting after one or two seasons' growth; or it may be sown at once where the crop is intended to stand. In the latter case the drills should be two feet six inches or three feet apart, or they may be made in beds five feet wide, with two-feet alleys between the beds, putting three drills in a bed; in either case the plants should be thinned out, so that they may stand about a foot asunder in the rows. Except on strong tenacious soils, which are easily injured by treading when gathering the crop in wet weather, we recommend sowing in consecutive lines. Where the land is clayey, the ground should be set out in beds, the soil dug out of the alleys to the depth of twelve or fifteen inches, and placed on the top of the beds; there will then be little danger from lack of surface-drainage. The Asparagus is very impatient of stagnant moisture about its crowns during the winter, and, on strong soils, trenches should be made of moderate size, say three inches deep and wide, and filled with sharp sand previous to sowing. The plants will probably not make as much progress in the first season as if they had been sown in the soil, but the roots will soon extend beyond the sand, and in after-years this will prevent water lodging about the crowns and rotting them.

Transplanting and after-Management.—The plants, if sown with the intention of transplanting, after one or two seasons in the seed-lines, should be encouraged by an occasional soaking of manure-water, during the growing season; and a liberal dressing of rich manure should be spread between the rows in winter. Transplant in April, when the ground is in good working order. If the ground has been properly prepared, set a line and take out a trench sufficiently wide and deep to allow of spreading the roots, and cover the crowns about two inches. On strong, heavy, imperfectly-drained soils, place sand about the roots and over the crowns, as recommended under the head of Sowing. The roots should be carefully taken up, avoiding all cutting or injury; and any that are decaying should be rejected. During the growing season keep the ground free from weeds, and the surface free and open by frequent hoeings; a soaking of manure-water may be given with advantage when the weather is droughty. Clear off the haum in autumn, when it will part from the crowns by a slight pull, and apply a dressing of well-decayed manure, which may be lightly forked in between the lines, at once or in spring. During the second and following growing seasons too much manure-water can hardly be given, and on light, sandy soils a sprinkling of salt applied two or three times in the course of the summer will be of great service, particularly where manure-water cannot be used freely. The plants should not be cut for use until they become strong and throw up fine grass, and cutting should not be continued very late in the season. The ground between the lines must be liberally manured every autumn.

Grayson's Covent Garden Giant Asparagus, per oz. 3*d.* per lb. 2*s.*

214.—ASPARAGUS (French Method).

The French practice with asparagus is to dig a trench five feet wide and the length of the bed, laying aside the best of the soil for surface-use. On the bottom of the trench is laid, first, six inches of rich stable manure; above it, eight inches of turf; again, six inches of well-rotted dung, and then eight inches of the reserved soil sifted; over this six inches of thoroughly decomposed manure, and six inches more of the soil thrown aside in making the trench, well mixed together by digging. The beds thus formed are five feet wide, with alleys between, two feet wide. The roots are planted in the beds in rows 18 inches apart, and 18 inches apart in the rows; a handful of fine mould is placed under each plant, over which the roots are carefully spread, the crown being an inch and a half below the surface; a spadeful of fine sand is now thrown over the crown, and the operation is completed. In order to procure an early supply of this delicious vegetable, they prepare a moderately warm hotbed, on which six inches of rich mould is laid, and a sufficient number of asparagus from an old bed planted. Over this they lay a few inches of the same soil, covering the whole with sufficient litter to keep out the frost, or by mats over the frame. The plants will soon start into growth. A little liquid manure applied occasionally will keep up a vigorous growth, and the plants, if properly managed, will be ready to cut by Christmas.

215.—ASPARAGUS IN FRAMES.

Early Asparagus is forced in the following manner with the most satisfactory results. In an ordinary melon-pit, about the beginning of February, a quantity of stable dung is set to work, by turning and shaking, to sweeten and regulate the heat. By the middle of the month, as much of this is thrown into the pit as will fill it to within a foot of the glass. Two days afterwards, this is covered with a layer of three inches of mellow soil. On a mild day previous to this, a quantity of asparagus-roots should have been grubbed up from an old bed (these make the best plants for forcing), and placed ready. As soon as the fermenting material has arrived at a safe temperature—about 80°, these roots should be packed thickly together on the three inches of soil, and more soil thrown on them, just sufficient to cover them, without increasing the weight too suddenly or too much. This precaution is necessary, because the addition of ten inches of earth would cause a rapid sinking and proportionate rise in the temperature of the dung, to the injury of the roots. Four days after planting, sufficient earth should be put on to cover the crowns about six inches. In ten days the crowns will begin to appear.

216.—ASPARAGUS-CUTTING.

Cutting Asparagus is an operation of some delicacy. It should be cut with a saw-edged knife, having a straight, narrow, tapering blade, about six or eight inches long, and an inch broad at the haft, rounding off at the point. When the shoots are fit to cut, the knife is to be slipped perpendicularly close to the shoot, cutting, or rather sawing, it off slantingly three or four inches below the surface, taking care not to touch any young shoot coming out of the same crown.

217.—BEANS.

Beans, like peas, can be sown in October, where the soil is light, well-drained, and well sheltered; where the ground is heavy, they may be raised in a pit or frame by sowing three in a 4-inch pot, and planted out in March; but if the soil is cold, and no conveniences are at hand for starting in pots they may be sown in the following manner:—Let the ground be laid in ridges 3 feet wide and 15 or 16 inches high, ranging east and west; on the south side of each ridge draw a drill halfway between the top and bottom, in which sow the beans about 3 inches apart: by this means they will be above the wet, catch every ray of sunshine, and will be stronger than if raised under glass and planted out. Peas may be managed in the same way. When about 10 inches high, level the top of each ridge to the row of beans behind it: they will not require earthing up again. If sown in October, a succession may be sown in January, in the same manner; and so on once a month till June: they do not bear well if sown after that. Those sown on level ground should have some earth drawn up to the roots when three or four inches high: this induces them to emit fresh roots. They are sown in rows about 4 feet apart, which leaves room for a row of brocoli, spinach, or lettuce between; but those who are not limited as to space had better allow 5 or 6 feet from row to row. On light soils the usual method is to stretch a line along where they have to be sown, and dib holes 4 inches deep, planting a row each side of the line, 4 inches apart, zigzag fashion; but in wet soils it is better to drill them in, laying boards along the row to stand on, so as to avoid clodding the ground by treading on it. The sort usually grown for first crop is the early mazagan; but the early long-pod is equally early and prolific, and larger; so is the prolific long-pod, for main crop. The monarch long-pod is also very good. The green, and the hang-down long-pods are excellent beans. The royal dwarf is a good bearer.

218.—BROCOLI.

All the varieties of brocoli require a deep, rich soil, and the ground should be trenched to a depth of at least two feet, incorporating, as the work proceeds, abundance of rich manure. Indeed to obtain fine large heads too much manure can hardly be used.

The early varieties, such as Purple Cape, Grange's White Cape, Walcheren, &c., should be sown from the middle of April to the middle of May, according to locality, and a second sowing of similar kinds should be made about a fortnight afterwards. These will succeed the cauliflowers, and will carry the supply on to Christmas. Two or three sowings of Snow's Winter White, put in from the beginning of April to the middle of May, will keep up the supply until the sprouting varieties are ready, and these again till the spring kinds come in. Sow the Purple Sprouting and Lee's new White Sprouting early in March; and those intended to furnish the spring supply or main crop at the latter end of April or early in May. When the plants are sufficiently strong, and before they are drawn by growing too closely together, transplant them into nursery beds or lines, allowing about four inches intermediate space. This will insure strong stocky plants, and will also induce the formation of an extra quantity of roots. When well established they may at any time be taken from the nursery beds and planted in permanent situations.

Division I.—For cutting during the autumn and early winter months.

Grange's White Cape,—the best White Cape variety for succeeding the cauliflower.

Early Purple Cape,—very useful, may be cut from August to December.

Early Purple Cape,—new large-headed.

Walcheren,—a very valuable variety for cutting in September and October.

Dancer's Late Pink Cape,—a valuable succession to the Purple Cape.

Division II.—For cutting during the winter months.

Snow's Winter White,—fine heads may be cut from this variety in November, December, and January.

Covent Garden,—the variety supplied to the Covent Garden market in winter.

Division III.—For cutting in March and April.

Adam's Earliest White, — the earliest spring brocoli, sometimes ready for use in February.

Imperial Early White,—a very superior variety, a fine succession to Adam's.

219.—COTTAGER'S KALE.

This is a variety of the tall cavalier cabbage which was raised at Sherburn Castle, Oxfordshire, from Brussels sprouts. Crossed with one of the varieties of kale, it was submitted to the Horticultural Society in the spring of 1858, and is said to be the most tender of all the greens, and of exquisite flavour. It stands four feet high when full-grown, and should be allowed an equal space to grow in, being clothed to the ground with immense rosette-like shoots of a bluish-green tint, which, when boiled, become a delicate green. The seed should be sown late in March or early in April, and the plants should have a rich deep soil assigned to them.

220.—ENDIVE.

Trench the ground to a depth of two feet, mixing a very liberal dressing of rich and thoroughly-decayed manure. For crops intended to stand the winter, a light, dry, and rather poor soil is best, and they should be planted in a sheltered situation.

Make the first sowing about the middle of May on a bed of well-pulverized rich soil, scattering the seed thinly, and covering it lightly. For the main crop sow in the middle of June, and again about the middle of July. Plants to stand the winter should be sown early in August. When the plants are about two inches high, transplant into nursery beds upon rich well-prepared soil, taking special care not to injure the roots, as this, as well as want of water in hot, dry weather, very often causes them to run to seed. In dry weather supply them liberally with water.

Planting and After-management.—When the plants are about four inches high, transplant, lifting them carefully with as much soil as can be kept about their roots. Place them in drills about three inches deep and 12 to 14 inches apart, and leave about the same distance between the plants. Give a liberal supply of water immediately after planting, and as often as may be requisite to keep the soil moist.

Green Curled, extra fine French,—very superior variety.

Batavian Green,—smooth broad leaves.

Batavian White,—large and very superior.

White curled,—very useful.

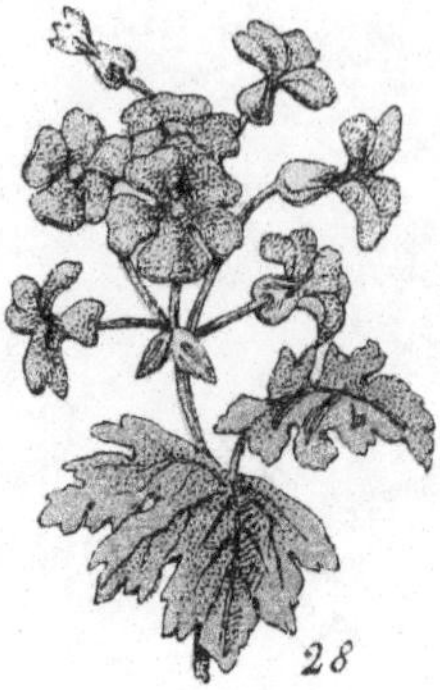
28

29

30

31

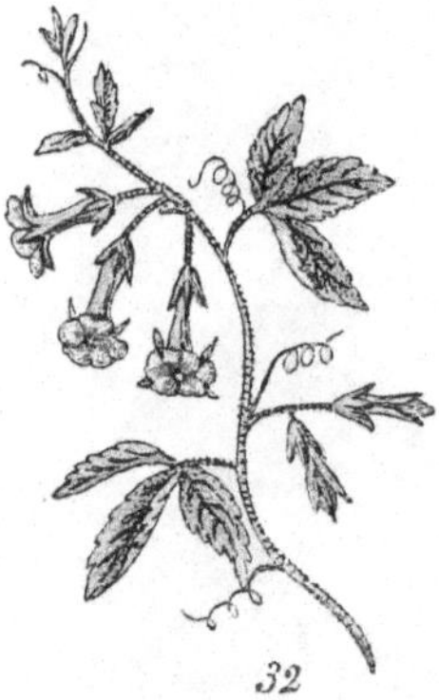
32

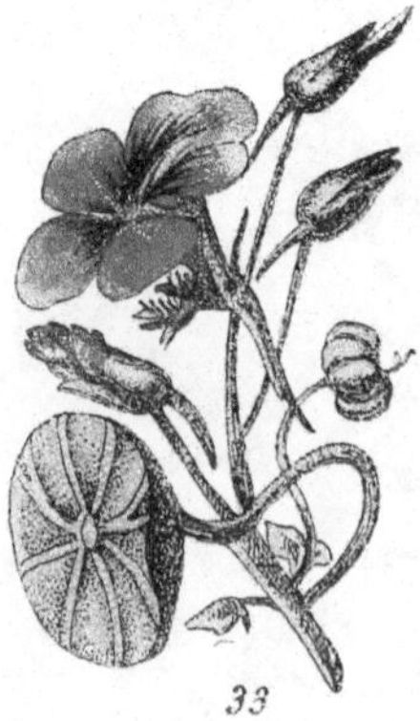
33

34

35

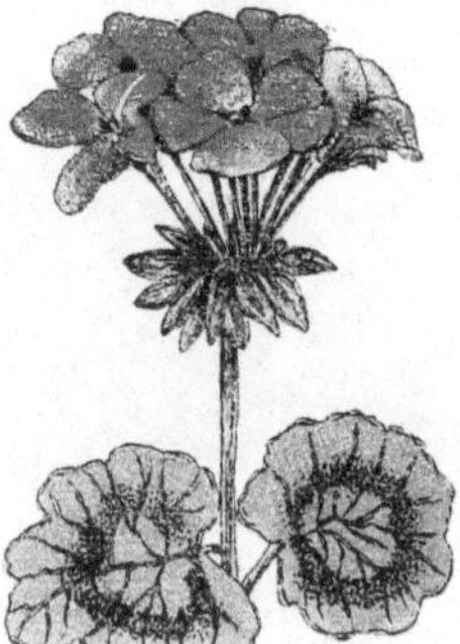
36

Pl. IV.

OUR COLOURED PLATE.—No. 4.

28. *Primula.*—A very large number of specific varieties is included under this genus, which embraces the common Primrose of our hedges and the most choice and tender greenhouse sorts. They are all beautiful; those which are marked in the gardener's catalogues as Primula Sinensis are especially deserving the notice of everyone who takes any interest in greenhouse plants; they are very showy and of easy cultivation, and, if good seed be procured, fine flowers may be had from November until the end of January, and even longer by successional sowing. The plants are raised from seed, and as soon as they appear above ground they must be kept in a growing state; they do not require artificial heat, except to start the seed.

29. *Quaking Grass.*—This is one of the garden grasses, its botanical name is Briza; the seed should be sown in March. It is a great pity that a few more varieties of this very extensive genus, Gramineæ, are not brought into cultivation; the grasses form most elegant bouquets for table ornament, and are extensively used for this purpose on the continent.

30. *Ranunculus.*—Among florists' flowers Ranunculuses are said to rank next to the Tulips, but we must decidedly agree with those who give a preference to them. They are far less formal, and certainly not inferior in brilliancy of colouring. Many of the named varieties are magnificent flowers. The tubers require a deep rich soil, and to be planted in a somewhat moist situation. The Turban varieties should be planted from October to February for a succession of blooms, and the Persian varieties from January to the end of March. Two inches deep to five inches apart is the usual mode in forming a bed.

31. *Rhododendron.*—Of American plants we always regard the Rhododendron to be chief. The sorts are so various and the heads of blossom so magnificent that few things can be compared to them for clumps upon lawns and borders. They rejoice in a peaty soil, but it is a great mistake

to imagine that peat is an essential to their proper growth—any light vegetable soil of a sandy character will answer well enough if peat is not at hand. We mention this because many persons seem to think that they cannot grow Rhododendrons as they have no peat in their neighbourhood.

32. *Tacsonia.*—These lovely climbers are allied to the Passion-flowers; most of them require a greenhouse or conservatory, and when planted in a suitable soil under glass their growth is in general so rapid, that a single plant will cover a large house in one season. To flower freely, they require sharp pruning.

33. *Tropæolum.*—The common Nastertium belongs to this genus, and is known as T. Majus. Many species have tuberous roots, and among these there are several showy climbers which do well against trellis work.

34. *Verbena.*—For summer bedding few plants are more popular than the Verbena. It is of easy cultivation and very useful for cut flowers.

35. *Zinia.*—These pretty bedding annuals are to be seen with both double and single flowers—our illustration is one of the former. As they require support, they are best grown in mixed beds.

36. *Zonale Geranium.*—Pretty foliage is a very great addition to a pretty flower, and most persons must admit that Geraniums with zoned leaves are greatly to be preferred to those without zones; the named sorts are extremely numerous, but in many cases it is very hard to mark any difference between them.

221.—FENNEL.

Fennel may be raised from seed in April or May. The seed should be covered lightly with fine mould, and, when the plants are strong enough, they may be set out in a bed about a foot apart. A good bed of fennel will last for years; but to insure fine leaves, the flower-stalks should always be cut off as soon as they appear, so as never to ripen seed.

222.—FRENCH BEANS.

These delight in a deep, friable, and rich soil; and where the land is of a strong, tenacious character, it should be trenched and ridged as early in autumn as possible, well intermixing a liberal allowance of manure. Upon very strong soils a good dressing of leaf-mould may be added with advantage.

The seeds being liable to rot if sown early in wet, cold soil, the first crop had better be planted in boxes or pans. Place these in a cold frame, or under the shelter of a south wall, and protect them from frost. When the plants are in the rough leaf, and the weather considered safe, transplant in rows about 2 feet 6 inches apart on a warm and sheltered border. Transplanting induces early fertility, and may be practised with advantage even where the plants are raised in the open border. Sow for the principal crops early in May, June, and July: and on light dry soils in warm localities a small quantity may be sown towards the end of July. Cover the seeds with about three inches of soil.

When the plants are about four inches high, ridge the soil neatly up on either side. This will prevent their being blown about by rough winds: and while there is any danger of frost, the early crops should be sheltered by well-furnished branches of evergreens, stuck into the soil in a slanting direction on each side of the rows. Keep the ground between the lines well stirred and free from weeds. During dry, hot weather, if the supply threatens to fall short, an occasional soaking of water will be of service. The following are the best varieties. Cost 1*s.* 6*d.* per quart:—

Canterbury White,—very prolific, well known, 1 ft.
Chinese Long-podded,—exceedingly productive, free-cropping variety, 1 ft.
Dun or Cream-coloured,—much esteemed for its earliness and free cropping, 1 ft.
Fulmer's Early Forcing,—a fine variety for forcing, very productive, 1 ft.
Mohawk, or Early Six Weeks,—a very good variety, 1 ft.
Negro Long-podded,—very fine cropper, with long pods, 1 ft.

223.—GOURDS.

All vegetables of this class which produce an immense amount of food, may be profitably and easily cultivated by attending to the following directions:—The seed should be sown in April or May, in pots or pans of rich, light soil, and raised in a warm frame. As soon as possible, the young plants should be potted off, and hardened in a cold frame for planting out in the end of May or early in June. Mr. James Cuthill tells us that marrows contain a rich sugary and farinaceous matter, and are a most excellent and nutritious article

10

of diet when dressed in the following manner:—Cut the marrows into short pieces, take out all the pith and seeds, and boil them in plenty of water with a little salt. When well boiled, scrape out all the marrow, put it between two dishes, and squeeze out all the water; then mash it well, adding salt, pepper, and a little butter. It is then a dish fit for any table. The cultivation Mr. Cuthill recommends is to sow the seed about the first week in May in the open ground, in a warm corner, transplanted to moderately rich land. "I can grow," he adds, "twenty tons of the marrows to the acre easily; and when ripe, they can be stowed away anywhere, and will keep good for a great length of time. In addition to their utility as a vegetable for the table, they form a most excellent and economical article when boiled for fattening pigs."

224.—LEEKS.

Leeks, for the main crop, are usually sown in April, about the same time as onions. Some gardeners sow them with a small sowing of onions, the latter being drawn young for salading, and the leeks being left on the bed or planted out. Some sow them in drills 18 inches or even two feet apart, and thin them to a foot or so apart in the row, planting the thinning, at the same distance. This gives room to draw earth up to them for the purpose of blanching the root and stem. Sow very shallow, tread, and rake, provided the ground admits of it: thin before the plants interfere with each other, and water in dry weather. This crop delights in a light rich soil, and in moist seasons grows very large. The London Flag is the sort most usually grown; but the Scotch or Musselburg is esteemed by many, as growing larger.

225.—SPINACH.

For the summer crop sow early in March, and at intervals of three weeks or a fortnight, until the middle of July, in quantities according to the demand. The round-seeded varieties are the best for summer crops. The winter crop should be sown from the middle of August to the beginning of September. The prickly-seeded is the hardiest, and should be partly used for this crop. All the crops should be sown in drills, from 1 to 2 inches deep, and from 12 to 18 inches apart, scattering the seed thinly, and covering with the finest of the soil. When the ground is dry the drills should be well soaked with water before sowing. If the seed is steeped for twenty-four hours before sowing, it will germinate sooner; but this should be done only when the ground is hot and dry.

After-management.—The winter crop should be thinned as soon as the plants are strong enough to draw, so as to leave them about 9 inches apart in the row; but the summer crops soon run to seed, and need not be thinned to a greater distance in the line than 3 inches. Some growers recommend a liberal use of manure-water for the summer crop, and this doubtless increases the size of the leaves; but it must not be depended upon to prevent the plants running to seed for more than a few days; and while the weather is hot a succession should be provided for, by making frequent sowings. Keep the ground between the lines free from weeds, and in an open state by frequent deep hoeings. The perpetual Spinach or Spinach Beet is a most valuable variety in small gardens.

226.—PUMPKINS.

These are used, when young, as a vegetable. When ripe they form a valuable esculent for soups and "pumpkin pies" in winter. The young shoots in summer are an excellent substitute for asparagus.

The varieties enumerated below constitute a portion of that magnificent collection of edible and ornamental gourds which was awarded the three principal prizes at the Great International Gourd Show, held at the Royal Horticultural Gardens.

Big Ben of Westminster, Blondin, Dr. Lindley, Emperor, Giant's Punch-bowl, Mammoth, Monster, Portmanteau de Naples, Victor Emmanuel.

227.—RHUBARB.

Directions for the cultivation of rhubarb will be found in the monthly calendars. It will grow without forcing; but is far better forced. The best kinds for early forcing are the Prince Albert and Linnæan, which force with less heat than most other kinds. If rhubarb be forced on the ground where it grows, nothing more is required than to cover with large pots and stable manure,—by this method it is blanched; but when forced in a frame, or otherwise, it is unnecessary to exclude the light, as there is no advantage in blanching it. Rhubarb may be planted at any time of the year, although mild weather in autumn or early spring is best: it should be planted on a clear open spot on good soil, which should be well trenched 3 feet deep. The plants should be not less than 4 feet apart; or, where it is intended to take up some every year for forcing, a distance of 3 feet will be sufficient. Before planting, a good substance of very rotten manure should be worked into the soil. When the plants are to be increased, it is merely necessary to take up large roots and divide them with a spade: every piece that has a crown to it will grow; and as it grows very quickly, this is a good method of propagating it. To insure fine rhubarb, a large dressing of well-rotted manure should be dug in about the roots, as soon as you have finished pulling the leaves. It is not right to wait till the winter before the plants are dressed.

228.—SAVOY.

In all respects the treatment is the same as with cabbages, removing the plants to a nursery-bed when 2 inches high, selecting the strongest plants first. When planted out permanently, they should stand 2 feet apart in the rows and 20 inches between the plants; but it is not unusual to plant them between standing crops of peas or other less permanent crops, whose place they thus occupy when removed.

229.—TURNIPS.

Preparation of the Land.—A somewhat light, sandy, but deep rich soil, is most suitable for turnips, and is indeed essential to secure bulbs of mild and delicate flavour. If the summer crops sustain any check during their growth, they are apt to be stringy and high-flavoured. Sow a small breadth of the early Dutch, for the chance of a crop, upon a south border, or in a warm, sheltered situation, early in March; and as this sowing is liable to run to seed soon, put in a small quantity of the same variety about the middle of the month, and again early in April, sowing a small breadth of the American Strap-leaf at the same time; afterwards sow at intervals of three weeks or a month till July, and for a winter supply from the beginning to the middle of August. On light warm soils, in favourable localities, useful-sized bulbs may be obtained from sowings made early in September. The Orange Jelly is one of the best varieties for autumn sowing; but if a white-fleshed turnip is required, use the red-top American Stone. All the sowings should be made in shallow drills from 12 to 18 inches apart, regulating the distance by the size of bulbs which may be most esteemed: 12 inches will be sufficient for the early and late sowings. Scatter the seed very thinly and evenly, and cover it lightly with the finest of the soil. In summer, when the ground is dry, the drills should be well watered before sowing, and if the seed is steeped in water for twenty-four hours, this will hasten germination.

After-management.—Thin out the plants as soon as they are sufficiently strong to draw, so that they may stand from 6 to 9 inches apart in the row. If fly makes its appearance—and this is generally very troublesome during summer in warm localities, dust the plants over with quicklime early in the morning, while the leaves are moist with dew. Repeat this operation as often as may be necessary. Keep the surface of the ground open and free from weeds by frequent stirring with the hoe.

230.—SCARLET RUNNERS.

These may be planted at any time from April to late in July. The seed should be dropped about 4 inches apart, and if a line be selected along the two sides of a walk in the kitchen-garden, a very pretty shady avenue may be made. Plant stakes 7 or 8 feet high in the row where the beans are; set two or three stakes to the yard, and bend them over at the top to form arches. In the spaces between the stakes place pea-sticks, to which the runners may at first be trained. The stakes should be tied together by wands arranged longitudinally, one along the top and one half way up each side. When this framework becomes covered with scarlet runners, a very pleasant shady walk will be formed. With a little care in manuring and watering, the runners may be kept green and in bearing till killed by the autumn frosts. The runners will blossom and bear much more freely if the old beans are all removed, and they are not allowed to ripen seed. A mixture of the white Dutch runner with the scarlet runner gives to the avenue a very pretty effect.

231.—SCORZONERA.

This is sown in the same manner as salsafy, and, by some, is much esteemed. To have it large, it should remain over the second season. It seldom grows large enough for use the first year, but is none the worse for remaining two or even three years before using.

232.—SAGE.

This useful garden herb is a salvia, which is a very extensive genus in botany. All the kinds should be grown in a light rich soil, and are propagated by cuttings, the division of roots, and seed.

233.—SALSAFY.

This may be sown about the end of April or beginning of May. It is best to sow this seed in drills 15 inches apart, or thereabouts, and thin to 6 inches in a row.

234.—PARSLEY.

Full crops of parsley should be sown in the spring along the edges of one of the borders. In order to grow this useful herb in perfection, it is necessary that the roots and stem should be kept in a perfectly dry state: this is indispensable to the health and freshness of the plant. In preparing the beds, therefore, remove the soil to the depth of six or eight inches, and fill in the bottom with the same depth of stones, brick-rubbish, and similar loose material. Over this prepare the bed of light rich soil, which will thus be raised considerably above the level of the ground, the bed being raked smooth and level. Towards the end of May, sow some seed of the most curly variety, either in shallow drills, slightly covered with fine soil, or thin broadcast raked in. If the weather continue dry, water frequently: in five or six weeks the plants will have appeared; when large enough, thin them out, so that they may be four or five inches apart. By the end of autumn they will be large and vigorous plants. At this time, drive a row of stakes or hoops into the ground, on each side of the bed, so as to form arches strong enough to support a covering of mats, which should be laid over them as soon as frosty or wet weather threatens to set in. During intense frosts, increase the protection, removing it on fine days, and removing it entirely in mild weather. The soil should be kept dry, and all decayed leaves carefully removed: in this manner this useful vegetable may be available all the winter.

235.—PARSNIPS.

Preparation of the Soil.—Parsnips succeed best in a deep, free, rich soil, and as the application of fresh manure tends to the production of forked and badly-formed roots, ground in high condition (having been heavily manured for the previous crop) should be selected. If manure *must* be applied, let it be well decomposed, or use guano. The ground should be trenched 2 feet 6 inches, and ridged up as long as possible before sowing.

Time and Method of Sowing.—Sow in lines 15 to 18 inches apart, as early in spring as the ground can be found in fair working condition, scattering the seeds thinly, and covering them half an inch to one inch with the finest of the soil.

After-management.—When the plants are about two or three inches high, thin them out, leaving six or eight inches between them. Keep the ground free from weeds, and the surface open by frequent deep stirrings with the hoe. Towards the end of November take up the roots, and, after cutting off the tops, &c., either store them in damp sand, in a cellar, or pit, as is done with potatoes. The roots, being hardy, would be quite safe in the ground.

Sutton's "Student."—This variety has been ennobled from the wild parsnip of Great Britain, and is considered an important acquisition: the flavour is very superior, and the roots are clean and handsome. We confidently recommend it.
Hollow Crown Improved,—the most useful for main crop.
Jersey,—a large valuable sort.
Chervil Parsnip.

236.—SEAKALE.

This useful vegetable may be grown from offsets and from seed.

Preparation of the Soil.—To grow seakale in the highest perfection, the ground must be deep and rich, and should be trenched two or three feet deep, working in a very liberal dressing of the richest manure. In strong tenacious soils, two or three inches of sharp sand or finely-sifted ashes, well mixed in, will be of service. The readiest method of propagation is by offsets, but it may be raised from seed.

Time and Manner of Sowing.—Sow in lines, as early in April as the ground is ready. Scatter the seed thinly, and cover it with about an inch of free soil. The distance which should be allowed between the lines will depend upon whether the plants are to be transplanted the following season, or used for forcing, or are to be cut where they grow. In the former case the rows should be 15 to 18 inches apart, and the plants thinned out to a foot apart; in the latter, 2 feet 6 inches should be left between the rows, and from 12 to 15 inches between the plants. In peculiarly favourable soils, roots of a useful size for forcing may be obtained the first season, but in most cases it would be advisable to sow in lines 12 or 15 inches apart, and transplant in the following spring, allowing 2 feet 6 inches between the rows, and 1 foot 3 inches between the plants.

Cultivation.—Destroy weeds as they appear, and keep the surface open by frequent deep stirrings. During the growing season give a liberal soaking of manure-water when the ground is dry, or a light sprinkling of salt, washed in with clear water, which will answer equally well. Keep the plants to one crown, or shoot, cutting off others, and removing seed-shoots as soon as they appear. After the second season's growth, the roots will be in the best possible condition for taking up and forcing, or for producing a crop of fine large kale where they stand.

Seed 4*d.* per packet; 6*d.* per ounce.

237.—LETTUCES.

Preparation of the Soil.—A rather strong and highly-enriched loamy soil is best; but ordinary garden soil, with plenty of old rotten manure will produce very fine lettuces.

Time and Manner of Sowing.—For an early crop sow under glass in February, and transplant on a well-prepared bed, in some sheltered corner, in April. For successional crops sow, in beds of well-pulverized soil, early in March, and at intervals of about a fortnight until the end of July. The crop intended to stand the winter should be sown in the second week of August and first week of September, using some approved hardy sort. Make the surface of the beds fine, sow thinly, and cover the seeds lightly with fine soil. The plants for the main summer crops may be transplanted with advantage into nursery lines, in beds of light rich soil, and if not transplanted, they must be thinned out in the seed-bed early, so as to afford them ample space to grow strong and stocky. A north border is a good situation in which to plant during the summer months, as the plants are less exposed to the sun, grow stronger, are more succulent and crisp, and are longer before they run to seed. Keep the surface of the ground loose and open by frequent hoeings. Some varieties require tying up in order to get them properly blanched, and this should be done when the plants are a fair size, and a week or ten days before they are wanted for use, selecting first the strongest plants, continuing to do this as they are wanted. Where lettuce is wanted for the winter, a portion of the plants from the sowing made at the end of July should be covered with glass. By thus protecting them from frost and wet and giving air freely, they will, if not too large, keep in good condition for a long time. The following are the best varieties:—

Cos Lettuces.

Covent Garden Giant White,—the best white Cos lettuce known; it is less affected by hot weather than other varieties, and does not readily run to seed.

Covent Garden Giant Brown, — large, crisp, and fine-flavoured.

Butler's Fine Summer White,—very fine variety.

Carter's Giant White,—very excellent.

Carter's Giant Brown,—very fine.

Dunnett's Giant Black-seeded Bath,—a most desirable variety.

Paris White,—a very superior variety.

Cabbage Lettuces.

Covent Garden Winter,—the best winter cabbage lettuce.

Drumhead,—the best summer cabbage lettuce.

238.—TOMATO.

An admirable vegetable by itself—it enters largely into a great number of our best and most wholesome sauces. It may be cooked and brought on to table like other vegetables in several different ways. Moreover, those who have analyzed its properties say that the tomato is singularly wholesome, and very useful, especially in cases of bad digestion—however, it is not appreciated or cultivated as it ought to be. There is, undoubtedly some little difficulty in our climate in fruiting and ripening tomatoes to perfection; but the following directions, if attended to, will generally be found to succeed:—Sow the seeds in pots in very rich light mould in March or April, and place them in a cucumber-frame, or other gentle heat. When the second leaf appears, re-pot the plants either singly or at most two or three together, keeping them near the glass and well watered. In May remove them to a cold frame for the purpose of hardening them before they are planted out, which should be done as soon as the fear of spring frosts is over, and the earlier the better. The best situation for tomato-plants is against a south wall fully exposed to the sun. The plants should be well watered with liquid manure to keep up a rapid growth. As soon as the blossom-buds appear, watering should cease. Stop the shoots by nipping off the tops, and throw out all those sprays that show little signs of fruit, exposing the young fruit as much as possible to the sun and air, only watering to prevent a check in case of very severe drought, of which the state of the plant will be the best index. In a very dull, wet, cold autumn, even with the greatest care, the fruit will sometimes not ripen as it ought; but in this case it may frequently be made fit for use by cutting off the branches on which full-grown fruit is found, and hanging them in a warm dry greenhouse or elsewhere to soften and ripen; a cool oven may be used advantageously to effect this.

239.—MUSTARD.

If a supply is required in winter, or when the weather is too cold for the seed to vegetate out of doors, sow in shallow boxes or pans, placing these in a warm house or pit. During the heat of summer a shady border will be the most suitable situation. Make the surface of the soil fine, level, and smooth, then water it and sow the seed very thickly. Press it gently into the soil, but avoid covering it with soil, else the earth and sand will adhere to the leaves and be with difficulty removed by washing. Exclude the sun's rays, and keep the seeds moist by coverings; but these must be removed as soon as the seeds have fairly germinated. To furnish a regular supply, sow at intervals of a few days, and never allow the plants to get too old before being cut for use.

240.—MUSHROOMS.

Mushrooms may easily be had at any season of the year by adopting an artificial process, and spawning, with Milltrack or other artificial spawn, a bed made after the following manner:—The best situation for the artificial growth of mushrooms is a cellar or underground tool-house, or any other place where the atmosphere is of that close, damp, foggy character which is always so peculiarly favourable to the growth of fungi. The antechamber or passage to an ice-house is an excellent place for a mushroom-bed, and is frequently made use of for this purpose: any shed, however, whether underground or not, may be made available; and, indeed, with a little more care, mushrooms may be grown in the open air, without any roof to cover them at all; but a cellar or underground hole has a decided preference. The foundation of the bed must be well-rotted manure from the horse-yard, which has been sweetened by being turned over two or three times: it may have a little good loam mixed with it, in the proportion of about two barrows of loam to twelve of manure. The bed is best made on a gentle slope, and the manure should be well and firmly beaten down with a spade. When the heat has fallen to about 75°, the spawn may be put in. This artificial spawn, which is usually made up in cakes, must be broken up into pieces about two inches square, and placed all over the bed, upon the surface of the manure, about ten or twelve inches apart. A covering of one inch, or one and a half inch, of good garden loam is then to be placed all over the bed, and the surface again beaten firm with a spade. The whole must then be covered well over with straw or other material, to exclude all light. The growth of the mushrooms will, of course, depend somewhat on the state of the atmosphere; but in a temperature of 45° to 55° they will usually begin to appear in about six weeks. Little or no water should be given to the bed until the mushrooms begin to come up, as its own moisture and heat ought to be sufficient to start the spawn; but as soon as mushrooms appear, a plentiful supply of water may be given, and it will be found that a little common salt, or, better still, saltpetre, will have a great effect upon the crop. It is essential that the surface of the bed be kept quite dark. If the bed be made in the open air, it may be necessary, after a time, to give to the spawn a fresh start, by placing a lining of hot manure around it; but on all occasions great care must be taken that the heat of the bed is not so excessive as to burn up the spawn. This, however, can never happen at a temperature of 75°; and when a bed is above this, no spawn should ever be inserted.

241.—ONIONS.

Selection and Preparation of the Land.—A rather strong, deep, and rich loamy soil is most suitable for this crop: where very large bulbs are desired, soil of this character is indispensable. Onions grown in a strong soil are much less liable to be attacked by the fly or maggot than in light, dry, sandy soils. The ground should be heavily dressed with rich well-rotted manure, trenched deeply, and ridged up early in autumn. If the soil is light and sandy, cow manure will be most suitable.

Time and Method of Sowing.—The main crop should be sown as early as the ground may be in working condition, and whether this occurs in February or early in March, a favourable opportunity for putting in the seed should not be suffered to pass. After levelling down the ridges, if the soil is light, tread the ground regularly and closely over, then rake and well pulverize the surface, making it as fine as possible. Set out the ground in four-feet beds, with alleys a foot wide between; draw drills half an inch to one inch deep, six inches from each alley, and nine inches apart. Sow the seeds thinly and regularly, and cover with the soil displaced in making the drills, or, where this is too lumpy, with other fine soil. A sowing should also be made about the middle of August, to furnish a supply of young onions during winter, and bulbs for use in summer before the main crop is ready. Where small bulbs, such as are used for pickling, are required, sow the Silver-skinned thickly early in May, upon the poorest soil, and in the driest situation at command, and thin out very sparingly.

White Spanish (true Reading), — the mildest in flavour, and most useful for main crop.

Deptford, or Brown Spanish,—similar to the above, but brown; a useful and good keeping variety.

Brown Globe,—a hardy useful kind.

White Globe,—a mild-flavoured good keeping variety.

Giant Madeira,—grows to a great size, and particularly mild-flavoured.

Blood red,—a very useful hardy kind

James's Long-keeping, — keeps longer than any other variety.
Silver-skinned,—the best for pickling.
Tripoli Large flat Italian, — the best variety for autumn sowing.
Tripoli Large Globe, — very fine for autumn sowing.
White Lisbon, — the variety sown in autumn by market gardeners for spring onions.
The Giant Rocca,—this is the largest onion grown, and very valuable for exhibition purposes.

242.—RADISH.

Preparation of the Soil.—Deep, light, and rich soil—not made so, however, by the application of manure—should be chosen. It should be carefully dug or forked deeply, making the surface soil fine and level before sowing.

Time and Manner of Sowing.—Sow early in January, on a warm sheltered border, and at intervals of three weeks until May; afterwards every fortnight during the summer, and at longer intervals when the weather becomes cold in autumn. Radishes are often sown much too thickly, and this causes the roots to be small, hard, stringy, and disagreeably hot in flavour. Sow broadcast, in beds of convenient size, and cover the seeds evenly and lightly with fine soil. The early sowing will require to be protected from frost by a covering of litter, but this must be removed every mild day, as soon as the plants appear above ground. When the weather is hot and the ground dry, well water before sowing; and some days before drawing, water the beds well, and keep the soil moist until the crop is finished. With the convenience of a frame and a little fermenting material, a supply may be obtained considerably earlier than in the open border, and with much less trouble. If grown in this way, cover the manure with 6 inches of light rich soil, and sow when there is no risk of the bed overheating. Give air on every favourable occasion, so as to secure stocky growth, and cover up at night when frosty. The Spanish varieties should be sown in drills, about a foot apart, and thinned out when sufficiently strong to draw, so as to stand from 4 to 6 inches apart in the rows. For a winter supply of these, sow from the middle of July to the middle of September.

243.—RAMPION.

This is a campanula, and very good for salad or for cooking. Sow at intervals, to be regulated by the size at which the roots may be most esteemed. For winter use sow in April or early in May, in lines a foot apart, covering the seeds very lightly with fine soil, and thin the plants out, so that they may stand 4 inches apart in the lines.

Seed, 2*s.* per ounce.

244.—VEGETABLE MARROW.

Preparation of the Soil.—These require a very deep, light, rich soil, and if planted in the open ground, a sheltered and warm situation. Dig pits 2 feet wide and deep, and fill with well-prepared fermenting manure, and cover about a foot deep with soil. The pits should not be less than 10 feet apart, and should be prepared about a week before planting, so that the soil may be properly warmed by the heat from the manure. The tops of compost-heaps, and hills of decaying leaves, manure, &c., will, however, afford the best possible situation for their growth.

Sowing and Preparation of the Plants.—Sow early in April, in a pot or pan, filled with light soil, covering the seed about half an inch; place in gentle heat, and as soon as the plants are sufficiently strong to handle, pot them off into 7-inch pots, putting two plants in each, and replace them near the glass in the warmth. When well established, remove to a cold frame, and gradually prepare for planting out, by a freer exposure to air, &c.

Planting and After-management.—Towards the end of May, or as soon as the weather is warm and appears to be settled, and the plants ready, plant them out, and protect them for a time by hand-glasses, or other means, and attend to watering until the roots get hold of the soil. Train and regulate the shoots, so as to prevent them from growing too closely together, and stop them if necessary, to forward the growth of the fruit. Do not allow the plants to feel the want of water at the roots, but if planted in suitable situations, watering will seldom be necessary.

The most approved varieties of the vegetable marrow are the following; each variety 4*d.* per packet, or the collection for 3*s.* 6*d.*

Barrel-shaped,—2 ft. long.
Cluster,—habit compact and bush-like.
Cream-colour,—large, 1½ ft. long.
Custard, — a small handsome variety, particularly delicate in flavour.
Clear yellow,—10 inches long.
Egg-shaped,—1½ ft. long.
Green and orange,—1½ ft. long.
Melon-shaped,—large and extra fine.
Oval,—10 inches long.
Pear-shaped,—10 inches.
,, beautifully striped.
Small,—white.
,, miniature, very pretty, 6 inches long.
Warted,—oblong, very handsome, 1 ft. long.
Warted,—large, cream, 14 inches long.

FRUIT GROWING.

245.—ORCHARDS.

Though, from the introduction of dwarf trees, upon which, in a good kitchen-garden, as much fruit may be grown as will be required for the consumption of a family, an orchard is not now so necessary as it was some years ago, still we are not inclined to see this useful appendage of the country-house wholly neglected. A piece of pasture where the soil is good may be very profitably employed as an orchard. It will yield both a crop of fruit and a crop of grass, and if the former be not required for consumption, there is at all times a ready sale for it. Apples, pears, and cherries are the fruits properly cultivated in orchards; but plums, walnuts, and filberts are not unfrequently considered as orchard fruit, and in cases where there is only one orchard, all these fruits may, with advantage, be included in it.

Upon the nature of the site and soil best suited for an orchard, Abercrombie observes, "Land sloping to the east or south is better than a level; a sheltered hollow, not liable to floods, is better than an upland with the same aspect, and yet a gentle rising, backed by sufficient shelter, or the base of a hill, is eligible. A good loam, in which the constituents of a good soil predominate over those of a hot one, suits most fruit trees; the subsoil should be dry, and the depth of mould thirty inches or three feet. Before planting, drain, if necessary; trench to the depth of two feet, manure according to the defects of the soil, and give a winter and summer fallow; or cultivate the site for a year or two as a kitchen-garden, so that it may be deeply dug and receive a good annual dressing."

In forming an orchard, Dr. Lindley recommends, and we quite agree with him, the early transplanting of the different trees. "They cannot," he says, "be removed from the nursery too soon after the wood has become ripe and the leaves have fallen off, for between this time and the winter many of them will make fresh roots, and be prepared to push forth their young shoots with more vigour in the spring than those whose transplanting has been deferred to a late period of the season." All young trees should be carefully staked and protected from the wind, and, if a dry spring should succeed the autumn of their planting, they will require to be watered, or, what is better, to have manure laid round their roots and be watered through it. Pruning and training are necessary; but, as a general rule, the knife should be avoided, if it is possible to bring the tree into a good shape without it. The permanent trees, which, if it is intended to lay the orchard down in grass, must be standards and half-standards, with from four to six feet of clear stem, should be planted in rows, from 30 to 40 feet apart, and in what is termed the *quincunx* style, thus:—

*   *   *   *   *   *

*   *   *   *   *

In selecting apples and pears for planting, and, indeed, all fruits that admit of sorts, it is of the greatest importance to take into consideration not only soil but climate. Very little good is gained by selecting the best varieties if they, or any of them, are not suited to the locality. Disappointment too often follows want of judgment in this respect. Whoever intends to plant an orchard, especially of apples and pears, should ascertain, in the first instance, what sorts flourish best in his part of the country. He should then select the best of these, and introduce such other sorts as, from their resemblance to them, may seem likely to answer. The following lists are taken from the fourth volume of the "Floral World." The sorts are undoubtedly all good and the descriptions accurate; but it would be quite wrong to imagine that they are all adapted to every locality.

Twenty-four Orchard Apples.—*K.* kitchen purposes; *D.* dessert.

Alfreston, K.—Large, round, skin light orange next the sun, greenish-yellow in the shade; flesh yellowish, crisp, sharply acid. November to March. A fruitful variety and showy grower.

Bedfordshire Foundling, K.—Very large, pale green when ripe; flesh yellowish, acid. November to April. A handsome kitchen apple, but rather fitful in productiveness.

Bess Pool, K., D.—Large, conical, handsome, yellow suffused with red next the sun; flesh white, sugary, vinous. Good from November to May. Rarely fails to give a good crop, and fit for any purpose.

Blenheim Orange, K., D.—Very large, ovate, yellowish, red next the sun; flesh yellow, sugary. November to June.

Court of Wick, D.—Medium size, very handsome, greenish-yellow, orange, and russety; juicy, high-flavoured.

Dumelow Seedling, K. — Large, round, yellow, and light red; flesh yellow, first-rate. November to March. Also known as Wellington and Normanton Wonder.

Devonshire Quarrenden, D.—This is the famous "sack apple" of the western counties. Medium size, deep crimson; flesh greenish-white, juicy, subacid. August.

Dutch Codlin, K.—Very large, conical, and ribbed, greenish-yellow, with light tinge of orange; flesh white and firm. A first-rate kitchen apple, always bears, and will keep till Christmas.

Fearn's Pippin, D., K.—Full medium size, round, and handsome, greenish-yellow, russety, and bright red; flesh greenish-white, sweet, and rich-flavoured. November to March.

Forge, K.—Medium, golden-yellow mottled with crimson, and dark red next the sun; flesh tender, juicy, and perfumed. Always bears well. September to February.

French Crab K., D.—Large, dark green, brownish next the sun; flesh green, firm, subacid. Bears immensely, and will keep any reasonable length of time.

Gooseberry Pippin, K.—Large, roundish, bright green; flesh greenish, tender, gooseberry flavour, which it retains till May or June, and it may be kept to the following August.

Hawthornden, K.—Large, ovate, yellowish-green, reddish blush next the sun; flesh white, juicy, almost good enough for dessert. This never fails to give a large crop; it is not a strong grower. The new Hawthornden is more robust, and produces a finer fruit, but is scarcely so prolific as the old. If but one apple-tree could be planted in the garden, we would have the Old Hawthornden in preference to any other. September to February.

Hanwell Souring, K.—Medium size, greenish yellow, red blush; flesh firm, crisp, acid. No orchard should be without it. November to April.

Kerry Pippin, D. — Small, pale yellow streaked with red; flesh yellow, firm, juicy, and sweet. First-rate in every respect. September to November.

Loudon Pippin, K. — Large, roundish angular, with five protuberances round the crown, deep yellow; flesh white, subacid. Very fruitful. October to February.

Melon, D.—Large, lemon-yellow and light crimson; flesh white, tender, juicy, vinous, perfumed. One of the best American apples, generally fruitful. December to February.

Nonpareil, Old, D. — Small, greenish-yellow, one of the hardiest; pale russet and brownish-red; flesh tender, juicy, rich. January to May.

Norfolk Bearer, K.—Large, green, yellowish, and crimson; flesh tender, brisk flavour. A prodigious bearer. December to February.

Northern Greening, K. — Medium, dull green, brownish-red; flesh greenish, subacid. First-rate. November to May.

Sturmer Pippin, D.—Medium, yellowish-green and brownish-red; flesh yellow, firm, sugary, and rich. January to June.

Skye-house Russet, D.—Small, roundish, greenish, aromatic, and highly-flavoured. One of the best dessert apples. November to March.

Winter Pearmain, K.. D.—Large, conical, handsome greenish-yellow and deep red; flesh juicy, sweet, and brisk-flavoured. October to April.

Yorkshire Greening, K.—Large, roundish, irregular, dark green striped with dull red; pleasantly acid. October to February.

From the same excellent authority, the "Floral World," we extract the following list and description of pears suitable for orchard cultivation.

Alexandre Bivort.—Medium size, melting, rich, and exquisite; prolific bush. December and January.

Bergamotte d'Esperen. — Medium, late, melting. Forms a handsome prolific pyramid or bush; but in wet or cold climates it requires a wall. March to May.

Beurré Brown. — Large and excellent. End of October.

Beurré d'Aboise. — A hardy variety of Brown Beurré.

Beurré d'Aremberg. — Medium, delicious, melting; forms a handsome prolific pyramid. Orpheline d'Engheim is a variety of this pear, equally good, less vigorous habit, a most prodigious bearer as a dwarf bush. December and January,

Beurré d'Amanlis.—Very large, melting; one of the best autumn pears, not particular as to soil. End of September.

Beurré Rance.—Large, late, melting, insipid from a wall; but on the quince, in the open grounds, its flavour is quite exquisite. Requires double working, and forms a better bush than a pyramid. March to May.

Beurré Easter.—Large, melting, perfumed, insipid from a wall; best on the quince, and forms a beautiful bush. January to May.

Beurré Goubaut.—Medium, melting, excellent; wonderfully prolific when worked on the quince; better as a bush than a pyramid. Middle of September.

Bon Chrétien (Williams's).—Large, perfumed, melting; should be gathered before it is ripe. September.

Broom Park (Knight's). — Medium size, melting pear; partakes of the flavour of the melon and pineapple; on the quince must be double-worked; a prolific bush. January.

Chaumontel.—Large, well-known, melting. December. This is the pear which grows so fine in Jersey and Guernsey.

Comte de Lamey.—Medium; one of our most delicious autumn pears. October.

Dr. Trousseau (Van Mons').—A new, excellent, and most delicious hardy melting pear, which succeeds well on the quince, and forms a prolific bush. December.

Duchesse d'Angoulême. — Very large and handsome, insipid from a wall; forms a fine pyramid. November.

Forelle, or Trout Pear.—Medium, a very handsome speckled pear, melting and good. Forms a prolific bush, or a pyramid of moderate growth. December.

Josephine de Malines. — Medium size, delicious melting pear, aromatic. On the hawthorn it forms a spreading, fruitful tree; succeeds well on the quince, but does not form a handsome pyramid; as a bush or espalier it is very prolific. February to May.

Louise Bonne of Jersey. — Large. Mr. Rivers says: "When cultivated on the quince stock, this is the most beautiful, as well as the most delicious, melting pear of the season. Every one possessing a garden of six square feet should plant a tree on the quince stock of this variety; it forms a most ornamental pyramid, and a compact fruitful bush." It is worthy of the orchard-house. October.

Marie Louise, or Marie Louise Nouvelle.—Large, melting, excellent; on the pear it forms a prolific pyramid, on the quince, double-worked, a prolific bush. October, November.

Monarch (Knight's).—Medium, excellent; forms a handsome pyramid on the pear. Deserves a wall, and may always be relied on. January till February.

Peach, or Poire Pêche.—Medium, early, melting, slight aroma, very juicy; a prolific bush. September.

Seckle.—Small, highly-flavoured, melting; bears profusely as a pyramid on the pear. October.

Urbaniste, or Louise d'Orléans.—Medium size, a delicious melting pear; succeeds well on the quince, and forms a handsome pyramid. November.

Winter Nelis.—Small, roundish, buttery and melting, rich and aromatic; an abundant bearer, and a beautiful bush. November to February.

Yat.—Medium or large, a hardy Dutch pear, melting and juicy, highly-perfumed; forms a prolific bush. September.

246.—ORCHARD-HOUSES.

Mr. Rivers, who is our best authority upon Orchard-houses, describes as a convenient form of house, a lean-to structure, 30 feet long and 12 feet 6 inches wide, made in the following simple manner: Six posts of yellow deal, 5 inches by 3 inches, or oak posts, 4 inches by 3 inches, and 9 feet 6 inches in length, are firmly fixed, and driven 2 feet into the ground, the lower ends being previously charred and coated with coal-tar. This is the back line of posts. Six other posts, exactly similar, but only 4 feet 6 inches long, are fixed 18 inches in the ground, forming the front posts of the house,—the one rising 3 feet and the other 7 feet 6 inches above the ground-level. Two posts at one end occupy the centre, and form the door-posts. On the six posts, both in back and front, a wall-plate is nailed to receive the rafters, one of which springs from each of the six front posts, resting on the corresponding back post.

The rafters are 14 feet long;—a 9-inch deal, 3 inches thick, will make four of them. On the upper side of each rafter is nailed a slip of ½-inch deal, 1¼ inch wide, which will leave half an inch on each side as rebate to receive the glass. The rafters so prepared are fixed in their place to the wall-plates by having a piece cut out at each end to correspond with the angle of the back and front plates. They are then firmly nailed, at back and front, by a strong spike-nail, leaving a space between each rafter of 5 feet, which is called a bay: this is filled up by smaller rafters or sash-bars, of a size proportioned to their length and the use they are to be put to,—vines trained to them requiring stronger bars. A piece of ¾-inch deal board, 6 inches wide, nailed along the top of each rafter, so as to be even with their upper edges, forms the ridge-board, leaving a groove to receive the upper end of the glass. A similar piece of inch deal, 6 inches wide, let in by sawing a corresponding piece out of each rafter, will receive the glass and carry off the water. The placing the glass is a very simple process: beginning at the top, a plate of glass, 20 inches wide, is laid in the groove, and fixed in its place by a brad driven into the rafter, a bed of putty being first laid; and so on till the whole is covered in,—open joints in the glass being rather advantageous than otherwise, if not too wide. No putty is used in the laps. The ends of the houses are fitted up to correspond with the roof, only that above the doorway a large sash is fitted in for ventilation. These sashes at each end, and the front or side sashes, are said by Mr. Rivers to be quite sufficient; indeed, he pronounces the ventilation perfect. Well-seasoned ¾-inch deal, planed and jointed, nailed outside the posts, forms the lower part of the house.

In the back wall, sliding shutters, 3 feet by 1 foot, will afford ventilation to the roof; and about 3 feet from the surface of the ground, two similar sliding shutters will ventilate the lower part of the house behind, and on a level with them. Ventilation is secured by sashes 2 feet 6 inches wide, and running the whole length of the house under the wall-plate: below these sashes the space is filled in with boarding, well painted. In summer it is impossible to give too much air. The house is now complete, except the door, which must open inwards for obvious reasons, and may be half glass, or otherwise, at the proprietor's discretion.

Within the house, a trench, 18 inches deep, is formed, to which two steps from the outside will lead. This leaves a platform or border on each side of 4 feet 9 inches; the back border requires to be raised 18 inches, and Mr. Rivers suggests that it would be improved by a second terrace behind the first, of 14 inches, supported by a 4-inch brick wall, so that the back row of trees need not be shaded while they are brought nearer to the glass.

Now, everything depends upon these borders; their surface loose and open, formed of old lime-rubbish and road-sand, mixed with manure, may be laid, 4 inches deep, the whole forked over, and well mixed with the soil 9 inches deep. The estimate for this house, as given by Mr. Rivers, is £28 5s. A handy person with his tools could probably do the whole for much less; at least, the material and glass, calculated at the prices in Montgomery's list of sash-bars and rafters, are under £15. Here is Mr. Rivers's description of a span-roofed house, which comprises the advantage of border as well as pot-cultivation:—Height at sides, 5 feet; height of ridge, 9 feet; width, 14 feet. The roof rests on oak posts 5 inches by 3 inches. The rafters are 20 inches apart; it is glazed with 16-oz. glass, in 20-inch squares. Under the eaves-boards, the sides, back and front are filled in with glass 15 inches deep, joined without putty. Under this is a ventilating-board on hinges, opening downwards; below this are ¾-inch boards, to the ground; the two ends are glazed to the same level as the side-lights; the doors with glass sash, opening inwards. Over the door, an angular space, 9 inches deep, is found sufficient for roof-ventilation; the rafters, 3-inch by 1½-inch stuff, are tied at the top with a light iron tie screwed to the rafters. No putty is placed in the laps of the glass, which serve every purpose of roof-ventilation found necessary in this house. The only ventilator is the shutter 1 foot deep on each side, and 2 feet 6 inches from the ground, and the angular opening over each door.

The borders in such a house need not be raised, nor the path sunk, except as a matter of choice; they should have a dressing of manure and sand, or manure and burnt soil, or any loose material well forked over, and mixed with a dressing to the depth of 6 inches, composed of the top spit of a pasture of tenacious loamy soil, which has been exposed to the air for the summer months, mixed with one-third of well-rotted manure, chopped up into lumps as big as an egg. In the border thus composed, two rows of trees may be placed; the front row 3 feet apart, the second being in the rear, zigzag-fashion, but half-way between, so that they are each 3 feet from stem to stem, and none shading the other. Such a house as this, without artificial heat, is intended for protection only, and not for forcing; but it would, Mr. Rivers thinks, grow oranges and camellias successfully, if the house could be heated in very severe weather so as to prevent it falling at any time below 26°. The most severe frost would not injure tea-scented roses so sheltered; but the house is essentially intended for the protection of fruit-trees, whether planted in the borders or in pots, and has the effect of bringing us, without artificial heat, to the temperature of Angers, in the south of France, where the royal muscadine grape usually ripens in the open air on the 25th of August.

The use Mr. Rivers proposes to make of the first structure is the culture of peaches, vines, and figs, in pots. Selecting a straight-stemmed maiden peach or nectarine, well furnished with lateral buds, and not more than 4 or 5 feet high, it is planted in an 11-inch pot, and each lateral shoot is cut in to two buds. As soon as the shoots have made three leaves, the third is pinched off, leaving two, not reckoning, however, one or two small leaves generally found at the base of each shoot. These pinched shoots soon put forth a fresh crop of buds, each of which, and all succeeding ones, must be pinched off to one leaf as soon as two or three leaves are formed.

"This incessant pinching off the shoots of a potted pyramid-tree, in the climate of an orchard-house, will, in one season, form a compact cypress-like tree, crowded with short fruit-spurs." In spring, these will require to be thinned, and every season the shoots will require to be pinched off as above described.

Dwarf pyramidal peach and nectarine trees may also be planted in the border, 2 feet apart, with excellent results. They require the same incessant pinching, and must be *lifted and replanted* annually in October ; but the span-roofed house is better adapted to the culture of trees planted in the borders.

247.—ORCHARD-HOUSES, FURNISHING OF.

Full directions for the management of the orchard-house will be found in our calendar every month; we need not, therefore, repeat them. It will be well, however, to give a list and description of such trees as may be introduced into an orchard-house without fear of disappointment, for it is hardly necessary to observe that all sorts of trees do not bear the confinement of glass, nor ripen their fruit under such circumstances equally well. All the sorts we shall mention are of course suited to the table; for no one would take the trouble which an orchard-house involves to grow in it any sorts of fruits which are fitted only for cookery.

Apples. — Ribstone Pippin. Blenheim Orange, Cox's Orange Pippin, Sturmer Pippin, Golden Reinette, Coe's Golden Drop, Melon, the Nonpareil, Margil, Newtown Pippin, Northern Spy.

Apricots.—Early Sardinian, St. Ambrose, Shipley's Early, Kaisha, Grosse Pêche, Royal Orange.

Cherries.—Archduke, Belle de Choisy, B. Magnifique, Bigarreau, B. Napoleon, Black Eagle, Coe's Late Carnation, Early Amber, Governor Wood, Kentish. Kirtland's Mary, Knight's Early Black, May Duke, Ohio Beauty, Royal Duke, Young's Black Tartarian.

Figs.—Early Violet, White Marseilles, Brown Turkey, White Ischia (for forcing).

Grapes.—Early Malingre, Prolific Sweet-water, Grove End Sweetwater, Muscat St. Laurent, Royal Muscadine, White Romain, Purple Fontainebleau, Black Hamburg, Chaplat, Muscat de Sarbelle.

Grapes (for forcing in the orchard-house). — Chasselas Musqué, Black Frontignan, Purple Constantia, White Frontignan, Bowood Muscat.

Nectarines.—Downton, Elruge, Early Navington, River's Orange, Violette Hâtive, Prince of Wales, Victoria, Pineapple.

The Orange Tribe.—Maltese Common Oval Orange, Maltese Blood Oval Orange, Citron, Lemon, Persian Lime, Silver Orange, St. Michael's, Tangerin, Egg Orange.

Peaches.—Acton Scott, Late Admirable, Barrington, Bellegarde, Chancellor, Claremont, Crawford's Early, Early Albert, E. Newington (Smith), E. Savoy. E. Victoria, E. York, French Mignonne, Galande, George the Fourth, Grimwood's Royal George, Grosse Mignonne, Malta, Millett's Mignonne, New Royal Charlotte, Noblesse, Red Magdalen, Royal George, Téton de Vénus, Violette Hâtive, Admirable Warburton.

Pears. — Doyenne d'Eté, Jargonelle, Duchesse d'Angoulême, Seckle, Louise Bonne of Jersey, Josephine de Malines, Passe Colmar, Williams's Bon Chrétien, Winter Nelis, Gansel's Bergamot, Cressane, Marie Louise, Beurré d'Amanlis, B. d'Aremberg, B. de Capiaumont, B. Clairgeau, B. Diel, B. Easter, B. Giffard, B. Goubault, B. Hardy.

Plums.—Angelina Burdett, Belgian Purple, Green-gage, Guthrie's Aunt Anne, Guthrie's Late Green, Huling's Superb, Ickworth, Impératrice, Jefferson's Kirk's New Blue, Mitchelson's Orleans, New Orleans, Oullen's Golden gage, Pond's Seedling, Prince Engelbert, Prince of Wales, Reine Claude de Bavay, Reine Claude Hâtive, Rivers's Early, Victoria, Washington, Woolston's Black-gage.

248.—APPLES.

The apple is a somewhat capricious fruit, some sorts affecting clay soils, while others do better in sandy loam, and even in well-drained peat soils. Apple-planting, therefore, requires some discrimination as well as observation as to the sorts most successfully grown in the locality. The following are among the most useful varieties.

Dessert Apples.

Early Harvest—ripens end of July.
Margaret,—early in August.
Calville rouge d'Eté,—middle of August.
Devonshire Quarrenden,—middle of August.
Barrowski,—end of August.
Emily Julian,—in August and September.
Summer Pippin,—beginning of September, but of short duration.
Monstrous Pippin,—September and October.
Oslin, a high-flavoured apple,—ripens in September.
King Pippin,—end of September.
Reinette blanche,—in October and November, in France.
Quatre Gouttes côtelées,—in October and November, in France.
Scarlet Crofton,—ripens with us in October.
Early Nonpareil,—in October, and keeps till March.
Potmaston Nonpareil,—in October, and keeps till March.
Court of Wick,—in use from October, and keeps till March.
Calville de St. Saveur,—ripens in November, in France.
Belle Fleur de Brabant,—in November, in France.
Reinette d'Angleterre,—in November, in France, and keeps till March.
Downton Pippin,—ripens in November, and keeps till January.
Golden Pippin —popularly supposed to be extinct, grows vigorously on a warm soil and in sheltered situations, ripening in November, and keeping till March.
Reinette dorée (Golden Reinette),—ripens in November, and keeps till April.
Ribston Pippin,—in November; keeps till May.
Ross Nonpareil,—in December; keeps till February.
Cornish Gilliflower,—in December; keeps till February.
Queen of the Reinettes,—in December; keeps till February.
Reinette du Canada,—in January and February.
Royale d'Angleterre,—in January, and keeps till March.
Wyken Pippin,—in January; keeps till March.
Old Nonpareil,—in November; keeps till May: a fine high-flavoured dessert apple.
Lamb's Pearmain,—in December; keeps till June.
Reinette Franche,—in use from February to July and August.

Cooking Apples.

Keswick Codling,—fit for use in July and August.
Monk's Codling,—fit for use from July till February.
Golden Winter Pearmain,—in use from October till January, as a kitchen as well as dessert apple.
Beauty of Kent,—in use from October till February.
Bedfordshire Foundling,—in use from January till March.
Winter Pearmain,—in use from November till April.
Winter Majetin,—in use from January to June.
Norfolk Beefing,—keeps till June.
Gooseberry Apple,—ripe in January, and keeps till June or July.

The modern system of dwarfing fruit-trees, by which space is so much economized, is produced by a special course of pruning, commencing a year from grafting, when the apple-tree should be pruned back, leaving about eight buds on the shoots. In the second year the head will exhibit eight or ten shoots, and a selection must now be made of five or six, which shall give a cup-like form to the head, removing all shoots crossing each other, or which interfere with that form; thus leaving the head hollow in the centre, with a shapely head externally, shortening back the shoots retained to two-thirds or less, according as the buds are placed, and leaving all of nearly the same size. In the course of the summer's growth the tree will be assisted by pinching off the leading shoots where there is a tendency to overthrow the balancing of the head. At the third year's pruning the same process of thinning and cutting back will be required, after which the tree can hardly go wrong. The shoots retained should be short-jointed and well-ripened; and in shortening, cut back to a healthy, sound-looking, and well-placed bud. After the third year, littl or no shortening back will be required, especially where root-pruning is practiced; the tree should now develop itself in fruiting stems, which will subdue the tendency to throw out gross or barren shoots.

Large standard trees in their prime only require pruning once in two or three years. At these intervals cross-growing or exhausted shoots, especially those in the centre of the tree, require thinning out, bearing in mind that the best fruit grows at the extremities of the branches, and that these branches must be kept under control.

249.—PEARS.

These are best grown dwarf. The varieties of the Pear are very numerous, there being upwards of 500 varieties known. Of these 500 the following is a selected list of tabie pears:—

Beurré Giffart,—ripe in July; suitable for a standard.
Citron des Carmes,—ripe in July.
Epargne,—in July and August, suited for espalier on an east or west aspect. Does not make a good pyramid.
Beurré d'Amanlis,—a standard; August and September.
Professeur du Breuil,—suitedfor espalier and east or west aspect.
Beurré d'Angleterre,—grafted on a stock as a standard; ripe in September.
Louise Bonne de Jersey,—an espalier on free stock; east aspect; ripe in October.
Beurré Gris,—an espalier on free stock; east and west aspect; ripe in October.
Beurré de Capiaumont,—espalier on free stock; east and west aspect; ripe in October and November.
Duchesse d'Angoulême.—as an espalier on east, west, or north aspect; ripe in October and November.
Bon Chrétien Napole,—espalier on free stock; east or west aspect; ripe in October and November.
Beurré Beaumont,—as a standard; ripe in August.
Jargonelle,—ripe in August.
Bon Chrétien, Williams's,—a standard, grafted on a free stock; ripe in August and September.
Seckle,— ripe in October.
Beurré Diel,—an espalier on east, west, or north aspect; ripe in November and December.
Beurré passe Colmar,—espalier on free stock, for east, west, or north aspect; ripe in November and December.
Althorpe Crassane,—ripe in November.
Hacon's Incomparable,—in December and January.
Glou Morceau,—in December and January.
Knight's Monarch,—in January.
Easter Beurré, — in March; keep till March.
Beurré de Rance,—espalier for east, west, or south aspect; ripe in February and March.
Ne plus Meuris,—in March and April.

250.—PLUMS.

The most useful varieties for small gardens are, for early fruiting, the Goliath, Greengage, Victoria, and Rivers's Prolific; for late fruiting, Purple-gage, Magnum Bonum, Coe's Golden Drop, and Damson. The Reine Claude de Bavy is an excellent plum upon a warm wall or under glass.

251.—PEACHES.

In the calendar of the different months will be found full instructions for the growing and management of peach-trees, whether out of door or in the orchard-house, and lists of the best sorts.

252.—NECTARINES.

Full directions for the treatment of Nectarine-trees will be found under the head of Fruit-Garden in the calendar for each month. It will be necessary here merely to mention that the following will be found very useful varieties, and form a good succession in ripening:—Fairchild's Nectarine, Violette hâtive Elruge, Murrey, and Red Roman.

253.—APRICOTS.

Apricots, as most other fruit-trees, flourish best in a good sound loam. For planting, prepare the soil about a yard deep, and manure with rotten leaves—one part of leaves to four or five of soil. Place a substratum of brick or other imperishable material below each tree. The apricot, when in a healthy state, produces more natural spurs than most other trees, and although some kinds will blossom and bear fruit on the young wood, yet the chief dependence for a crop of fine fruit must be on the true spurs. In pruning, stop all leading shoots, and pinch off to a few buds all shoots not required to fill up vacant places on the wall. Thin partially all fruit where it is thickly set, but preserve the final thinning until the fruit has stoned. The apricot, and especially the finest of them (the Moor Park), is subject to a sudden paralysis: first a branch, then a side dies away, until scarce a vestige of the tree is left; and this generally occurs on fine sunny days in spring and early summer, when the sap-vessels are young, and the sap is easily exuded by a few sunny days. In this state a frost occurs, the sap-vessels are burst by the thawing of the frozen fluid, and the whole economy of the plant deranged. Under these circumstances, which are so often occurring, the injured limb having consumed the sap, can draw no further supply: it yields to the solar influence, languishes and dies. Such is briefly Mr. Henry Bailey's diagnosis of the disease under which this delicate tree suffers. The remedy is to retard, or rather prevent, premature vegetation, and when that can no more be done, to provide protection; for this he recommends netting made of sedge, of about four-inch mesh, to envelop the main branches.

254.—MELONS.

The culture of the Melon is very similar to that of the Cucumber. The preparation of the manure, making the bed, raising the plants, the stopping and setting, are the same; but the soil in which they are finally planted should be trodden down rather firmly: and as the fruit appears all nearly about the same time, it is advisable to have the melons swell off as nearly as possible together; otherwise, the most forward will take the lead, and become much larger than the other. Two melons on a plant are as much as can be expected to do well; but never more than three should be allowed to remain: pinch off all the rest, and every other unnecessary growth. It is important that the plants be not allowed to ramble after the fruit has begun to swell; for this will require the whole strength of the plant. The fruit takes some four or five weeks, occasionally more, from the time of setting to the time of ripening, which is indicated by the stalk appearing to separate from the fruit. They should be cut and used on the day this takes place, or very soon after.

Both cucumbers and melons, it is known, are sometimes grown in pits heated with hot water. The superiority of this plan is so fully established, that none would be troubled with dung-beds after having tried it. The diminution of labour, the cleanliness and comfort, and last, but not least, the ornamental appearance of the suspended fruit, have a decided advantage over the many inconveniences attached to the management of hotbeds.

255.—BUSH FRUIT.

Under this title are included gooseberries, the different kinds of currants—red, white, and black,—and raspberries; fruits so extremely useful that they are to be found in every garden, and are grown extensively for markets in the neighbourhood of every large town. It too frequently happens that bush fruit, from the readiness with which it yields a crop, is left to take care of itself; but the quantity and quality of the fruit produced will be found to depend very materially upon the good management of the bushes. We shall give general directions for the different sorts, but further information will be found under the different heads.

256.—GOOSEBERRY.

Though the gooseberry will grow on the poorest soil, it will not produce fine fruit unless planted in a deep, rich soil, and treated generously. Though hardy, it requires moderate shelter, and though rejoicing in moisture, it will not flourish in undrained land. Cuttings should be planted any time from October to March. Select for the purpose shoots of a medium size, not root-suckers, about a foot or more in length. Cut the base of the shoot square; no fruit canes should ever be planted with slanting heels; after this, remove with a knife every bud from the base to within two inches of the top. If the cuttings are fifteen inches long, and four heads are left at the top, the future stem will be a foot high, which will be ample for a useful tree. The lower buds are removed in order to secure a clean stem and prevent the formation of suckers. Plant the cuttings in the shade four inches deep, and fix the earth firmly about them. During summer, young growing shoots strike readily under a hand-glass on a shady border, and a season may frequently be saved in this way. The first season's growth of cuttings put in in autumn should be very little interfered with. If any pruning is requisite, it is best done by rubbing off buds and by pinching in shoots which would interfere with the proper shape of the bush. At the end of the season, cut back all leading shoots to two-thirds of their length, so as to cause them to break next spring and form well-shaped bushes. At first, it is frequently desirable to plant cuttings only a few inches apart, and after the second year's growth to plant them out finally about six feet apart. Each bush would then have about eight leading shoots to form a head, and must be kept in shape and order by yearly prunings. If large fruit is required, it is not desirable to shorten the shoots except they grow too vigorously and incline too much downwards. Weak and superfluous shoots should be removed, and this is best done by taking them off as close as possible to the old ones, and removing all bottom buds, so as to prevent the formation of too many young shoots. The trees may be trained in many ways: sometimes the form of a fan or an espalier hedge is adopted, which has the advantage of being easily netted if birds are troublesome. The cup and funnel shapes are especially suited for the production of fine fruit, as air may be admitted to the centre of the trees. The best plan of protecting these from birds is by encircling the bush with wire netting, and covering the top with a piece of string netting, which can be removed when the fruit is to be gathered. The ordinary bush form can be protected in the same way.

With regard to the selection of sorts, we cannot do better than give the ex-experience of so good a judge as Mr. Shirley Hibberd. The best of the old varieties, he tells us, still hold their ground. There are none equal to the Champagne for flavour. The Red Champagne is of the same quality, differing only in colour. The old Rough is the best for preserving, and Warrington is unequalled as a profitable late gooseberry. For early work, take Golden Drop, Ostrich, and Early Green Hairy. For the latest crop and for retarding, the best are Warrington, white; Viper, yellow; Pitmaston, green; and Coe's Late Red. The most profitable sorts are Keen's Seedling and Warrington, red; Globe and Husbandman, yellow; Profit and Glenton, green; Eagle and Wellington Glory, white. For large exhibition berries, the following are a few of the best established sorts:—Red: Companion, Slaughterman, Conquering Hero, and Dan's Mistake. Yellow: Leader, Leveller, Goldfinder, Peru, Catherina. Green: Thumper, Gretna Green, Rough Green, General, and Turnout. White: Snowdrop, Antagonist, and Lady Leicester. The Lancashire gooseberries, which are generally distinguished by long drooping branches, bear the largest fruit. Seedlings have been shown at Manchester varying in weight from 20 dwt. to 26 dwt., and, we believe, even beyond this. Such fruit however, is generally produced at the expense of the crop. When fruit is to be gathered green, it is most profitable to keep the bush as thick in shoots as possible: for ripening fine fruit, the more open the bush the better.

257.—CURRANTS.

Red and white, may be pruned and treated in the same manner; black currants require a different treatment. In managing the cuttings, proceed as directed for gooseberries. Plant out the second year, when the cuttings have about 8 inches of stem and about five leading shoots. The pruning of both red and white currants is very different from that of gooseberries. When the requisite number of branches has been produced, so as to form a uniform bush, the greater part of the young shoots should be taken off annually, leaving only those that may be required for new branches, and shortening these to 4 or 6 inches with a clean cut just close to a bud. In pruning off the superfluous lateral shoots, take hold of each branch at its extremity with the left hand, and with the knife in the right hand, remove every fresh lateral up the stem, leaving to each a short spur of a quarter or half an inch in length—from these spurs the bunches of fruit are produced. As the bush increases in age, it will be necessary to remove all old mossy wood, and also to thin out the spurs when they have become too crowded. Of late years, great improvement has been made in both red and white currants. Visitors to Covent Garden market frequently express surprise as to the size of the bunches and the berries. These currants are not only peculiar sorts, but very great pains are taken in the cultivation of them. To grow fine currants, make the plantation in an open sunny position, on a stiff, well-manured loam; plant the bushes 5 feet apart each way, and every autumn trench in a good dressing of half-rotten manure in such a way as not to injure the roots of the trees. At autumn-pruning all the young shoots must be cut in to 2 inches. The sorts which produce the largest fruit are White Blanche, with amber-coloured berries, and White Dutch. Of red currants, Cherry is the largest; La Fertile and Knight's Large Red are also excellent varieties.

258.—BLACK CURRANTS.

The cultivation of the black currant is almost the same as the gooseberry, and the pruning is the same, only not so severe, as the black currant does not form so many young shoots. All dead and unproductive wood should be removed each year, and the shoots thinned so that light and air may freely enter the bush. The best varieties are the Naples Black and Ogden's Black; both of which, under good culture, are profuse bearers, and very large. Black currants are best left to grow as bushes; they do not thrive well trained to walls, or as espaliers.

259.—RASPBERRIES.

Raspberries flourish in any good rich loam, and will grow to perfection in a dark unctuous soil. As a general rule, raspberries do well where black currants flourish, and neither of these are very productive where cherry-trees thrive best. Before planting, the ground should be well trenched and manured; for though the roots lie near the surface, it is well to induce them to strike downwards in the event of a dry season. The second or third week in October is the best period for planting. Strong canes should be selected, and great advantage is gained if they be taken up with soil upon their roots. They may be put in singly, in rows, or in bunches of three canes each. In this latter case, it is desirable to cut the canes of different heights; the strongest may be 4 feet, the second 3 feet, and the third 2 feet. Staking will be necessary before the plants begin to grow in the spring, and great care should be taken that the ground is not trodden in wet weather. The pruning of raspberries is an easy matter. In June the bushes should be gone over, and all suckers removed, except about six of the strongest. These, at a later period, may be reduced to four, and if the parent plant be weak, two or three will be sufficient. There is great benefit in cutting the canes of different heights, for as the top buds grow strongest, the young fruit-bearing shoots are more equally divided, and enjoy more air and light. The ground in which raspberries are grown should not be broken up, but have a top dressing of good rotten manure yearly. The most useful varieties of the raspberry are the Red Antwerp, Fastolf, Prince of Wales, and Vice-President, the Yellow Antwerp, and Large-fruited Monthly. By a little management, raspberries may be made to bear a crop of fruit during autumn. There are sorts called Double-bearing Raspberries, but the result is greatly due to pruning. For late bearing, as soon as root suckers show themselves in June, the old canes should be cut away entirely, so as to prevent summer fruiting; and encouragement given during July and August to such suckers as show blossom-buds, for these will bear fruit in autumn. Autumn-bearing raspberries must be kept thin, or they will not prove successful. The canes for this purpose should be planted in single rows, and not in threes, as recommended for summer-fruiting. They should stand about 1 foot apart.

260.—STRAWBERRIES.

Good sorts are Keen's Seedling, Oscar, the British Queen, Dr. Hogg, Alice Maud, Eleanor, Sir Harry, the Black Prince, Goliath, and Carolina Superba.

261.—STRAWBERRY-BEDS.

The proper time to make new strawberry-beds is the month of August; but if space of ground cannot then be had, or the time spared, it is an excellent plan to take the runners at that time, and set them only a few inches apart in peat soil, on a north border, where they will soon make good root, and become strong plants. In the early spring they should be taken up separate, with a ball of earth, by means of a trowel, and planted a proper distance from each other in the bed intended for them.

An American writer, Hiram W. Buckley, New York, gives the following novel method for forming new strawberry-beds:—"Whenever I wish to make a new strawberry-bed, I allow the old bed to run into mass. Then, at any time during the autumn or early spring, I line strips about 7 inches wide and 20 inches apart through the length of it, and cut them into squares. These squares I take up with a spade, about 3 inches deep, and set them 18 inches apart in furrows previously made 2 feet apart, in a new bed. As the roots are not disturbed, they bear a full crop the first season, and this pays well for the heavy labour of removing so much earth. I make no account of the large number of roots required, since they spread so rapidly as to cover the ground in a single season, even when runners are kept back till after fruiting. The trenches made in the old bed are filled with rich earth and manure, and the bed is soon covered with plants again, when other trenches may be made and filled; thus renewing the old bed by degrees, as occasion requires." It is of so recent a date that this American plan cannot have been tested in this country. It seems, however, very likely to answer, and to be a saving of time in the formation of new beds. Still it must be remembered that when fine fruit is required, the plants should be kept single—quantity undoubtedly is gained by allowing the beds to run into mass.

262.—FIGS IN THE OPEN AIR.

Almost any well-drained soil will suit the fig-tree. Care, however, must be taken that it is not too rich, for if so, the tree will not produce fruit. Three sorts of figs are usually grown,—the Brown Turkey, Brunswick, and Black Ischia, all require a wall and a sunny situation. The best mode of training is perpendicular. Fix to the wall as many permanent leaders as are required at from 10 to 15 inches apart; get rid of all unnecessary wood by disbudding, and stop the fruit-bearing shoots at the end of August or beginning of September, according to the habit of the tree and the nature of the season. This operation is performed by merely pinching off or squeezing flat the terminal growing-point. This stopping, the object of which is to induce the formation of fruit for the ensuing season, is a matter of much nicety. A too early stopding with most trees will cause a too early development of fruit, the consequence of which will be that it will not stand through the frost of winter. The fruit for next year must not be much larger than a pea when winter sets in.

263.—FIGS UNDER GLASS.

Instructions for the proper treatment of these will be found in the calendar for each month.

264.—CHERRIES.

The Cherry delights in a light rich soil and open situation. Waterloo and May Duke are among the earliest varieties, and the late Duke and Morello among the latest. Other good sorts are Bigarreau, Black Eagle, Black Tartarien, White Heart and Kentish; the last is much admired for cooking purposes.

265.—FILBERTS.

The trees may be introduced into orchards, shrubberies, plantations, or hedgerows. Planted close to each other, they form valuable screens or shelter in exposed situations. A filbert-walk is a great addition to any garden. Filberts are not merely ornamental, but profitable. They will thrive almost anywhere, and are much improved by pruning. Webb's exhibition Cobb filberts are the best.

Plants may be had of Messrs. Barr and Sugden, Covent-garden, at 50*s*. per hundred, or 9*s*. per dozen.

266.—GRAPES.

It is certain that our moist and cloudy climate is not favourable to the ripening of the grape; its cultivation in the open air, therefore, requires great care; and in many seasons the most skilful management will fail to bring it to perfection. Nevertheless, the graceful trailing habit and beautiful foliage of the vine render it highly ornamental on the walls of a house; and for this it is worth cultivating, with the prospect of some fruit in favourable summers. It is also certain that in former days vineyards of considerable extent were cultivated, some remains of which are still found in Gloucestershire.

The vine is propagated by cuttings and by layering. Cuttings made early in March or the latter end of February may be planted about the middle of March. The cuttings must be shoots of last year, shortened to about 12 inches, or three joints each; and if they have an inch or so of last year's wood at the bottom, it will be an advantage. They may be planted either in nursery rows until rooted, or planted at once where they are to remain, observing in the latter case to plant them in a slanting direction, and so deep that only one eye or joint is above ground, and that close to the surface.

Vines are propagated by layering of the shoot of last year, or of a part of the branch, laying them about four or five inches deep and covering them with soil, leaving about three eyes above the ground; they are also layered in large pots, either by drawing the branch through the drainage-hole and filling the pot with soil, or by bending the branch and sinking it four or five inches in the soil and pegging it down there; it may then either be grown as a potted vine, or, when fully rooted, transferred to its permanent place on the wall or vine border; in the latter case the soil of the border should be dug out for three or four feet, as directed for other wall trees, a solid concrete bottom formed, with thorough drainage to carry off the water, and the border filled in again, first with bones and other animal remains. then with lime rubbish where that is available, and the surface with good loamy soil. In this soil the vine should be planted, the roots being previously trimmed and spread out horizontally, so as to radiate in a half circle from the crown of the stem. Under such an arrangement as this the vine comes rapidly into bearing.

When the vine is approaching a bearing state, and the leaves have fallen, a general regulation of the shoots becomes necessary. In every part of the tree, a proper supply of last year's shoots, both lateral and terminal, should be encouraged, these being the principal bearers to produce next year's fruit. All irregular and superabundant shoots should be cut out, and with them all of the former year's bearers, which are either too close to each other or which are too long for their respective places. Where it is not desirable to cut out the branch entirely, prune it back to some eligible lateral shoot, to form a terminal or leading branch. Cut out also all naked old wood. The last summer's shoots thus left will in spring project from every eye or bud young shoots, which produce the grapes the same summer. The general rule is to shorten the shoots to three, four, five, or six eyes or joints in length, according to their strength, and cutting them back from half an inch to about a quarter of an inch at every eye, the strongest branches being limited to five or six joints, except where it is required to cover a vacant space on the wall. When left longer, the vines become crowded, in the following summer, with useless shoots, and the fruit is smaller in consequence. This pruning should be performed early in spring, even as early as February: in pruning at a later period, when the sap has begun to ascend, the wound is apt to bleed when the thick branches have been cut off. A second pruning should be performed about the middle of May, when the grapes are formed and the shoot has attained a length of two or three feet: at this time pinch off the shoot about six inches above the fruit and nail it to the wall in such a way that the fruit may be in contact with it. About midsummer a third pruning should take place, when all the branches should be gone over, and the fruitless ones, not required for next year's wood, removed. A vigorous vine will require a fourth and final pruning in August, when the long shoots from the previous stoppings must be shortened back again, and all leaves lying too much over the bunches of fruit removed; taking care to prune, however, in such a manner that there is always a succession of young branches advancing from the lower part of the stem properly furnished with bearers, as well as a sufficient supply of young wood to replace the old as it becomes unserviceable. The pruning finished, let the branches be nailed or tied neatly to the wall or trellis, laying them regularly six, eight, or ten inches apart. Vine-pruning may be performed any time during the winter months, when the weather permits; but the sooner the work is done the better. The young shoots of last year produce shoots themselves the ensuing summer; and these are the fruit-bearers, which are to be trained horizontally or upright according to the design of the tree.

In May the vines will shoot vigorously, producing, besides bearing and succession shorts, others which must be cut away, and bearing and other useful branches nailed or tied up close to the wall before they get entangled with each other; and all weak and straggling shoots, especially those rising from the old wood, should be cleared away. Much of this summer pruning may be effected by pinching off the young shoots with the finger and thumb while they are young and tender. This should be continued during June and July. Many small shoots rise, one mostly from every eye of the same summer's main shoots laid in a month or two ago; these must be displaced, in order to admit all the air possible to the advancing fruit. All new shoots whatever should now be rubbed off as they appear, except where they are required to cover the wall. In August, even these must be rubbed off, being utterly valueless even for that purpose. During this month, the fruit itself requires attention. Where the branches are entangled, or in confusion, let them be regulated so that every branch may hang in its proper position. All the shoots that have fruit hanging on them, or which are ranging out of bounds, may be stopped, and where the grapes are too much shaded during August and September, remove a few of the leaves which intercept the light and heat. They should now have all possible aid of the sun to enrich their flavour. It will be necessary now to protect them from birds, wasps, &c., by bagging the best bunches in gauze or paper bags In October the bunches are ripe to bursting, and ready to gather, preparatory to a new year of growth and decay. Bear in mind that success depends on well-ripened wood—a short-jointed branch, ripened under an August sun, being a fruitful bearer of highly-flavoured fruit, and for this purpose a light porous earth is preferable to more tenacious clay soils. When the bunches of grapes are formed, pinch off the leading point of the growing shoot one joint above that from which the bunch proceeds. This is done to check the tendency of shoots to overlap one another. After the young points have been stopped, each joint below the stopping will put forth a side-shoot. These are termed lateral shoots. While this close stopping limits the extension of the tree, the size of the berry is much increased. This stopping is continued till the stoning period commences. This process occupies six or eight weeks, during which the growth of the fruit remains stationary, and the leading shoots may be suffered to push wherever they may.

During the swelling of the berry, the fruit begins to acquire flavour, and the buds plumpness and firmness. Henceforth they must have all the sunlight possible. To obtain this, all the lateral spray and others which shade the larger leaves must be stript away, leaving the larger leaves exposed to the sun; for the fruit receives its flavour through the agency of the leaves.

Pruning varies with the fancies of the operator. *Spur-pruning* consists in carrying up one leading shoot to the whole extent of the house or wall, either at one year's growth, or two or three, leaving spurs or lateral shoots to develop themselves at regular intervals on the stem. This is usually the result of three years growth, the cane being allowed to make a third of the length the first year, a second third the second year, and the remaining third during the third year. There will thus be five branches the first year, ten the second year, and fifteen the third year. The subsequent pruning is confined to pruning each of the laterals back to the last eye at the base of the shoot.

Long-rod pruning consists in establishing a stump with three strong branches or collars, from each of which, in its turn, a shoots springs, which, by a regular system of pruning, is worked in successive lengths, the one running the whole length of the rafter, the second half the length, and the third, recently pruned back, is to produce the renewal-shoot.

Where the object is to cover a wall or house, the leading shoots are carried almost at random, the pruner selecting those which suit him, without heeding much, so long as they are short-jointed and strong, shortening back the renewal-shoots, according to the space they are to occupy, from three to six or eight eyes.

PESTS OF THE GARDEN.

267.—ANTS, TO DESTROY.

Place an inverted garden-pot over the nest, and the ants will work into it. Remove the pot in a day or two by placing a spade underneath it; then plunge it, with its contents, into boiling water, and repeat the process if necessary. Ants may be expelled from any particular plant by sprinkling it well with sulphur; they may also be kept away from wall-fruit, and other fruit while ripening, by drawing a broad band with chalk along the wall near the ground, and round the stem of the trees.

268.—APHIDES.

Aphides, or plant-lice, and their congeners, are indicated by an unhealthy appearance in plants; the leaves and young shoots curl up, and multitudes of ants, which seem to feed on their secretions, are seen about the stems. The remedy is repeatedly syringing the leaves and stems with tobacco or lime-water, or with gas-tar water when that can be obtained; but plants should be carefully examined in May, and the winged parent of the *Psilla Pyra*, and its congeners, destroyed before they have deposited their eggs. Lady-birds *(Coccinelidæ)* render great service in destroying myriads of aphides.

269.—WIREWORMS.

If any bed or favourite plant suffers much from wireworm, a good trap may be made by placing small potatoes with a hole in them just under the surface of the ground, at different intervals. The wireworms will, in general, prefer this to any other food, and a daily examination will serve to entrap a great many of them.

270.—MILDEW.

Syringe the plant upon which the mildew has begun to make its appearance, with a strong decoction of green leaves of the elder; or use in the same way a solution of nitre, made in the proportion of one ounce of nitre to one gallon of water. A mixture of soapsuds and sulphur will, in many cases, answer the same purpose.

A composition made by Mr. Bell, of Exchange-street, Norwich, is the most valuable of all remedies for mildew. It may be safely used anywhere, and is indispensable for wall fruit-trees.

271.—SNAILS.

To prevent snails crawling up walls and trees, they must be looked for, picked off by the hand, and killed. Make a thick paste with train oil and soot, and daub the bottom of the wall with it: this will form an effectual barrier, over which no snails will attempt to pass.

272.—WASPS.

No class of insects give more annoyance in the garden than wasps: the sweetest and ripest of the wall-fruit become their prey, unless precaution is taken to prevent it. Towards the end of autumn, that is in September and October, every specimen of the common wasp should be sought for and destroyed; those individuals still left being females in search of a quiet corner in which to deposit their eggs and pass the winter. Each of these, if left undisturbed, is destined to become the foundress of a fresh nest. One of the best traps for catching wasps is formed by placing a couple of hand-lights on the top of each other, making a small hole at the apex of the lower one, attracting them into it by placing ripe half-eaten fruit beneath the under one. Another mode of dealing with wasps is as follows:—Having found the nest, rinse well with spirits of turpentine a common wine bottle, and while the inside of the bottle is wet, thrust the neck of it into the chief hole of the nest, carefully stopping up all other holes with clods of moist earth, to prevent the egress of the wasps. This, of course, is best and most effectually done in the evening. The fumes of the turpentine will first stupify, and then destroy the wasps: in a few days the nest may be dug up. All persons interested in gardens should be careful, as we have just observed; to destroy the large female wasps, which may be seen about singly late in autumn, and on fine days in the early spring; these, as we have already stated, are looking out for a wintering-place for their eggs.

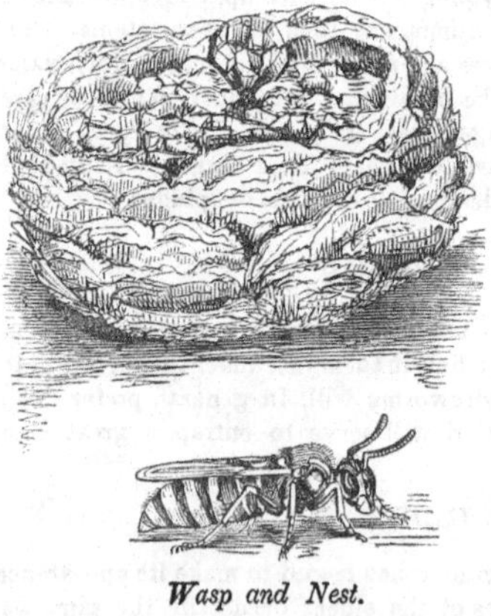
Wasp and Nest.

273.—SAW-FLIES.

Among the saw-flies, so called from the females possessing a saw-like apparatus at the extremity of the body, *Cladius difformis*, which is very destructive in gardens, measures a sixth of an inch in length, black and shining in body, with dirty yellowish-white legs. It feeds upon the leaves of various kinds of roses; the caterpillars are found feeding on them in the beginning of July, remaining in the pupa state a fortnight or three weeks, when they appear as perfect insects.

274.—MOLES.

These troublesome intruders may be driven out of the garden by placing the green leaves of the common elder in their subterranean paths, for the smell of these is so offensive to them, that they will not come near it; or they may be poisoned by placing in their paths worms, which, for some time, have been left in a place with a small quantity of carbonate of barytes.

275.—MOSS ON FRUIT-TREES.

Wash the branches of the trees wherever moss appears, with strong lime-water: strong brine will also answer the same purpose.

276.—MOSS ON GRAVEL WALKS.

Sprinkle the walks and yards over with refuse salt, but be careful to keep the salt from box-hedgings and the sides of the grass. This sprinkling should be done in dewy or damp weather, but not during rain.

277.—MOSS ON LAWNS.

All remedies are useless until the lawn is well drained; when this is done, rake the grass with a sharp-toothed rake in different directions to drag out the moss, and roll with a very heavy roller in wet weather. Nitrate of soda, at the rate of one and a half to two cwt. per acre, should be sown in the spring, over the mossy grass. Very fine coal-ashes, also, may with great benefit be spread over those parts of the lawn where moss abounds, especially if done in wet weather, or before a soaking rain.

278.—GREEN-FLY.

Fumigate with tobacco the plant infected, and syringe it well afterwards with clean water, or, if it is not possible to fumigate, wash the plant with strong tobacco-water, by means of a soft brush.

279.—INSECTS.

As all insects are produced from eggs, and as a natural instinct enables the mother to place the eggs in a spot where they will not only be safe, but where the young grub will find food to support itself until its first transformation takes place, a knowledge of the habits of the more destructive species is absolutely necessary to the gardener; the most effective remedy being to destroy the egg; for the caterpillar or larva state is that most destructive to vegetation. In this state the name of caterpillar is applicable to lepidopterous insects or moths, and butterflies, and some of the Hymenoptera, or bees. Grubs are the larvæ of beetles. generally with three pair of feet, strong jaws, and fat, mis-shapen bodies; maggots are the larvæ of flies, moving along the ground by the muscular action of the rings of the body; the larvæ of bees and ants being also generally called maggots.

When the larvæ of these creatures have exhausted the food near which the provident care of the mother has placed them, they are generally prepared for their second transformation—viz., the pupa or chrysalis state: winding themselves in their cocoons, they bury themselves in the earth, or in some other obscure place, and emerge in a few hours in forms as various as were their larvæ, the beetles with rudimentary feet, which are developed in their perfect state; the butterflies naked, suspended by the tail, or attached to the branch of some tree or wall; the moths enveloped in a bag or cocoon, which they have spun round themselves, as in a shroud; the flies and two-winged insects, smooth oval substances, are fixed to the plants or trees which have supported the larvæ. At length their last metamorphosis occurs: the caterpillar becomes a moth or butterfly gaily painted in its garb of summer; the grub becomes a beetle, with its diaphanous-coloured, hard, shining shell; the maggots develop themselves in thousands of shapes, floating and humming in the air—the two-winged insects, or Diptera.

All the mischief, however, has been done, so far as the garden is concerned and the gardener has only to look forward, as he ever must, to the next season. The insects humming and buzzing around him are short-lived: one object of their creation has been attained; they have performed, so far, their office of scavengers; their next is, to perpetuate their species; and the object of the gardener must be to circumvent them here, by destroying their eggs as they are deposited.

280.—INSECTS ON ROSES.

There is no class of flowers so much exposed to the depredation of insects as roses, and no remedy can be applied to their depredations without a precise knowledge of their habits and different states of transition. St. Pierre, when he had studied the economy of the different insects which infest the rose-tree for thirty years, still found something new to note. Moths, beetles, and gall-flies, and other insects hardly known to the initiated, seem to unite their forces in order to attack the queen of flowers. During June and July, the rose-beetle (*Cetonia aurata*) may be seen wheeling round the rose-tree, with its low hum, its wing-cases and elytra erect, instead of being extended from the body. It feeds upon pollen and honey, and in doing so bites off the anthers of the flowers, while its larvæ feed upon decaying wood and vegetable matter, burying themselves in the ground like the cockchafer.

Among the moths, the bell-moth (*Argyrotoza Bergmanniana*) is distinguished by the rich golden yellow of its breast and fore-wings, slightly clouded with orange, and bars of purple-brown with silvery scales. This moth, in the caterpillar state, is very destructive round London to roses. The moths deposit their eggs in the summer in the incipient buds, and they commence their operations on the leaves as soon as they appear, attaching them back to back by their silk-like thread-fibres. Round these leaves others grow in distorted shape, while the caterpillar revels on its core, "a worm i' the bud," devouring the petals of the flower as well as the leaf. When disturbed, the caterpillar drops down, suspended by a thin web which it spins, and by which it is able, when the danger disappears, to resume its former position. The only method of destroying these insects is by sharply pinching the buds where they are suspected to be in the early spring: this will relieve the plant, and enable it to throw out fresh leaves. If allowed to arrive at maturity, the moths should be destroyed as soon as they appear, and before they can deposit their eggs. The ashy-white bell-moth (*Spilonota aquana*) is another moth of the *Totricidæ*, which has been reared from the leaves of the rose, and of habit similar to the preceding. The yellow-tail moth, which has usually been found on the oak, the elm, and the blackthorn, Mr. Westwood has also found on the Scotch rose in his garden, feeding upon the petals, and afterwards attacking the leaves. This moth appears at the end of July, and the caterpillar (which is thickly clothed with long black hairs) feeds also on the pear.

281.—INSECTS ON PEAR-TREES.

Pear-trees are subject to the attacks of several species of lepidotera, saw-flies, and aphides. Among the lepidotora, the beautiful moth, *Zeuzera pyrena*, with its antennæ feathered on each side, is furnished with an elongated telescope-like ovipositor, with which the female deposits the eggs to a considerable depth in the crevices of the bark of the tree. The perfect insect appears in July, and the caterpillars in August, when they immediately burrow into the wood of the tree. In September they moult, and in the following June they are full-grown. Sparrows are the gardener's best ally in destroying this insect in the perfect state. Several other small lepidotera are injurious to the pear: *Argyromyges scitella*, one of the Tincidæ, deposits its eggs on the uuder surface of the leaves towards the end of May. The young larvæ penetrate the under cuticle, and feed on the fleshy parenchyma, leaving the surfaces untouched, giving the leaf a flabby and blistered appearance. The chaumontelle is said to be particularly subject to the ravages of this creature, especially in the beginning of autumn.

282.—SLUGS.

Of slugs there are several varieties, but the most destructive in gardens are the small white and small black slugs, which bury themselves in the ground or under leaves, and come out in the night-time to feed. To destroy these, take fresh lime in a powdered state, put it into a coarse bag, and after nightfall or before sunrise dust the ground where slugs are about: every slug touched with the smallest particle of the lime will die at once. If the weather be wet, the power of the lime will soon be destroyed: but if the ground be strewed in the evening with fresh cabbage-leaves, the slugs will hide under these, and may be destroyed in the morning.

283.—RED SPIDERS.

The free use of the syringe, and an occasional washing of walls, &c., with lime or sulphur, is the best method of getting rid of these troublesome pests.

284.—MICE.

There appear to be three sorts of mice, all doing more or less injury to gardens,—the common house-mouse, and two descriptions of field mice, the short- and long-tailed. They are all very destructive to newly-sown peas and beans, also to crocuses and other bulbs. To preserve peas and beans from injury by mice, let them be well saturated with a solution of bitter aloes before they are sown, or, having soaked them in salad oil, let them be rolled in powdered resin, which will answer the same purpose. Chopped furze, also, may with great advantage be placed in the drills over the seeds. The most effectual remedies, however, are poison and micetraps: of the latter, that usually termed the figure-of-four trap, formed with three pieces of stick and a tile, is perhaps the most simple and efficacious.

285.—DRESSING TO DESTROY INSECTS' EGGS.

An excellent Dressing to destroy the Eggs, &c., of Insects that infest the bark of trees and old walls.—Take ½ lb. of tobacco, ½ lb. of sulphur, ¼ peck of lime; stir these ingredients well together in three or four gallons of water; leave them to settle, and syringe the trees and walls well with the clear liquid. More water may be added when the first is used up.

286.—FIGURE-OF-FOUR TRAP.

This trap, which is formed with a common brick, supported by three pieces of wood placed in the shape of the figure 4, is very efficacious in the destruction of garden mice.

287.—DANDELIONS, TO KILL.

Cut the tops off in the spring, and place a pinch of salt, or a little gas-tar, on the fresh wound.

288.—DAISIES ON LAWN.

To clear a lawn of daisies, there is nothing equal to the continued use of the daisy-fork. With this ingenious little tool several square yards of apparently the most hopeless grass can be cleared in a few days. The fork should be used in moist weather, and the grass well rolled afterwards.

289.—DAISY-FORK.

This is a useful little tool for removing daisy roots from lawns: it is fitted with two iron prongs, and acts as a lever by means of a long or short handle. Price, 1*s*. 6*d*. to 3*s*. 6*d*. each.

290.—DAISY-RAKE.

The daisy-rake is for removing the flowers by drawing it over the grass. Price, 5*s*. to 7*s*. each.

291.—EARWIGS.

To get rid of earwigs, place pieces, about four or six inches long, of the hollow stems of any plant, in an horizontal position, in different parts of the trees on which earwigs abound. The earwigs will congregate in these, and may be shaken out into boiling water and destroyed. Very small garden-pots containing a little dry moss may be inverted on the top of a stick, and in this way will form a good trap when placed among flowering plants.

292.—EARWIG-TRAP.

The amateur's registered earwig-trap is to be obtained of Messrs. Barr & Sugden, price 7*s*. 6*d*. per dozen. Those who cultivate dahlias, roses, and such other plants as are subject to the depredations of earwigs, &c., will find this trap invaluable.

No. 1 engraving represents the trap as it appears fixed on a stake; No. 2 shows the section with the chamber (*from which there is no escape*), where the insects retire after their nocturnal depredations. The trap is then inverted, the inner cone removed, and the culprits dealt with as may seem good in the eyes of their captors.

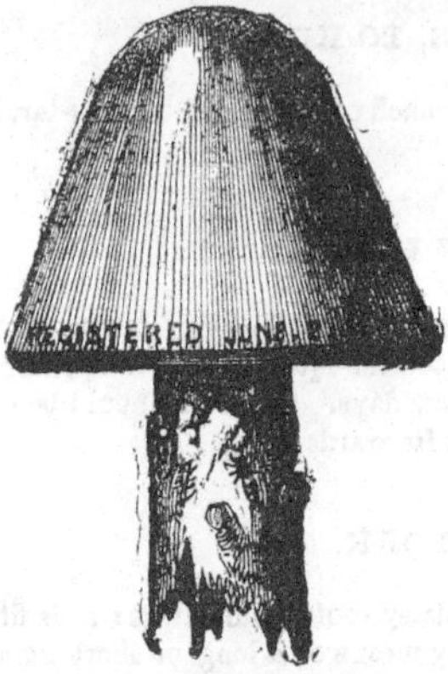

No. 1.

No. 2.

293.—GOOSEBERRY CATERPILLARS.

These pests have, of late years, become exceedingly prevalent, and in different parts of the country the gooseberry has been nearly destroyed by them. They come principally from a saw-fly, which lays its eggs in rows along the under-ribs of the leaves, and after having committed its ravage, falls to the ground, where it lives in the pupa state till the following season. The bushes should be carefully looked over once a week to watch the hatching of the eggs, when the infected leaves may be picked off. To prevent the fly from settling, the bushes should be dusted over with hellebore powder, or watered with a strong decoction of the Digitalis, or common foxglove. If the caterpillar has begun its ravages, the ground beneath the bush should be sprinkled with new lime, and a double-barrelled gun fired two or three times under it to shake the caterpillars down into it. The most effectual preventive, however, is to remove the top soil from under the bush during winter time, and destroy the grubs in it by mixing it with salt or soot: the parings so mixed may be buried or entirely removed, and new soil placed round the roots instead of it. Layers of bark from the tan-yard, when used as a covering of the soil underneath the bushes, have been found very useful in destroying the insect in its chrysalis state. One of our very best practical gardeners, Mr. Ogle, at Erridge Castle, Kent, says, "In the autumn or winter, when digging between the bushes, sow the whole ground over with *fresh-slaked* lime, using a liberal supply of lime, more particularly round the stems and about the roots of the bushes, forking the ground over. About the middle or latter end of March repeat the application, more especially round the roots, and rake the ground in, repeating the operation in two or three weeks. Few caterpillars will suruive this treatment."

294.—GUMMING.

To get rid of gumming in fruit-trees, scrape the gum clear away, wash well the place where it has accumulated, and stop it with a compost of horse-dung, clay, and tar.

MONTHLY CALENDAR.

295.—JANUARY.

January is the first month of our year, and the second of winter. The average temperature is 39° during the day and 32° during the night, and the mean temperature, during an average of many years, does not fall below the freezing-point; severe frosts, and frosts of long continuance, occurring in January, are therefore exceptional occurrences in our climate.

The Gardener's attention must now be concentrated on the future. All arrears of labour must at once be discharged. Nothing tends more to mar the success of gardening operations than dragging through the necessary work three weeks or a month behind the time proper for its performance. The peculiar fickleness of our climate renders gardening precarious and difficult enough with every advantage of judgment and foresight. It will be well, therefore, to bear in mind that the work can only be done "weather permitting." For instance, it is impossible to dig, plant, or sow when the frost has set its strong seal upon the earth. It is bad practice to dig in snow, and worse than useless to attempt anything on the surface of the ground when an excess of moisture has converted it into mud. It may thus occasionally happen that a part or the whole of the work prescribed for one month, may have to be deferred to another, and thus a double portion fall upon one or any of the winter or spring months. In such cases, extra labour must be employed, or diverted from other departments, until the whole of the work indicated is completed. There is a difference of several weeks in the climate of different parts of the country; operations that should be performed at once in the South, may thus generally be deferred for several weeks in the North. The term flower-garden seems almost a misomer at this season. But now is the time to consider what can be done to prepare the garden for the return of flowers.

The moment that flower-beds are cleared of their summer occupants, they should be dug up as roughly as possible. But rough-digging, while it can never present a smooth, may always exhibit an even surface, and in this way is not unsightly. It would be difficult to say whether the mechanical or chemical influence in enriching the quality of the soil is the most important. Certainly both are of the highest value, and their influence will be powerful, or the reverse, in exact ratio to the quantity of fresh surface exposed to atmospheric influence. Hence the importance of rough-digging, of forking over ground in frosty weather; resulting in that finely pulverized, mellow, genial soil in spring, in which plants delight to grow.

Next in importance to drainage, trenching, and manuring, and often of greater moment than any or all of them put together, is the frequent digging, forking, and scarifying of the surface; and from December to April are the months specially adapted for those operations.

Shrubberies on poor soils would be much benefited by manuring during this month. The usual practice of raking every weed and leaf off the surface, and cruelly disrooting the plants by a deep winter or spring digging, may be designated barbarous. When once shrubberies are properly established on good soil, no rake should ever cross their surface; and every leaf that falls upon them should be merely dug in, any time from December to April,—the earlier the better. Leaves are Nature's means of maintaining the fertility of the soil; and whenever or wherever art removes them, without applying a substitute, the soil rapidly tends towards sterility.

All newly-planted shrubs and trees should have their roots protected during the first winter with long litter. When placed close together in nursery lines, plants shelter and protect each other, and the massiveness of their tops, and possibly their summer leaves, shield their roots from the frost. Their condition is widely different when placed thinly, in newly-formed shrubberies. Hence the propriety, and in many instances the necessity, of what is termed mulching, that is, covering the surface with some good non-conducting material. The next point of most importance in planting trees or shrubs, especially of large size, is to secure firmly the top to a strong stake, or by any other method, so as to keep it immovable, for when trees, both top and root, are the sport of every fresh breeze, the probability is, that after the roots have made a feeble effort to grow, and been forcibly wrenched from the soil, they will perish.

During January, plant crocuses and any other hardy bulbs for succession: the main crops should have been planted in October or November.

Place old verbenas that are to be used for propagation into a gentle heat, and prepare a slight hotbed for cuttings. All bedding-plants will bear a much stronger heat while they are striking in the spring than in the autumn: verbenas will root in a week, placed in a close pit, with a bottom-heat of from 80° to 90°.

Place scarce varieties of dahlias in heat, for the purpose of securing plenty of cuttings. Proceed with potting-off singly all cuttings in store pots, using 48-sized pots for geraniums, and large 60 for verbenas, &c., where abundance of space is available. Where this is not the case, the potting-off must be deferred till another month.

Prepare a good stock of soil, clean pots, sticks, labels, stakes, &c., in bad weather, so that there may be no hindrance during the busy se son. No soil is better for the majority of bedding-plants than equal parts of loam and leaf mould, and a sixth part of sand.

Keep all cuttings free from green-fly, thrip, and other insects; cover up, water, and give air with caution and judgment, and ever bear in mind that the beauty of next summer's garden is dependent not only upon the quantity, but the health and the cleanliness of your stock.

In the reserve garden, in very severe weather, autumn-sown annuals should be protected by having some boughs stuck among them, or by being covered with mats, canvas, &c. Dig and ridge all vacant ground, and get the beds intended for the main sowing of hardy annuals prepared. Give beds intended for choice ranunculuses a liberal dressing of two-years-old cow-dung, and lay them up rough, ready for planting next month. Protect beds of hyacinths and tulips during severe weather, as they are often injured when coming through the soil.

In the rose-garden, proceed with planting and pruning hardy roses, and protect tea and China roses with boughs, covering the roots with old tan, cinder-ashes, &c. Procure stocks for budding; liberally manure all the ground occupied with roses, and take care that their roots are not injured during the process of digging. Florists' flowers require attention. Water auriculas, polyanthuses, pinks, carnations, &c., in frames, with care; remove early blooms from polyanthuses; examine pinks in beds, and if any of the plants are heaved up by the frost, press them firmly down in the soil.

Collect materials, and form new rock-works. Stone is the only really legitimate basis for scenery of this descrcription; but as this is not to be found everywhere, and as the taste for rock-works is universal, we must submit to have rocks of clinkers, débris of pottery, bricks, cement, chalk, concrete, and timber.

A collection of hardy heaths should be found near the rockery. Sweep up leaves; roll grass and gravel; remove all litter of dead or dying plants; and let the impress of neatness and the stamp of order be everywhere apparent.

Kitchen-Garden.—The work to be done in the kitchen-garden in January depends altogether on the weather. In open frosty weather no opportunity should be lost for wheeling manure on the vacant ground. All the refuse about the grounds should be collected and added to the manure-heap, and that burned or charred which will not readily decompose.

Now the forethought of the gardener may be exhibited. He has to lay down his plan of operations for the year, or at least for the next three months; and on his judgment in doing this much of the successful cultivation depends. If he cover too much ground with early crops in these three months, notonly will great waste arise, but he will have forestalled the space required for the main crops in April, May, and June, when some of the most important crops are to be sown. He should make his calculations, so as to secure a constant succession of the various products as they are required, leaving little or nothing to run to waste. It is a good practice, in going through the orchard, bush-fruit, and trees generally, to cut off all spare wood at this season, and to tie it up in bundles ready to be used for peasticks and other purposes.

The crops to be got into the ground this month are peas and beans, in the open ground or in cold frames; also in frames, radishes, lettuces (the black seeded cos does well if sown early), Walcheren or early Cape brocoli, cauliflower; on a slight hotbed, early horn-carrot and potatoes: of course, these will be earlier and better for the assistance of a slight hotbed of two feet or so in height. A little parsley sown now on a slight hotbed will be useful for planting out early. A little celery for an early supply, and a little cabbage also, should these be scarce, so as to fill up the main crop, if thinned out by severe frosts.

At this season it is necessary to be provided with mats or litter to cover the glass, in case of sharp frost; for, though most of these crops are hardy, yet, when young and growing, they are not unlikely to be cut off by frosts.

They are also much strengthened and hardened by exposure to the air in mild weather. A warm shower, if it can be had, is beneficial; but too much wet is injurious.

Early peas may be got in any time that the weather permits. Where the ground is tolerably porous and well drained, and a warm border, well sheltered on the north, is available, nothing more is required than to sow in rows, five, six, or more feet apart, the rows running north and south: for dwarf peas, five feet will suffice. About London it is the custom to sow spinach between the rows of peas, the spinach coming off in time to be replaced with brocoli before the peas are over; it is better however, to sow spinach apart, and leave the spaces between the peas till the time for planting potatoes, French beans; and other open ground crops.

Lettuces may be sown in a warm border under a south wall or fence: they are better sown in a frame at this season, if one can be spared, or even under a hand-light. Wanting either, it is advisable to cover the seed, when sown, with straw or light litter, taking it off sometimes to give a dusting with lime, in case any slugs may be harboured; the ground should be well dug on, spit deep, a dressing of manure being turned in, as lettuces require a rich soil to grow them to advantage.

For the purpose of growing very early potatoes, nothing is more suitble than a broad roomy melon-pit. About the beginning of January, let some middling-sixed tubers be laid in a warm and moderately dry place, well exposed to the light: here they will make short plump shoots by the time the bed is ready. Prepare a quantity of dung sufficient to make a bed three feet six inches in depth. When the bed is in order, lay on three inches of soil, and plsce in the potatoes fifteen inches apart, covering them with six inches more of soil. Some seed of the scarlet short-top radish may be scattered over the surface. As these begin to grow, give abundance of fresh air in mild weather, so that neither potatoes nor radishes be drawn up; and as they come up, remove the radishes which happen to be immediately about the crowns of the potatoes;

earthing up will not be required. The radishes will be ready in March; the potatoes early in May. Sow mustard and cress.

Plants under Glass.—Without protection of some kind during the winter months, no collection of plants can be kept together; but when mere protection is all that is sought, it is easily obtained; a trench two feet deep, dug in the ground, if the soil is dry, and a drain at hand to carry off surface-water, will suffice, if covered with frames, straw, hurdles, or other efficient covering; for it is ascertained by numerous, experiments, that the earth at two feet deep is warmer by 2° or 3° than the surrounding air in winter. A vacant frame, a cold pit, a greenhouse, or a conservatory, will also any of them serve the purpose. On the other hand, where plants of a warmer climate are to be forced into early bloom, artificial heat must be applied, not only to keep out the cold, but to assimulate their native climate and atmosphere.

Forcing-houses.—The routine business here during the month commences in earnest in January; a few plants of all kinds for ornamenting the house and conservatory should be introduced and started gradually; Indian azaleas, bulbs, roses, and lilacs, if already somewhat advanced, should have others brought forward to succeed them. Towards the end of the month, a good stock of pinks sweet-williams, and lilies of the valley, should be started in pits or frames. The temperature of the forcing-house should not be suffered to fall below 50°; and as the days lengthen, the temperature should be increased 4° or 5° until it attains a minimum temperature of 60°, and a maximum of 70°, by artificial heat, and an increase of 10° by sun-heat; giving air daily, even if for a short time only, and keeping the atmosphere always moist and genial by syringing or watering the pipes and flags.

Greenhouse. With the opening year and the lengthening days the busy season in the greenhouse commences; plants of all kinds begin to move, and most of them may be assisted with a little heat. Soft-wooded plants may be stimulated by it; and, when they begin to grow, moved into the larger pots in which they are to flower; while those which are more advanced and showing bloom, may be introduced into a warmer place. Many cinerarias are now in bloom, and may be removed to the window or conservatory; those reserved for blooming in May and June should still be kept in cold pits or frames, taking care to guard them from severe frosty weather, and especially from moisture. If large cinerarias are required, shift a few into larger pots, and pinch off the tops to produce a bushy head, tying or pegging down the side-shoots to keep them open: keep them moderately supplied with moisture, and give air on every possible occasion.

Fuchsias may be started, and large early-flowering specimens produced by cutting down the old plants and shaking the roots out of the old soil as soon as they have broken, re-potting them in a good rich compost, with sufficient drainage. Strike cuttings for bedding-plants as soon as the shoots are long enough. Calceolarias require great attention, as to watering. Remove all decaying leaves as they appear, peg down the shoots to the soil, that they may root up the st ms, and thus strengthen the plant. As seedlings advance, shift them into larger pots, and prick off those sown for late blooming. In potting, use a compost of light turfy loam, well-decomposed manure and leaf-mould, and a liberal portion of silver-sand, with an ample drainage of potsherds and charcoal, and keep them free from insects.

Pelargoniums which are strongly rooted may be shifted into larger pots, with stronger soil, using silver-sand freely, and taking care that the pots are clean and dry and the drainage good. Stop some of the plants required for succession, remove decaying leaves, and thin out weak shoots. Stake and tie out the shoots of those sufficiently advanced, to admit air to the centre. In plants of dwarf habit, peg the shoots down to the edge of the pot to encourage foliage. Stir and top-dress the soil from time to time, if required: a watering once or twice with lime-water and soot imparts a rich dark colour to the foliage, and destroys worms in the soil.

Where early flowers are required, and a stove or hot-house or other forcing convenience is at hand, remove a few plants, such as Admiral Napier, *Alba multiflora*, Amas, Jenny Lind, or any more recent early-flowering sort, for forcing.

Should frost appear, or the weather prove damp, light the fires in the afternoon, and shut up the house before the sun disappears, keeping the heat as low as is consistent with keeping out frost and dispelling damp. Water those plants which have become dry, but water them copiously. The fancy varieties, being the most delicate, should be kept in the warmest parts of the house, and their foliage thinned out occasionally. Use fumigation, to prevent the appearance of the green-fly.

A constant and ample supply of compost, well turned and thoroughly dry, should now be prepared for spring potting, and pots washed and dried for use, when wanted.

It is usual, where circumstances permit, to grow hard-wooded plants, such as heaths, azaleas, camelias, and others of similar habit, in a separate house; and some cultivators go so far as to recommend those having limited accommodation to confine their culture to one family, contending that it is better to have a houseful of finely-grown heaths, geraniums, or camellias, as the case may be, than a miscellaneous collection of indifferently-cultivated plants.

Care should be taken that hard-wooded plants do not suffer from the absence of moisture at the roots. After severe frosts, when fires have been used, evaporation by the sides of the pots is very great, while the surface seems to be moist enough: this should be seen to. To camellias and other plants of similar habit advancing into bloom, occasional doses of manure-water in a tepid state should be given, and the plants syringed with tepid water every other day, until the flowers begin to expand.

Among the hard-plants, the same remarks respecting heat are applicable; a temperature of 40° should be aimed at during the night, rising a little by natural causes during the day. Air should be given from above or by means of ventilators, without exposing the plants to cold draughts, and moisture encouraged by sprinkling the floor, flues, and pipes, when warm, with water.

With heaths, guard against mildew, and water moderately. Avoid artificial heat, if possible, but keep out frost; if heat becomes necessary, remove such plants from its influence as are required for later flowering'

Camellias should be advancing into full bloom, and the young expanding buds should be protected from cold currents of air, but without much fire-heat.

Azaleas should be growing freely, if they were shifted and promoted to a warm place last month. To get early-flowering plants, some of the more advanced specimens should be introduced to greater heat, while others may be retarded for a succession, to supply the conservatory or window cases. *Indica alba* is a fine early-flowering variety.

The conservatory, being only a more ornamental variety of the greenhouse, the same directions apply to it. Being generally a lofty building, the conservatory is not so well calculated for growing plants unless they be of a climbing habit, when they may be displayed to great advantage, Every thing should now look fresh and healthy. Acacias should be advancing into bloom. Camellias are either out or advancing rapidly into bloom: to promote this, see that they do not want for water. If there is a stove in the establishment, many orchids hyacinths, arums, tulips, and other bulbs, with heaths, epacridæ, from the greenhouse; and if only a frame is available, cinerarias, violets, and mignonette, will render the conservatory both gay and fragrant.

To preserve all these in bloom for the longest possible period is now the object. Keep the atmosphere moist and genial, but not wet; water regularly when necessary, especially the bulbs, giving as much water, of the same temperature as the house, as they assimilate; keep the temperature about 40°, rising a few degrees from sun-heat during the day, ventilating daily, if only for a short time, but avoiding cold draughts of air.

Vines, where they form a feature in the cultivation, are usually treated on some principle of succession, either by dividing the house by partitions, or by having a succession of houses. Supposing the plants to have been started in October, they would break last month with a temperature in houses of about 70°. This should now be the point aimed at, the minimum being 60° during the night. The habitual heat, however, should be regulated by the state of the external border. If the heat is failing, then fresh heating materials must be applied; for on that depends the result. Later sections may follow for succession, beginning at a lower temperature, and increasing the heat gradually as the vines break and advance. Vines in pots, if started in October and exposed to regular heat, will now be setting their fruit, and may be pushed on vigorously; for the roots being entirely under control, there is less danger of the plants being injured by over-forcing. Fresh plants should also be forwarded. This may be done by plunging them into a hot-bed and frame, and adding linings to keep up the heat until they break, when the heat of the vinery will be found sufficient. The plants showing fruit may be assisted by occasional applications manure, water, and bright weather.

FEBRUARY.

The mean temperature of February is nearly 2° higher than January, and the average number of frosty nights is about eleven. Less rain falls this month than in any other, and hoar-frosts at this season generally precede it.

Flower-Garden and Shrubbery.—Where the beds are filled with shrubs in winter, they should be hoed deeply several times during the month, to expose a fresh surface to the air. Beds occupied with crocuses and snowdrops should have the surface broken with a rake occasionally, before the plants appear, Borders similarly furnished require the same treatment. This not only imparts additional neatness, but, by breaking the crust, enables the plants to appear more easily and speedily, and in dry weather it considerably modifies the power of the frost. Beds planted with herbaceous plants, as well as herbaceous borders, would be benefited by similar treatment, provided they were dug early in November. Finish digging among herbaceous plants, circumscribing, dividing, rearranging, and replanting all where necessary, during mild weather. The old-fashioned way of arranging these according to their height is still the most effective.

The great business of propagating tender plants for furnishing the flower-garden must now be vigorously prosecuted. Stock must be taken; calculations made, judgment and foresight exercised, and activity displayed, if the garden is to be liberally filled next May. For the last few months, the great object has been safely to keep what we have. During the next three, the plants we have must be used to furnish what is required to fill the garden in May. If the bedding system is to maintain its ground, the garden must be filled with *flowering plants* by the end of that month. To effect it, verbenas must be planted four inches apart, and geraniums from six to eight inches. Measure the superficies of your beds; calculate at these distances, and increase your stock accordingly. With the exception of calceolarias, and probably geraniums, nearly all other bedding plants grow and flower as well, if not better, when propagated in the spring than in the autumn. Geraniums grow equally well; but autumn-struck cuttings flower more freely, and certainly two, three, or four-year-old plants flower more freely than any cuttings whatever. Boxes of geranium-roots that have been stored in cellars through the winter may now be brought out into the light of day, and, if they have been carefully managed, the whole surface will be alive with buds and shoots.

Sweep and roll turf and gravel; finish laying turf; top-dress, turn, renew, and relay the edgings of walks; and let cleanliness and neatness compensate as far as possible for the absence of floral beauty.

Shrubberies.—Push forward the digging and clearing of shrubberies. The great point in the management of shrubberies, however, is so to plant, prune, and train the shrubs, as to render these operations unnecessary. The raw edges and masses of bare soil, that render digging and cleaning an injurious necessity, mar the beauty aud grandeur of masses of shrubs. As a rule, their branches should sweep the edges of the turf: and the culture of herbaceous plants should never be atempted among them.

Reserve Garden.—If the weather continues open, the following hardy annuals should be sown during the month:

Alyssum calycinum (Sweet Alyssum).
Bartonia aurea.
Calandrinia speciosa.
Calliopsis Drummondii.
,, bicolor atrosanguinea.
Chrysanthemum coronarium.
Collinsia bicolor.
,, grandiflora.
Erysimum Peroffskianum.
Eschscholtzia californica.
,, crocea.
,, ,, alba.
Eutoca Manglesii.
,, viscida.
Gilia tricolor.
,, ,, alba.
,, ,, rosea.
Iberis coronaria (Candytuft).
,, umbellata.
,, ,, alba.
,, odorata.
Leptosiphon androsaceum.
,, flore albo.
,, densiflorum.
Limnanthes grandiflora.
Nemophila atomaria.
,, insignis.
,, discoidalis.
,, maculata.
Schizanthus porrigens.
,, pinnatus.
,, ,, Priestii.
Silene pendula.
Sphenogyne speciosa.
Viscaria oculata.

Florists' Flowers.—*Auriculas*, *Carnations*, and other florists' flowers in frames, must be carefully attended to. During dull weather, the lights should be tilted up, either at the sides or the top, in preference to pushing them down or drawing them up. This mode of ventilation will prevent the fog or moisture in the air from being deposited on the leaves, a point of the first importance during damp or frost. Never water unless absolutely necessary.

Top-dress auriculas and pinks with soil, and re-pot all pansies that are intended to be bloomed in pots. Eight-inch pots are those most generally used for this purpose, but the size must depend upon the strength of the plants. To insure a good bloom, they should not be shifted into larger pots after this period.

Kitchen-Garden.—The operations in the kitchen-garden this month depend very much on the weather. In mild open weather, a sowing of radishes is made, and to protect them from birds and frost, cover lightly with straw or fern, uncovering the beds occasionally in mild weather. Netting stretched over the beds will admit light and air, and exclude the birds; white worsted will keep them off for a day or two; but they soon get used to it.

Dwarf peas may always be grown advantageously where sticks are an object, and these may be sown closer together. The Bishop dwarf, long-podded, is a good cropper.

Broad Beans of any sort may be got in for succession.

Cabbages.—Look over the rows of cabbages, and see if any are eaten by vermin. Dust with lime when the ground is wet, or early in the morning, to destroy slugs. It is a good plan at this time of the year to let a few ducks into the kitchen-garden for half an hour or so every morning; they will destroy immense quantities of slugs, snails, worms, and grubs. Replace all the plants that have been destroyed by frost or otherwise, and draw earth up to the stems. Sow under hand-glasses a little cabbage, of some quick-heading kind, as early York or Eastham: they will follow those which have stood the winter, and be very useful in July, August, and September. Some Brussels sprouts may be grown; also purple Cape and Walches and brocoli for autumn use.

Parsley may be sown in drills or broadcast, or as edgings or between dwarf or short-lived crops. The seed should be but slightly covered, trodden or pressed in, according to the state of the soil, and raked evenly. It takes several weeks to germinate at this season of the year.

Carrots of the Short-horn sort, if sown on a warm border now, will come into use in May. Sow rather thickly, and thin to two inches apart.

Onions for salading may be sown on a warm border. A small sowing of leeks may be made at the same time and in the same manner, but not quite so thick. A top-dressing of soot once a fortnight, or even oftener if the weather be rainy will have a very great effect upon the onion crop, and will prove an effectual remedy against the maggot at the root, which so often destroys the entire crop, especially on highly-manured land.

Red Beet, if sown now, will be very useful late in the summer. Sow in drills nine or ten inches apart, or broadcast, and thin. to the same distance. The white beet may be sown for the leaves, which are eaten like spinach in summer.

Early Potatoes may be planted on a south border, or under a wall having a sunny aspect. At this time it is well to plant middling-sized tubers whole. The early tops are apt to get cut off by spring frosts; but they bear none the less for it, aud they may be recovered if not too severely frozen, by watering with cold water before the sun is up.

Salading and Potherbs.—Lettuces should be sown now for succession; mustard-and-cress also, under hand-glasses. American cress, which is much the same as watercress, may also be sown on a sunny border: it is very useful for salading, and easily cultivated. Chervil may be sown about the end of this month, and also other potherbs, as savory, marjoram, coriander, and hyssop.

Fruit-Garden.—Prune and dress all bush fruit, if not already done.

Apples and Pears.—Finish pruning all fruit-trees this month, whether standards, espaliers, dwarf-bushes, pyramids, pillars, or trained on walls. In pruning these, the main object is to produce short fruiting-spurs, so that all vigorous shoots should be shortened; but the stronger the shoots the less they should be cut; for too close cutting throws them into the production of wood and leaf, and not fruit. Figs on walls should scarcely be cut at all, and no trees should be cut in frosty weather.

Peaches, Apricots and Nectarines, on walls, ought to be unnailed and pruned this month. Thin out the shoots till they lie about six inches from each other, and shorten or not according to the strength of the tree or shoot: if very strong, shorten but little; but if a weak shoot terminates still more weakly, cut back to a double bud—that is, one leaf-bud between too flower-buds: prune neatly in this way, and tack the branches up again with fresh nails and shreds. The trees should be gone over in this way every winter; it is important also that the shoots be thinned out and disbudded in summer.

Flowers under Glass.—In the conservatory camelias, arums, epacrises, *Salvia splendens*, Chinese primroses, a few heaths, lachenalias, and perhaps forced lilacs, azaleas, rhododendrons, with hyacinths, narcissuses, crocuses, and other bulbs, will now be in flower or coming into flower-bud. Keep a night temperature of from 40° to 45°, allowing a rise of 10° with sun-heat. Unless during very severe frost or cutting winds, give air daily, if only for an hour at noon, to change the atmosphere of the house-damp. Prune, and destroy scale and other insects on climbers and other permanent plants. All plants should be carefully examined before they are introduced into this house, in order to prevent an importation of insects. Smoking with tobacco, and other insect-destroying processes, are not only very disagreeable, but are most inimical to the beauty and longevity of the flowers. Examine, water, and top-dress if necessary, any of the borders. Remove all plants back to their respective quarters as soon as their flowers fade, and introduce fresh supplies from forcing-pits or stove; let no dead leaf or flower, or any dirt of any description, be allowed to mar the sense of delight which the conservatory should ever inspire.

In the greenhouse more air may be given, and 5° less heat will suffice for this house than for the conservatory. Now is a good time to examine and clean the whole stock of plants.

Towards the end of the month several species of greenhouse-plants, such as kalosanthes, baronias, chorozemas, Dillwynias, pimeleas, and azaleas, might be shifted into larger pots. Most of these thrive well in good fibrous peat and a little loam, liberally intermixed with sharp silver-sand and charcoal. One of the chief things to be attended to before placing any plant, and especially any hard-wooded plant, into a larger pot, is to see that the old ball is in a nice healthy growing state. The extremities of the roots should also be carefully untwisted or unwound, to induce them to start at once in the fresh soil. The new soil must also be pressed firmly into the pots, or the water will pass through it, instead of penetrating through the old mass of roots. Many hard-wooded plants are destroyed through inattention to these points. At this season of the year green-fly often attacks pimeleas, leschenaultias, and other plants: let it be destroyed at once, by fumigating with tobacco-smoke.

Fancy Pelargoniums will bear a temperature of 5° or 10° more than the other varieties. They should be potted in lighter soil, and even more carefully watered, as altogether they are more tender. The shoots will now require thinning and training.

Calceolarias and Cinerarias enjoy a temperature of from 45° to 55°. Well-rooted plants of the former should be shifted into the compost recommended last month; and plants of the latter, for very late flowering, may also be shifted. The earliest cinerarias will now be opening their flowers in the conservatory, and a succession coming on. Green-flies are particularly fond of these plants: they must be destroyed on their first appearance.

Fuchsias, after repotting, thrive best if plunged in a gentle bottom-heat. Water carefully until fresh roots are emitted; shade in bright sunshine to prevent flagging.

Forcing-Pit or House.—This is an indispensable adjunct to a well-kept conservatory, and should now be occupied with bulbs for succession. Rhododendrons, azaleas, Ghent and Indian (most of which, especially the ***Indica alba***, force admirably), roses, lilacs, Anne Boleyn white and other pinks, true carnations, cloves, &c., maintain at a genial growing temperature of 55° to 65°; on very cold nights, however, it may fall 5° or 10° with impunity.

Plant-Stove.—Maintain a temperature of from 60° to 65° fire-heat. Start the first batch of achimenes, gesnerias, gloxinias, &c. Prune plants of allamanda, dipladenia, clerodendron &c. Pot *Glorissa superba* during the month: it thrives best plunged in a brisk bottom-heat. Many ferns, begonias and other plants, should also be potted, and started into fresh growth. Some of the dendrobiums, Stanhopeas, and maxillarias should now be watered and pushed into flower. Prepare plenty of good peat and loam, broken potsherds, charcoal, &c., &c., for a general potting of all plants that require it.

Fruit under Glass.—In the present day this includes almost all known fruits. Vines in houses, started in October, will now be swelling their fruit. Thin in time, and maintain a steady growing temperature of 65°.

Those started in January will show their bunches this month, and a temperature from 55° to 60° will be suitable. Some prefer leaving the disbudding until the bunches show, so as to leave the best. This is safe practice, and the buds up to this stage do not much exhaust the vine.

In many places the first or second house will be started this month. See that all loose bark is removed, that the vines are thoroughly cleaned with soap-and-water, and painted over with a thick coating of equal parts of sulphur, soot, lime, and cowdung, made into a paste with strong soap-suds, previous to starting them. Begin with a temperature of 45°, and slowly and gradually increase it during the month 10° or 16°. Maintain a genial atmosphere in all the houses, by sprinkling the paths, syringing, &c.; and give as much air as the weather will permit, allowing a rise of 10° or 15° during sunshine. Grapes grown in pots require the same general treatment as those planted out. It will very much hasten the ripening of the fruit, if the pots are maintained in a steady bottom-heat of from 70° to 80°.

Pines.—From 65° to 70° should be the minimum temperature during the month: the bottom-heat may range from 5° to 10° higher. During dull weather weather a dry atmosphere must be preserved. The plants should be carefully examined previous to watering, and this operation performed so as to prevent the water getting into the axils of the leaves. Plants swelling their fruit should be placed at the warmest end of the house, and those intended for autumn or winter-fruiting kept steadily growing, carefully guarding against any sudden check. Succession-plants in pots must be kept rather dry, and the linings and coverings carefully attended to.

Peaches.—In their early stages these are very impatient of heat. Begin with a temperature of 40°, and gradually rise to 50°. This should not be much exceeded until the fruit is set; then, by gradual ascent, from 5° to 10° may be added; this is the maximum of fire-heat for peaches until the period of stoning is over. Syringe twice daily in bright weather, except when the trees are in flower. The borders should have a good soaking, if dry, before forcing commences. Give as much air as the weather will permit at all times. Unless bees make their appearance, the trees should often be gently shaken when in flower and the pollen distributed by a camel-hair pencil, to insure the fructification of the blossoms.

Figs will bear a higher temperature than peaches, and may be started at 50°. The terminal buds of the young shoot should be removed, to insure a good crop. Maintain a moist atmosphere, and water copiously when necessary.

Orchard-houses should stand open night and day, unless during severe frosts. Plants in pots must not, however, be allowed to become too dry. The trees would be benefited by being painted over with a similar composition to that recommended for vines. This would tend to prevent the attacks of insects, kill all moss and fungi, and render the buds safe from the ravages of birds. which often play sad havoc with trees in orchard-houses.

During dull weather care must be exercised not to force the fruit-houses too rapidly, or elongated, spongy growth will be made at the expense of strength and fruitfulness.

Hotbed and Frame Cultivation.—Cucumbers in full growth require every attention. See that the heat of the beds does not fall below 70°: apply fresh linings as soon as this is the case. Attend to stopping and setting; allow no more than two or three cucumbers to grow at the same time on one plant; admit air in sunny weather, but not enough to produce a draught; give all the light possible, but cover at night with mats and straw, and add fresh earth if required. Water of the same temperature as the bed is absolutely necessary. Always water over the leaves, as well as at the roots, about twice or three times a week, which is as often as the plants are likely to want it. In watering forcing-beds use a fine rose, that the surface of the soil be not beaten down in the process; and take care that the water is of a temperature nearly equal to that of the bed —rather above than below: if liquid manure is used, it should be rather weak. For asparagus and seakale a little salt, about a teaspoonful to the gallon, may be advantageous; it is better to water effectually at once, than to water little and often, because the latter is apt to keep the surface slimy and soddened.

Frame Potatoes.—February is also a good time to put some potatoes in a little heat. An excellent plan is to pare the soil off an old cucumber, or melon-bed; add three inches of fresh earth, then set the potatoes fifteen inches apart, and cover with five or six inches more earth; put on the lights, and then give a good lining of prepared dung: this will cause heat; the potatoes will root into the dung of the old bed and be very fine: give them plenty of air, but never allow them to get frosted.

French Beans may still be sown: they may be placed in an old hotbed fresh lined. As the season advances, they will require less heat, but will not do out of doors yet. Radishes, if sown now on a slight hotbed, will come in much earlier than those in cold frames.

Now is the time for making a hotbed for sowing ridge cucumbers, vegetable marrows, tomatoes, capsicums, and such plants.

Early Carrots may be sown, for succession, on a slight hotbed; and very dwarf peas also, which may be treated in the same way as French beans, also mustard, cress, and lettuces, for succession.

MARCH (Monthly Calendar).

This is generally a busy, but by no means a genial month. Trees, plants and shrubs that may have borne the rigours of winter with impunity, often succumb beneath the chilling blasts of March. The increased temperature during this month is chiefly observable during the day: it is still very variable, advancing, as it were, by starts: but the mean temperature of the month is about six degrees higher than February; the thermometer ranges from 28° to 63°, including the night and day temperature, the mean maximum being 49° 9′; and the mean minimum 40° 49′.

Flower-Garden and Shrubberies.—Magnolias, delicate roses, and other scarcely hardy plants on walls, should receive some shelter from the stern bite of March frosts and winds. Care must be exercised not to keep them too close and warm, or the remedy will prove more disastrous than the evil. For walls, nothing answers better than a thin layer of straw, covered over with a mat, and kept as dry as possible. Protection against the exciting energy of the sun's rays during this month is almost of equal importance to warding off the effects of extreme cold. The later in the season tender plants can be kept in a dormant state the better, and nothing secures this object more effectually than a thin covering of dry non-conducting material such as straw. The utmost caution must be exercised in removing protection. Uncover by degrees.

Grass lawns must be frequently swept and rolled; gravel walks turned fresh-gravelled, raked, rolled, and swept; edgings cut, planted, or altered; and all planting, pruning, and digging finished at soon as possible. This is also a good season to remove plantains and daisies from the turf, and to sow grass-seeds for new lawns. Fork over flower-beds on frosty mornings, to expose a fresh surface to the atmosphere, and provide a finely-pulverized soil for the roots of bedding-plants. Stir the surface by flat-hoeing or deep-raking among borders of annuals and bulbs. Complete pruning and training clematises, jasmines, bignonias, and other creepers on trellises. Remove all prunings and winter rubbish, to be either rotted or charred, and see that the entire garden has a neat appearance.

This is the proper month for planting all the hardy gladioli. If they were taken up in November and kept in a proper temperature, they will now be starting, and should be planted at once.

In addition to the sowing of the annuals named in February, the following should be at once sown, either in the reserve-garden or on beds, or in rows where they are intended to flower:—

Adonis Flos.
Calandrinia.
Calliopsis.
Campanula Lorei.
Centauria.
Chrysanthemum.
Clarkia.
Collinsia.
Convolvulus minor.
Erysimum.
Larkspurs.
Linum.
Love lies Bleeding.
Lupines.
Malope.
Œnothera.
Poppies of different kinds.
Saponaria.
Sweet Peas.
Venus's Looking-glass.
Veronica.
Virginian Stock.
Viscaria.

Half-hardy annuals, which require a frame, are perhaps better left till April.

Proceed with the potting-off of all bedding-plants; keep them close for a fortnight after potting. No place is so good for them as a frame.

This is also the best month for increasing dahlias by cuttings. If the old stools were placed in a warm pit or house last month, cuttings three or four inches long may now be secured. Cut them off close to the stem, if you can find as many as you want; if not, leave one or two eyes on the old stool, and in another week these eyes will furnish two, four, or six more cuttings.

Shrubberies.—Let all planting and alterations cease for this season at once. Top-dress rhododendron-beds with equal parts of cow-dung (thoroughly decayed) and leaf-mould. On poor soils this imparts a rich gloss to the foliage, and causes luxuriant growth. Where such material is not procurable, a thick layer of leaves may be *pointed* in with excellent results. Finish digging and clearing. Attend to staking, tying, mulching all newly or recently-planted trees and shrubs, before the March winds tear them half up by the roots. Choice specimens, recently moved, would be much benefited by a copious syringing with the engine, on the evenings of dry, pinching days, to check perspiration and husband the scanty juices of the plants. The usual routine of sweeping and rolling turf and gravel must be assiduously attended to: if the weather is mild, the grass must have a first mowing during the month.

Reserve Garden.—Stir the soil among winter-sown annnals; transplant them to their blooming quarters in the flower-garden. Sow ten-week stocks on a sunny bed for succession to those raised in heat. Prepare a piece of ground for sowing anemone-seed: Hortensis, Coronaria, and Rectifolia are the most useful varieties for shrubberies. Rub the seed clean in sand; sow in shallow drills nine inches apart, and cover with fine sifted leaf-mould and sand. Get ground in readiness for a general sowing of all biennials and perennials next month. The oftener it is forked over, the more thoroughly pulverized it will be; consequently, the better adapted for raising seeds of every description. Protect seeds from birds, which are often most destructive just as the seeds are vegetating. Enrich the hooped beds, designed for the temporary protection of bedding-plants next month, with a liberal dressing of manure.

Rose-Garden.—Finish planting all hardy roses, if bloom is expected this season. The excited state of the shoots from a mild winter must not make you impatient to finish pruning. The more excited they are the greater the necessity for delay, as the expenditure of the sap in the terminal buds will preserve the buds near the base of the shoots longer in a dormant state; and it is upon these buds we are dependent for next year's blossom. In pruning roses, every bit of old wood, loose bark, &c., should be carefully removed, as it is exactly amid such *debris* that the larvæ of caterpillars, aphides, &c., are deposited.

Florists' Flowers.—As the power of the sun increases, if the weather continues mild, auriculas might have the benefit of warm showers. The lights should be drawn off daily on fine days. Avoid cutting draughts. See that the plants have plenty of water, as they will now be throwing up their flower-stems. The material in which they are plunged may be sprinkled, to keep up a moist, genial atmosphere. Plants intended for show should have seven pips as level as possible, round and well-shaped· any ill-shaped small pips may be cut off. Cover up securely from frost, and shade for a few hours on bright days: take off offsets, insert them in a close frame, and water with care until rooted.

Carnations and picotees should now, if the weather is mild, be placed in their blooming-pots, and sheltered under glass during bad weather: they should be potted firmly, care being taken to keep the soil out of the axils of the leaves. Pinks in pots or open borders should be top-dressed with a mixture of the fine loamy soil and half-rotten manure.

Kitchen-Garden.—During this month most of the principal crops must be got in. Hitherto, warm and sheltered spots and borders have been appropriated, while the larger quarters have been dug up into ridges, and exposed to atmospheric influences. Now the whole garden is to be cropped upon a carefully-considered plan, so that no crop of the same character should follow on the same spot.

Seakale still requires some covering, but less than last month, blanching being now the main object of it; and sand, ashes, or leaves, will effect the object. When the kale is past blanching, its use does not end: the leaves may be eaten all through the summer and autumn while they are green, merely dressing them in the same way as winter greens. It will be found a very profitable crop for cottagers: it grows well in shady places, is not particular as to soil, and will stand a cold, bleak climate. A top-dressing of very rotten dung, of any kind, may be improved by addition of a little salt,—about a pound to the barrowful of manure: wood ashes are also beneficial, and may be added in any quantity.

Celery.—It is too soon to sow the main crop of celery, but a little may be sown for early use. First sowings may be in seed-pans; but for the main crop shake together a small heap of stable-dung, just sufficient to give a slight heat; spread three inches of soil on it, sow the seed, and cover with a hand-glass. Seed sown in this month will be ready to transplant in April.

Jerusalem Artichokes should be planted.

Globe Artichokes will be making off-sets which should be taken off for propagation.

Cardoons are not so generally cultivated now as formerly, especially in small gardens, on account of the space they require. The seed is sown in March, in a warm sheltered spot, or under a hand-glass or frame. When large enough, they are planted eight or ten inches apart, in rich or well-manured soil. Then again they are planted in rows or trenches, after the manner of celery, only at a much greater distance from each other. During the autumn, earth up to blanch. The plant grows very large, after the manner of the globe artichoke.

Potatoes.—About the beginning of this month is the time to get in early potatoes. Some recommend planting them in October, placing them deep enough to be out of the reach of frost. In porous well-drained soils this answers admirably; but the advantage is not so great as to recommend it for general practice. To insure a good crop, the ground should be bastard-trenched in October or November, and left in ridges; in February levelled, and some thoroughly decomposed manure forked in. In March the frosts will have left it well pulverized, and ready to receive the sets. Some prefer middling-sized potatoes for setting, planting them whole, scooping out all the shoots except one or two; others choose large ones, cut in two or more, assuming that a large potato makes stronger shoots, capable of standing erect in full light of day.

Carrots may be sown now, but the main crop should be deferred till the first week in April. Such sorts as the intermediate may be sown in the four succeeding months: they will be useful to those who like to have this vegetable fresh from the ground. The ground should be deeply dug or bastard-trenched in autumn, left at first in a rough state, and when it has been well-frosted, stir and level it in January or February. In preparing the ground for carrots, no manure should be applied, as it induces them to fork and to become grub-eaten. A dressing of sand is advantageous.

Cabbages.—It is advisable to sow some cabbage-seed of a quick-hearting sort, to follow those raised in January, or that have stood the winter. They will be of great service in July and the following months. The Early York, Large York, Nonpareil, Matchless, indeed, any sort, will do.

Cauliflower-seed sown now will furnish plants to be set out in May and June: it may be sown in the open ground, or in a frame- or hand-glass. Sow on the surface, tread and rake, and protect with litter or netting.

Brocoli.—Such sorts as Walcheren, Purple Cape, or any sort that heads in autumn, should be sown at this time in the same manner as cabbage or cauliflower. They will be ready to plant out in May or June, and will be very useful at a time when summer crops are over, and winter crops not ready.

French Beans may be sown towards the end of this month, choosing an early dwarf sort; but the principal sowing should be deferred till next month: those sown this month should be in a border, sheltered from cold winds, but open to the sun.

Radishes may be sown thinly between the rows of the more enduring crops, such as onions.

Peas should be sown this month in succession, and coal-ashes scattered at the roots of those coming up, to prevent their destruction by slugs: sow a row of many-leaved spinach between the peas.

Fruit-Garden.—If there are any fruit-growers who still doubt the efficacy of protecting the blossoms of apricots, peaches, and other wall-fruit, this month will test their faith. Protection cannot be dispensed with in our uncertain climate. The best protection will probably be found in temporary wood copings, projecting ten or twelve inches from the wall, with canvas curtains attached, which can be readily removed in fine weather; next to the coping, worsted netting is, perhaps, the most efficient defence against severe weather, and causes the least obstruction to the necessary circulation of air, light, and rain. Those who have curtains will do well to use them, not only against frost, but against the extreme ardour of the noonday sun, to retard and strengthen their blossom. Generally speaking, pruning and nailing will be finished, but the trees should be washed with the garden-engine or syringe, using tepid water, with solution of sulphur and soot, or lime-wash, against scale and other insects. The apricot has a great tendency to die prematurely,—first a branch, then a side, until scarce a vestige of the tree remains; this generally occurs on fine sunny days in spring and early summer, and is supposed to arise from the sap-vessels being excited too early and rising too rapidly, so that they are in too watery a state to resist the severe frosts which sometimes follow.

Apples and Pears.—Pruning should now be finished, and this is the last month for planting until the autumn; the various operations of grafting and budding are now in full progress. This is especially the season for crown-grafting, where it is desired to use some vigorous old tree bearing an indifferent fruit. In this case, the grafts should be taken from the trees before the buds begin to swell.

The Culture of Flowers under Glass.—The Conservatory.—The interest and beauty of this house will now increase day by day. Let a minimum temperature of 45° be maintained, allowing for a rise of 10° from sun-heat, and give as much air as the state of the weather and the maintenance of a kindly genial atmosphere will permit. The less fire-heat used the longer the flowers continue in blossom; therefore, in very cold weather, suffer a depression of 5° from the above, rather than increase by artificial heat. Keep the heating apparatus cool in the morning if there is the slightest chance of bright sunshine. Nothing destroys flowers so fast as the sun shining upon a house while the pipes or flues are in operation.

Camellias in full flower must be liberally watered at the roots: during the expansion of a heavy crop of buds, the demand on their roots is very great. Clear weak manure-water will excite them gently; it must, however, be both clear and weak, or it will do harm rather than good, for camellias seem to have no power of assimilating gross food. The blossoms must on no account be rubbed, touched, or wetted: they show at once any bruise or spot of water on their clear and delicate petals. Two buds can scarcely be held in the hand at the same time without injury. In cutting the flowers, each should be placed separate in a basket divided into small compartments, or in pots filled with sand.

The Greenhouse—Proceed with the shifting of all plants requiring it. Free-growing plants, such as lischenaultias, boronias, &c., may be treated on what is termed the one-shift system, provided they are very healthy. They require turfy peat, well coloured with gritty silver-sand, and a fourth part of clean sweet leaf-mould. Dirty putrid water is certain death to hard-wooded plants.

Pelargoniums will require careful training. Remove every dead leaf, thin cut superfluous shoots, and keep the plants scrupulously clean.

Cinerarias.—Keep clean, remove decayed leaves, and throw away all but the most choice varieties as soon as they have finished flowering. Save the best sorts for seed or suckers, and sow seed at once for the earliest plants.

Calceolarias.—Thin out the worst of the crowded leaves; peg down the shoots to increase the strength of the plants, and sow seed for next year.

Forcing-pit.—Introduce fresh batches of azaleas, lilacs, rhododendrons, roses,

&c. Remove pinks as soon as they fairly show flower, to a cooler house. Hydrangeas introduced now will force well, and make useful plants for the conservatory. Part of the pit should now be devoted to sowing tender annuals in pans or boxes, a first sowing of balsams, amaranthus, egg-plants, mesembryanthemum, ipomœa, thunbergias, *Primula sinensis*, humea, &c.

Stove.—Keep a nice growing temperature of from 65° to 70°. If the sun continues very bright throughout the day, houses containing variegated plants will require shading for a few hours about noon. This will be the more necessary after repotting. Clerodendrums, allamandas, stephanotis, ixoras, &c., should be pushed forward in a sharp bottom-heat. They may receive a liberal shift, and be allowed to grow rather loosely for a time, to encourage a rapid extension of parts. Ferns should be thoroughly overhauled, examined, shifted into larger pots, or reduced, as circumstances may require: nice fibry peat, leaf-mould, sharp sand, and broken sandstone suit them well.

Now is the time to destroy scale.

Fruit-Culture under Glass.—March is a peculiarly trying month for forcing. The extreme changeableness of the weather, varying almost every hour, from the fiercest sunshine to the bitterest cold, renders the utmost attention necessary. This is the more essential, as at this season the young foliage and fruit are so easily injured. Perfect ventilation may be said to constitute the main feature of successful cultivation throughout the month. The powerful rays of the sun compel us to give air,—the keen withering wind forbids it. Both must be obeyed: a skilful balance, resulting in a genial atmosphere, must be struck between these contending forces. The moment that one ceases to act, the other must be checked. Hence the extraordinary attention required, and labour involved in ventilating houses during March.

Vineries.—The above remarks are peculiarly applicable to grape vines in the early stages of their growth. When fully expanded, the leaves will bear the strongest sun, and exposure to a cold air in the autumn, without inconvenience; but when young they are very easily injured. The earliest grapes may now be stoning. Don't attempt to hurry them during this process; for you will certainly weaken the vines. Stoning occurs when the grapes are about three parts grown, and often causes vexation and disappointment to young beginners. The grapes make no *visible* progress for six weeks or two months. They are, however, progressing within, forming their seeds, or stoning, as it is technically called. A temperature of 60° at night is enough until this work is completed.

The Pinery.—Many of the fruiting plants will now be showing flower. Maintain a minimum temperature of 70°, allowing a rise of 10° or 15° in the sun, and a rather dry atmosphere, until the blooming period is over. Drip, or two much water on the blossom, will prevent it setting. Unless it set, that pip will not swell, and one pip vacant in a pine destroys the beauty and symmetry of the finest fruit. Use water at 80° immediately after potting, to prevent the roots receiving a check from the cold soil, and maintaining a nice growing heat of 65° to 70°.

Peaches.—Guard against sudden or great variations of temperature, and cutting draughts, and syringe morning and evening as soon as the fruit is set. Begin to disbud the more forward woodbuds, leaving the strongest and best-placed shoots. This disbudding should be done very gradually; say at five or six periods, during the early stages of growth. Early peaches, after they are stoned, will bear a temperature of 70° with safety. They should be exposed to all the light and air possible.

Strawberries.—There is no better position for these plants during winter than the floor of an orchard-house, cool, dry, and free from frost, which preserves them in a healthy dormant state. They may now be looked over, top-dressed, raked, and plunged in a pit with a bottom-heat of 50°, giving air in sufficient quantity, dry and bright, to keep the top for another fortnight at 40° to 45°. This will secure a root-action in advance of the top; so that when the top moves, and the trusses appear, plenty of active roots may be ready to support them. The temperature in the pit may be raised from 45° to 55°; and this should not be much exceeded until the fruit are set. They will then bear ten degrees more heat during the ripening period. Plants may also be introduced upon shelves in vineries, &c., but a pit for themselves is the best place for them. For succession, introduce a fresh batch of plants every fortnight. Cuthill's Blackfriar is a useful early sort. Nothing, however, is better than Keen's Seedling for the early, and British Queen for the late crops.

Hotbed-and Frame-cultivation.—*Cucumbers.*—Where cucumbers have not already been started, it should be done now. Let the manure be shaken and turned over three or four times; on this everything depends; the heat lasts longer, and the plants are not exposed to violent and irregular action. When the bed is made, some gardeners recommend its being left a short time to settle, before putting on the frame and lights, in order to prevent violent heating and rapid sinking, from the additional weight of the frame; but if the bed has been well turned and beaten down in the process of making, this will hardly be necessary. If the frame is not put on at once, however, it is advisable to cover the bed with litter or mats, in case of heavy rains, which would reduce the temperature of it. After the frame is on, place about a bushel of loamy soil under the centre of each light; too much soil at once would induce too much heat. It is an old-fashioned but safe plan to thrust a pointed stick into the bed. By drawing it out occasionally, the temperature of the bed can be ascertained by feeling it: if more exactness be desired, a ground thermometer might be plunged into it. If the plants have been raised in a temporary bed, they may be planted five or six days after the bed is made: they will thus be ready to start into active growth at once. If no plants are ready, sow two seeds each in 3-inch pots, only half-filled with soil at first, and add fresh soil as the plants grow. The soil in which they are to grow should be rather coarse, and by no means sifted. The after-treatment is the same as that described in January.

Melons.—This is also a good time to make up hotbeds for melons to ripen in June and July. The soil should be put into the frame at once to the depth of 8 or 10 inches, and trodden or pressed rather firmly, if the dung has been carefully turned and the bed well beaten down in the making. Two plants should be planted under each light, the vines radiating from the centre; or place them further apart, and train the vines back and front, picking off all superfluous soil, and leaving only sufficient to nourish the fruit. Where a cucumber or melon-bed is in full operation, the other seeds may be sown in pots, and placed in them; and when up, repotted, and grown till the beds destined for them are ready—a great saving of time and dung, to maintain the heat; for any decline below the point of safety, which is about 70°, will check the growth of the plants, and throw them back considerably. This applies to the culture of cucumbers and melons, and all forcing plants; but, in the case of plants which are to be turned out later in the season, it is necessary to inure them, by a gradual decrease of heat in the frame, to the natural temperature of the air.

Vegetable Marrows may be sown thickly in pots, and placed in a cucumber or melon-frame. When up, they should be separated and planted out, two or three in a 4-inch pot, where they may either continue till their final planting-out, or be separated again, and potted singly, to prevent their getting pot-bound. At the end of March, or early in April, plant them out on a bed of manure of sufficient heat to start them, covering them with hand-glasses. In May plant them out, without any such stimulus, on ridges in the open ground. *Tomatoes* and *Capsicums* are raised in the same manner. They may be planted out under a south wall, or grown in pots, in frame, pit, or greenhouse, during the summer.

Asparagus.—Slight hotbeds should still be made for forcing *Potatoes*, *Asparagus*, *Seakale*, *French Beans*, *Strawberries*, and *Radishes*, or any of these may be sown or planted on an old bed; the old lining removed and fresh but prepared linings applied to give the necessary heat. If they are forced in a pit, let the dung be well worked, laid in carefully, levelled and beaten down, and filled high enough to allow for sinking. At this time of the year, no other heat than that supplied by the dung in the pit will be necessary; for late spring forcing, brick pits are preferable, on account of their cleanliness.

Salading.—For a supply of *Mint* or *Parsley*, some roots planted now in a hotbed will produce young shoots or leaves. Some roots of *Horseradish* and *Chicory* may be planted in the same way, and blanched by excluding the light. A succession of *Mustard-and-Cress* should be sown every week. *Radishes* may still be sown in frames, or in the open air.

APRIL.

Aspect of the Month.—The Aprilis of the Latins, from *aperire*, to unveil oneself, fairly lands us amidst the glories of spring. The opening buds and blossoms respond to the returning warmth of the sun, although in our northern and sea-girt climate there is, perhaps, little of that genial temperature which suggested the name of the month.

The variations in the temperature are still very great, the thermometer ranging from 75° to a degree or two below the freezing-point in the meridian of London; the mean maximum of an average of ten years being 57·82° in the atmosphere, and the mean minimum being 35·33°; the temperature being lowest at sunrise, there being, on an average of ten years, six frosty nights in the month. An unusual fall of rain in April is supposed to indicate a dry season for the harvest.

Flower-Garden and Shrubberies.—All seeds intended to flower during the summer should be sown during this month. In places where hardy annuals are extensively grown, another sowing might be made. It would be best to sow now where they are intended to remain. The modern system of furnishing the flower-garden has limited the use of annuals. In gardens, however, where the family may not be always at home, or where the proprietor is indifferent to more permanent and durable flowers, a very brilliant display may be made for several months with annuals.

It is now late for planting in shrubberies, and anything that has been left till this month must have copious watering. The holes dug for the roots should be well soaked with water before the plants are put in. This expedient will prove the safety of many a late-planted shrub. Rhododendrons and kindred American plants may still be set out either singly or in masses.

Gravel Walks and Lawns.—Walks should be broken up and turned, if not done in March; if turned then, roll twice a week at least. Lawns should now be mown once a week, and carefully, for nothing looks worse than the marks of the scythe on an otherwise smooth lawn. All gaps in box-edging should be made good, well-watered and trimmed. Place stakes to all such plants as require support, bearing in mind that "as the twig is bent the plant inclines;" fix the sticks firmly in the ground, bring the stalks to the stake, and tie them neatly but firmly to it, without galling the plant; remove all straggling, broken, or decayed shoots, and keep all clear of weeds, raking smooth with a small rake.

Rose-Garden.—All pruning and any planting not done last month must be finished early in this, and all recently planted trees copiously watered, the ground stirred, but left rough,—at least unraked. Beds for tea-scented roses should be prepared for planting towards the end of this month or in May. Good ordinary garden soil will produce roses large enough for ordinary purposes; but to grow them in perfection, a hole in the ground should be opened two feet square and a foot deep. This station should be filled with a compost consisting of two good-sized spadefuls of thoroughly rotted dung for each plant, mixing it well with the soil.

The season for planting may be any time from the fall of the leaf till the buds again begin to swell, in April or the beginning of May. After that there is danger of the tree dying off.

Florists' Flowers.—Polyanthuses require protection from sudden storms and cold winds. Auriculas are now in bloom, and require attention,—the trusses thinned, and deformed pips removed. Weak manure-water should be applied in the mornings, shading the plants afterwards from the sun. Seed should now be sown in shallow pans, and lightly covered with soil, and the pans placed in some gentle heat. Dahlia-roots may now be planted out in the beds, three or four inches deep, and five feet asunder; tulips protected from cold winds and frosty nights by netting thrown over hoops, and by mats, in severe weather, on the beds, leaving plenty of light and air. Ranunculuses require the soil to be loosened as they come up, and watering with weak manure-water; a watering with lime-water will destroy any worms in the beds.

Carnations in pots should have the surface stirred, and a little new compost added, and be watered with lime-water, to destroy worms. Sow seeds in pots or boxes during the month, place them in a west aspect, and cover them with a sheet of glass. Some fine hybrids have lately been raised, between the Anne Boleyn class and other varieties. *Pansies* will now be interesting: water the fresh-potted plants sparingly, until the roots reach the edge of the pots: top-dress the beds with rotten manure; look for and destroy black slugs; plant out seedlings, and put in cuttings. Hollyhocks kept in pots during the winter should be planted out, about six feet apart, in deep rich soil. Cuttings of choice sorts are more tender than seedlings, and would be safer with a little protection for six weeks after planting. Seed may also be sown now in the reserve garden for autumn flowering.

Tender Annuals sown last month should be pricked out three or four inches apart on a fresh hotbed, prepared as described in March; on this they will grow without interfering with each other for three or four weeks.

Half-hardy Annuals may now be sown in warm sunny borders. Give them the protection of hoops and mattings at night and in severe weather; sow hardier sorts in the beds and borders, in small patches, where they are to flower, observing that their position is to be regulated according to their height and colour. The mode of sowing is to form a shallow basin in the soil, such as might be made with the convex side of a breakfast saucer; in this hollow sow the seeds, and sift half an inch of fine earth over them. Thin out the patches as the plants begin to grow.

KITCHEN GARDEN.

Turnip.—A sowing of early Dutch turnip may be made in this month. This crop is very apt to run to seed instead of swelling at the root, if sown too early; but a great deal depends on the kind of soil—it does best on a rather retentive soil; but should be in an open, unshaded piece of ground, for it never does much good if shaded or overhung by trees. A dressing of soot at the time of sowing seems to benefit turnips greatly. The seed should be sown evenly but rather thin,—a small quantity of it will cover a large piece of ground, if regularly scattered. Broadcast sowing is preferable for this crop; but if sown in drills, let them be fifteen inches distant from each other, and the plants left not less than a foot apart in the rows.

Jerusalem Artichokes may still be planted where not previously done. Give four feet one way, and three or four feet the other. This should not be delayed after the first week this month.

Globe Artichokes.—The best method for propagating these is to take offsets from them in this month, and plant them, three feet apart, in a row, and the rows five feet apart. They will bear little or nothing the same season, but will produce abundantly the following. To keep them in good bearing, it is advisable to plant a fresh row every year, and remove one of the old ones. If they are protected with straw, fern, or leaves in winter, they bear rather earlier in summer. When left unprotected, they are killed to the ground, but break up strong in the spring. Before planting, the soil should be trenched three feet deep.

Parsley, Chervil, &c. should be sown for the purpose of keeping a stock of young plants to gather from, as young leaves are best.

Spinach may still be sown, and is often useful for colouring green pea-soup in the summer. Sow in a shady spot, if possible; it will last the longer.

Potatoes for the main crop must be got in this month. As the ground is more likely to be dry at this time, they may be dibbed in whole, which yields food for the young shoot till it can find its own. When potatoes are cut, it is best to expose them for a day or two, to render the surface callous. In planting them, let it be in rows two feet apart; or, if space be limited, allow three feet, which admits of planting later crops between, before they are taken up.

Carrots may still be sown; and those who know the sweetness and delicacy of the short-horn kinds, in their young state, will take care to have a supply of them. They may be sown till the latter end of July.

Cabbages.—The first week in this month is a good time for sowing the various sorts of *Brassicæ* for main crop, selecting the beginning of the month for the meridian of London, and a fortnight later north of Cheshire and Lancashire. If sown earlier, except for early use, they are apt to make superfluous growth. The treatment for all sorts is nearly the same. Let the seed-beds be open, and away from trees or other shelter, tolerably dry, but not parched, at the the time of sowing. Mark out for each sort its allotted space: give plenty of room,—at least a square rod; sow the seed broadcast regularly over the ground; tread it well in, unless the ground be wet and binding; in that case stand in the alleys, rake level, and pat the surface with a piece of flat board: this will press the seed in without hardening the ground. If dry enough to tread, rake the surface even. If the weather be dry, and continue so, it will be necessary to give the seed-bed a copious watering. When the seed is up, keep the beds moist, so as to promote vigorous growth; giving a liberal dusting of line, salt, or soot now and then, which will benefit the young plants, and prevent the attacks of the fly. When large enough to handle, thin them, and prick out those drawn, in nursery-beds five or six inches apart from each other.

Peas for late crops may be sown any time this month. The tall-growing sorts are best to sow now, and if sticks be plentiful, these should have the preference. If on good soil, or well mulched, the yield is far above all other sorts; the *Ne plus ultra* yields an enormous crop: sow them six or seven feet apart from row to row, or ten or twelve feet, where crops of cabbages can be sown between the rows. Any of the wrinkled or marrow peas do well sown at this time: earth-up, mulch, and stake those sufficiently advanced. Good peas for sowing at this time are Hair's Dwarf Green Mammoth, Knight's Dwarf Green, British Queen, Tall White Mammoth, *Ne plus ultra*, Knight's Tall Green. The first two grow about three feet high, the others about six feet.

Salading.—Lettuces should be sown for succession; the large drum-head, or Maltese, does well sown at this time: but Cos lettuces are generally preferred, as being most crisp: any sort will do sown at this season. *Chicory* is used both as salad in spring and also the roots as a vegetable: it should be sown late this month and the two following. Sow in shallow drills a foot apart, and thin to eight or ten inches in the row: they need not be disturbed again until taken up for use, or put in a frame to blanch the tops: in common with all crops, they must be kept clear of weeds. *Parsley* may be sown at any time: a principal sowing is usually made now. Some make sowings in shallow drills, eight or ten inches apart; but broadcast is preferable, at least if the ground is in condition to be trodden. Use the small hoe as soon as up, and thin out gradually, till the plants are ten inches or a foot apart. *Chervil* is sown and treated in a similar manner, and is much used in some families, See that a good curled sort is sown. *Marjoram*, of the sweet, or knotted kind, is usually sown this month, on a clear open spot: the seed is small, and should be sown on the surface, trodden, and raked evenly, and watered in dry weather.

The Fruit Garden.—Planting of all kinds, except in cases of absolute necessity, should now be over for the season. Should it still be necessary to plant, precautions must be taken to protect the tender roots, while they are yet foreign to the soil, both from frost and heat, by mulching with long stable manure, or, as some recommend, by placing a layer of pebbles over them, on a bed of sand, and covering this, during the spring months, with ferns, haulm or other rubbish. All winter pruning will now be completed. Peaches and nectarines are advancing towards blossem, and apricots, on a south wall, will be showing their bloom. These now require the greatest attention.

Apples and Pears, which bear their fruit on spurs, when cultivated in gardens are usually trained as espaliers, as pyramids, or dwarf. In the mature state they require care in selecting the shoots to be retained, preferring ripe, short-jointed, brownish shoots, shortening back these to a bud which will extend the growth of the tree, studying,—first, the productions of spurs; second, to keep the heart of the tree open; third, as the finest fruit is borne on the extremities of the branches, to keep these as compact as possible.

Vines are now pushing forth their young shoots in great numbers. At this season only those which are obviously useless, and especially those issuing from old wood, unless wanted for future years' rods, should be rubbed off with the finger and thumb close to the stem. The useless ones being disposed of, those left should be trained close to the wall at regular distances apart, so that all may enjoy the light, heat, and air.

Strawberries, which have been under mulch all the winter, should now be uncovered; the old foliage would be cut down in March, as directed; and after clearing away all weeds and useless runners, a spring dressing of half-decayed material from the cucumber-frame, mixed with soot and decayed leaves will be useful, watering frequently towards the end of the month.

Gooseberries and Currants, pruned in January and top-dressed in March, by removing an inch or two of soil, and replacing this with a compost of loam and decayed dung, in equal proportions, will require little attention till the fruit begins to form.

Flowers under Glass.—Conservatory.—While any probability of spring frosts remains, ventilation must be cautiously given, especially with newly-potted plants and tender flowers from the stove or forcing-house. As they begin to grow, air should be given whenever it can be done with safety. Where artificial heat is used, ventilation may be rendered safe by using extra firing. Camellias and other plants with large coriaceous leaves, if not perfectly clean, should be washed with sponge, and, if necessary, with soft soap, to eradicate insects; and a moist, genial heat maintained by sprinkling the floor, stage, and pipes. Boronias, Lechenaultias, Chorozemas and Tropæolums will now be fit to remove to the conservatory. Place them in as airy a situation as possible, maintaining a temperature of 45° to 50° at night, rising 10° or so from sun-heat.

It may be partly from contrast with the dormant state of plants out of doors but chiefly on account of the intrinsic beauty of its occupants, that the conservatory is so much more beautiful for the next three months than it ever is afterwards throughout the year. There is a delicacy and fragrance about spring flowers that never seem equalled afterwards.

In addition to the plants named last month, this perfume will now be enlivened by the lily of the valley, roses, sweetbriar, and violets. Each is exquisite alone; but, all combined and added to the odour of lilac, hyacinths, narcissus, and other spring flowers now filling with fragrance an artificial, partially, confined atmosphere, constitute a delicious odour.

Greenhouse Plants divide themselves into hard and soft-wooded plants. Among the former are Boronias, Hoveas, Acacias, and Chorozemas, Epacridæ, Genistas, and Pultenæas, which will now be coming into bloom, if well managed.

Azaleas will be coming forward, where there is a good stock of such plants as *A. lateritia*, *Gladstonesii*, Prince Albert, *præstantissima*, and others of similar habit: their bloom should be retarded by placing them on the shaded side of the house. Plants that have been forced should have the seed-vessels picked off, and shifted, if the pots are tolerably full of roots.

Heaths in full growth require an astonishing quantity of water at this season of the year; mere driblets are certain death to them.

Pelargoniums, and other soft-wooded plants, now growing rapidly, require every attention. Water carefully, so as to avoid any check in their growth, using manure-water occasionally, composed of equal parts of sheep-, cow-, and horse-dung, and a little lime.

Fuchsias.—This is still a good time to buy in plants. The variety now is endless; but many of them are almost the same, with a new name.

Orchids.—As soon as these begin to grow, a general potting should take place. The beautiful palm-like leaved cyrtopodiums should be shaken out and potted in a compost of equal parts loam, leaf-mould, turfy peat, sand, broken crocks, and charcoal. They are noble-looking plants. Bletias may be treated in the same manner, using more loam, however, for them and the beautiful dove-plant, *Peristeria elata*. Plunge Aerides, Vandas, &c., in water, when their flower stems appear, until they are thoroughly soaked. Shift into fresh baskets Longaras, Brassias, Cropegias, &c. Keep Oncidiums rather dry at present. The beautiful old *Goodyera discolor* will now be in full blossom. Its striking dark purple-veined leaves, and noble heads of pure white blossom, make it still a charming object. Clean all plants when in a dormant state, and secure a moist growing atmosphere of 70° to 75°. Mechanics and cottagers who have no glass may make their houses very gay by taking up a few patches of crocuses, violets, hepaticas, pinks, &c., carefully potting them, and placing them in a sunny window.

Fruit Culture under Glass.—The changeable temperature of the early spring months is a source of immense anxiety to the gardener. From cloud to sunshine, and from sunshine to storm; warm days succeeded by frosty nights, and cold winds by perfect calms, are constant occurrences, and keep the gardener and his assistants in a continued state of uncertainty. With every attention and skill, proper ventilation at this period is a work of great difficulty, from the fact that on the brightest days the air is often only a degree above the freezing point Unless provision is made for introducing the external air through a heated chamber, no *front* air should be admitted until the end of April or beginning of May. It is always dangerous to give direct air in front of vineries until the fruit is ripe. In the absence of some better means of partially heating the air admitted at the back or top of the house, before it reaches the plants, a close woollen net, or double or treble Nottingham netting, might be fixed over the ventilators or open spaces where the lights run down.

As soon as grapes begin to swell, a rise of 10° may take place, which may be continued until the first spot of colour appears. The minimum may then be from 60° to 65°, with a *little air constantly* in the house, never omitting to close it at night. Successional houses will now require great attention,—disbudding, thinning, and tying the shoots, &c. Raise the temperature through the different stages. Stop young shoots a joint beyond the bunches, excepting always the leading shoots on young vines. After a few stoppings, if the leaves become crowded, take the young wood off at the same point at every stopping, as two or three large leaves beyond the bunch are sufficient to supply its wants, and more useful than a number of small ones.

Pines.—Shift all the succession plants as soon as possible. It will facilitate this operation very much if one man places his arms carefully round the leaves and another slips a tie of soft matting round the plant, sufficiently tight to compress, without bruising, the leaves. This will render the plants manageable, and enable you to *pot* without *gloves*. If the plants have been properly kept during the winter, remove the crocks, gradually unwind the roots, take away as much of the old soil as possible, pull off from three to six inches of the bottom, place the plant two or three inches deeper in the new pot than it was in the old, as pines root up the stem, and have no permanent collar; press in the earth firmly, and the work is complete. Turfy loam, mixed with a little charcoal and broken bones, is the best compost; enrich this with manure water during the rapid growing and fruiting stages. If ferns or leaves be used for bottom heat, these will now require renewing. This work should proceed at the same time as the potting, so that the plants may at once be moved back to their proper quarters. Keep the plants level during the process of plunging the pots, and after two rows are plunged, cut the ties and arrange the leaves of the back row; plunge another row, then cut and arrange the second row, and so on throughout. A mild day must be chosen for these operations, as five or six hours' check from cold will often throw a whole pit of succession pines into premature fruit—one of the greatest calamities that can happen to a cultivator.

Figs.—A dry close atmosphere often causes the embryo fruit to drop. Dryness, or excessive moisture at the root, may produce the same results. When the fig is in full growth, the latter evil is almost impossible; but there appears to be but little demand upon the roots for moisture until the leaves are fully expanded. Maintain a temperature of 60°, and syringe the leaves daily.

Orchard-House.—Unless this is heated keep it constantly open when the outside temperature is above 32°. Success here depends upon retarding the trees as much as possible. If they start now, and we have a sharp frost in April, the chances are you will lose the crops. If a pipe runs round part or the whole of a house, it may now be allowed to move at a temperature of 40° to 45°. Place plums, apricots, and cherries in the coolest part, nearest the ventilators. See that the trees in pots and borders are well watered previous to starting, and give all the air possible to keep down the temperature during frosty weather.

Strawberries.—Give plenty of air when in bloom, maintaining a drier atmosphere during that process. After they are fairly set, they will bear a temperature of 70° to swell off; but 60° to ripen, with abundance of air, is quite enough. On shelves, place each pot in a pan, or within a second pot half-filled with rotten manure. Water with manure-water, syringe twice a day, and keep the plants clear of insects.

Hotbed and Frame Cultivation.—Cucumbers, in growing condition, require more air in the day-time as the sun acquires more power; healthy plants will bear the full light without shading: if they droop under its influence while air is given freely, something is wrong at the roots or collar, and fresh plants should be raised to supersede them, provided they do not recover. Air should be admitted in proportion to the weather; and as this varies every day, more or less, watchfulness and care are necessary. Peg down the bines, and pinch off shoots that are not wanted, and all shoots above the fruit; add fresh soil and fresh linings outside as required.

Melons should be syringed occasionally with water of a temperature rather higher than that of the bed: all shoots not wanted pinch off, so that the strength of the plant can go into the fruit. Fresh cucumbers and melons should be started for successions. The heat of the dung now lasts longer, and is not counteracted by severe frosts, and the sun begins to yield more heat; the days also are longer, the plants receive more light, and consequently are likely to be more stocky and short-jointed; the dung however, must be well prepared. Much time is saved by raising the plants in pots upon the fruiting-bed already going; if none are in operation, make a small bed with part of the dung, and cover it with a frame or hand-glass. Hotbeds at this time of the year are of the greatest importance in gardens where other appliances for raising plants are limited. The most tender plants may be raised from seed, and cuttings of almost all plants strike root most readily in them.

A melon-pit, divided into compartments of two or more lights each, will be useful at this time, and will answer most of the purposes to which frames are applicable. Vegetable marrows, ridge cucumbers, tomatoes, capsicums, chillies tea-plants, egg-plants, may be sown and raised with the aid of manure, managed as for melons. This is a good time to raise all these, or to pot them and plunge them in the dung, if already raised.

Seakale should, when cut, be removed; the roots planted in the open ground if required for increase.

Asparagus should be watered with weak liquid manure, but care should be taken not to overdo it: be rather sparing of stimulants than otherwise.

Potatoes may be tried by scraping away the earth near the collar. The largest tubers are generally near the surface, and may be removed without disturbing the plants, which should be left to perfect the smaller ones: water if required; but liquid manure is not necessary.

French beans that are flowering should receive plenty of light and air, and be kept tolerably dry overhead, and tied up to sticks if they hang over: keep the roots moderately moist; but, if allowed to root through the pots, they will require no other stimulus.

Strawberries will require plenty of water and a liberal supply of liquid manure. While the fruit is swelling, give, if possible, more heat and more air; if kept close, the fruit is apt to fog or mould without swelling. In damp weather, tilt the lights; in dry sunny weather, push them down.

Tender Annuals should now be sown in heat, and half hardy ones in cold-frames. Pot or prick off any that may be up. Balsams, cockscombs, and globe amaranths still require heat, and should be kept near the glass, to prevent being drawn up. Cuttings of all soft-wooded plants should be struck in great numbers for bedding out: these root and grow freely in hotbeds.

Salads may still be sown in cold-frames, and a good plan is to move the frames from place to place. merely using them to protect the seeds from birds: a frame placed over rhubarb will bring it on fast: lettuces, &c., may be urged on in the same way.

MAY.

The average temperature of this month is nearly 10° above April, but it presents even an increased variation in its extremes of heat and cold, which renders it very dangerous to the tender flowers and fruits of spring, which now, in consequence, require increased care in protecting them from cold frosty nights, and shading from the sun's heat. The maximum average of heat in May for the ten years ending in 1853 was 65·36°, the minimum 41·73°, and the average mean 53·54°.

Flower-Garden and Shrubbery.—When autumn-planting has been neglected, Portugal laurels, evergreen oaks, red cedars, arbor-vitæ, &c., &c., and hollies, have been found to take root more freely now than when planted earlier in the spring. Continue to prick off annuals raised in frames into small pots, and harden such as are established preparatory to turning them out into the open ground. As the season advances have everything ready, by hardening the plants, that they may experience no check by removal. Turn over and well work the soil to get it into a proper state for planting.

Large plants of some genera, as phloxes, asters, &c., generally throw up too many flowering-shoots: where such is the case, thin them out at once, so as to obtain not only fine heads of bloom, but increased strength to the remaining shoots, to enable them to need less assistance from stakes. Hollyhocks for late blooming may still be planted.

Where bedding-out is practised, this is a busy month. If an early display be wanted, they must be planted rather thicker, and need not be stopped; if they are to bloom not before a later period in the summer, plant somewhat thinner; the flower-buds also should be pinched off as they appear, till the plants have filled the beds.

Stake or peg down such plants as require it, as planting proceeds, or the wind will break many of them. Plant out in rich soil a good supply of stocks and asters for the autumn, and sow a succession of annuals for making up any vacancies which may occur, and likewise another sowing of mignonette in pots for rooms or for filling window-boxes.

Tender Annuals, such as cockscombs, balsams, amaranthus, egg-plants, and others, wanted early, should now be shifted to another hotbed previously prepared for them, either on the surface of the ground or in a trench of the size of the frame.

Half-hardy Annuals for beds and borders should now be planted out in the ground, others potted or pricked out on a slight hotbed, and those pricked out last month will be fit to transplant, having been gradually inured to the open air. For this purpose let them be taken up with the roots entire, and carefully planted with their ball of earth in the places where they are to remain. Ten week-stocks, mignonette, and China asters, may still be sown in a bed or border of rich soil; a gentle hotbed will bring them forward so as to flower a fortnight earlier.

Hardy Annuals.—Lupines, Adonis, lychnis, mignonette, and many others, may still be sown in beds or patches, where they are to flower, watering them after sowing and in dry weather. Perennials may be increased by cuttings of the young flower-stalks.

Grass Lawns and Gravel Walks should be in high order, the grass well mown once a week if possible, and kept clean and orderly; gravel walks kept free from weeds, well swept and frequently rolled, especially after heavy rains; borders, beds, and shrubberies free from weeds, and where vacancies in the beds occur, let them be supplied; let the earth be clean and well raked, and the edgings, whether of turf or box, be in perfect order.

Florists' Flowers.—Hyacinths and tulips, ranunculuses and anemones, are now in full bloom. The more valuable hyacinths and tulips should be planted in beds defended by hoops, which, in hailstorms and heavy frosts, are covered with mats. These protecting coverings may now be only kept at hand ready to throw on when their shelter is required. By this means the blooming season for these gems of the flower-garden may be prolonged for a fortnight or three weeks, and their brilliancy increased.

Spring crocus, snowdrops, crown imperials, and all other flowering bulbs, should also be taken up when the leaves decay. This should especially be practised in the case of bulbs which have remained in the ground two or three years and increased by offsets into large bunches.

Dahlias potted off last month, and hardened by exposure, may be planted out about the third week. If the pots are getting too small for the growing plants, it is better to re-pot them in larger pots than to plant out too early.

Auriculas going out of bloom should be placed in a shady place, if in pots, and receive shade from the sun, if in beds.

Carnations and *Picotees* in pots should have every assistance given them: sticks should be placed to support the stalks towards the end of the month, the plants watered in dry weather and kept clean, the soil occasionally stirred, and kept free from dead leaves, and a sprinkling of fine fresh soil added occasionally. All the side-stalks rising from the stem should be taken off, leaving none but the top buds; shading the pots from the mid-day sun. Pinks, as well as carnations and picotees in beds, require the same treatment.

Pansies may be planted for successional beds in a north border; for this spring seedlings may be used. Plants in bloom should be shaded at noon in sunny days, and well watered in the evenings. Blooms not required for seed should be cut off as they fade, and side-shoots taken off and struck.

Ranunculuses should have the soil pressed round the collar and watered when it becomes too dry.

Tulips beginning to show colour should be shaded by an awning, but not too soon; neither should it remain on after the sun has begun to decline. Watering round the beds will keep the bulbs cool, and protract the blooming season.

Phloxes, whether in pots or beds, should be watered occasionally with liquid manure.

Kitchen-garden.—Asparagus.—New plantations of asparagus may still be made, but it must be well watered, unless rain occurs. Sow asparagus-seed where it is to grow, and thin the plants to the proper distance. Beds that are in bearing should be kept clear of weeds, and the ground stirred occasionally, adding a sprinkling of salt, which improves the flavour. In cutting, use a rough-edged knife, and insert it close to the head to be cut, to avoid cutting others in the process.

Artichokes.—Stir the earth well about them, and reduce the shoots to three, and draw the earth well about the roots. The offsets taken off may be planted in threes, 4 feet apart one way and 5 another, giving a copious watering till they have taken root.

Seakale should be cleared of the litter used in forcing, and the ground

14—2

forked between the rows, keeping it clear of weeds till the following December, unless the season should prove a dry one, when one or two copious waterings should be given, especially to newly-raised plants, the roots of which are yet shallow. If the leaves are used, they must not be thinned too much.

Rhubarb roots may yet be divided and planted 4 feet apart: it is a good practice also to sow the seed, which may be done at this time. Sow broadcast, and leave the plants till the following spring, so as to judge of the earliest, for thinning is unnecessary till this is is ascertained Roots for forcing may be raised thus in abundance.

Horseradish.—Pinch out the tops where running to seed, and use the hoe freely all the season through—it will require little other attention the rest of the season.

Beans may still be sown; about the end of this month some will be in full bloom; pinch out the tops of such, to hasten the setting of the flowers.

Celery.—Prick out that sown in March, giving 6 inches distance from plant to plant; in order that they may get strong, let plenty of good rotten manure be worked into the soil. An excellent plan is to cover a hard surface with 4 inches of rotten dung, over this 3 inches of soil, which having trodden and raked even, prick out the young plants the same distance apart, and water plentifully; they will form a mass of fibres, and may be cut out with a trowel for planting in the trenches. A little shade will benefit them in sunny weather.

Cardoons may be treated in a similar manner; they may still be sown—they will grow large enough for every purpose; there is no advantage in having them over large.

French Beans.—These may be sown plentifully this month: they will be found exceedingly useful, as they follow the main crops of peas. Sow in drills 3 inches deep and 3 feet apart. Earth-up those that have made a pair of rough leaves, after thinning to 4 or 5 inches. They should have no manure, as it is likely to make them run all to haum.

Runner Beans may also be sown: being of a climbing habit and very quick growth, they must have plenty of room. Sow in rows 7 feet apart, or sow 10 or 12 feet from row to row, which will allow of planting ridge cucumbers between; drill them in 4 or 6 inches deep, or dibble them in clusters or circles, of five or six beans in each cluster; these being 6 feet apart, they may be grown with fewer sticks, and are ornamental. The Giant White, the Case-knife, and other varieties, have all the same habit; but that most usually grown is the scarlet runner, which is unsurpassed either for flavour or productiveness.

Nasturtiums are often grown for salad, and also for the seeds, which in the young state are useful for pickling. Sow in drills, the same as peas; they are ornamental as covering for rustic fences, hurdles, &c.

Peas.—To sow now, use such sorts as Knight's Dwarf and Tall Marrows, Mammoth, or British Queen, in good soils, and even in poor soils, if mulched with good sound manure, the latter sort yields immensely. Observe the same rule in sowing these as in scarlet runners, as regards distance. Earth-up and stick any that may be advancing, as they grow quicker now than in former months: this must be done in time, or they will fall over. Dwarf sorts will not require sticks, and are very useful on this account.

Carrots that are advancing should have the small hoe employed between them; nothing benefits them more than continually stirring the surface of the soil; thin them to the proper distance. Fresh sowings may still be made. Horn carrot sown now will be useful in the autumn; they should be sown thicker than larger sorts.

Onions may still be sown, more particularly for salading, for which purpose thin out the earliest sowings and clear from weeds: drenching the soil with liquid manure occasionally will benefit them. Give a dredging with soot occasionally.

Leeks.—Thin where forward enough, and plant the thinnings a foot apart, in rows two feet from each other; give liquid manure to those that remain, and stir the ground between.

Parsnips.—Thin out to a foot apart at least—18 inches is not too much.

Potatoes.—Continue to plant if desirable; no fear need be entertained of their doing well. Several good late sorts do as well planted this month as earlier. Earth-up those that are forward enough, but not too much; more earth than is just sufficient to cover the tubers is likely to prove injurious to the crop.

Turnips may do well sown now, if wet or showery weather occur; sow broadcast, tread the seed in, and rake soot in with it. This seed germinates very quickly at this time, especially if sown on fresh-dug ground. Such as are up should be hoed between and thinned out immediately; doing this early will be of great advantage to the crop—the oftener it is done the better.

Scorzoneras, *Salsafy*, and *Hamburg Parsley* may still be sown, the treatment of these being very much the same. Sow in drills 15 or 18 inches apart, and thin to about 9 inches when up. Their culture is very simple, merely requiring the hoe between them during the summer. It is as well not to give manure before sowing.

Lettuce.—Sow in drills a foot or more apart, especially on light ground; let as many as possible continue where sown. Those transplanted had better be in drills, for the greater facility of watering, an abundance of which they must have in dry weather, to insure crispness and milky flavour, which indicates a well-grown lettuce. The soil for lettuces cannot be too rich. The large-heading kinds of cabbage-lettuce are proper for this month, but cos lettuces do equally well. Tie up cos lettuces about a fortnight before using.

Endive.—The Batavian may be sown now; it may be used in the same manner as spinach; treat in the same way as lettuce.

Sow the white as a substitute for spinach, and also silver beet to be used as seakale; treat as the red.

Rampion.—Sow broadcast or in drills, and thin to about 10 inches. This being a very small seed the ground should be raked over before sowing, to prevent the seed getting buried too deeply; tread it in and rake afterwards.

Spinach may be sown: but as it is apt to run very quickly, it is advisable to sow on a north border. Give plenty of room; it is less likely to run than when crowded.

Chervil and *Parsley* sown now on a sunny border will be useful in winter. Sow either in drills or broadcast; tread the seed in before raking; thin out that which is sufficiently advanced to 9 inches; plant out the thinnings at the same distance—they curl better when planted out.

Radishes.—Sow for succession. These must be well protected from birds, as they are immoderately fond of pulling them up as they begin to grow. They must be well watered, to prevent their becoming hot and woody. A good retentive soil suits them best at this time.

Cress.—American, Normandy, and Australian cress, and corn-salad, to come in in August, should be sown now in shallow drills or broadcast, treading the seed firmly in before raking; they will all require copious watering.

Cabbage.—To hasten the hearting of those that have stood the winter, tie them in the same way as lettuces. Plant out early-sown ones, and sow again for succession.

Cauve Trouchuda is much esteemed in some families for the midrib of the leaf, which is used as seakale. Treat as Brussels sprouts. All the members of this group like a retentive soil, highly enriched with manure, which is best given in the form of mulch. They then fibre on the surface of the ground and grow luxuriantly.

Brussels Sprouts and *Borecole* may still be sown; treat in the same manner as brocoli.

Savoys.—Sow again this useful vegetable, since moderately-sized heads of good colour are better than large white ones, the result of too early sowing. The main point in their culture, in common with the rest of this group, is an open situation and plenty of room,—2 feet each way is not too much; they must also receive their final planting before they are drawn up in the seed-bed.

Cauliflower.—Plant out early-sown 18 inches apart. Those that have stood the winter should have liquid manure, or, at least, plenty of water, unless they were previously mulched, which prevents evaporation, and also feeds the plants. Break the centre leaves over any that may be heading.

Brocoli.—This being a good time for sowing late sorts, as Purple Sprouting Miller's Dwarf, &c., care should be taken to have a good supply of them: they are invaluable in the early spring time. Give them an open situation sow broadcast, each sort separately, and rather thinly. Walcheren sown now will be very useful in the autumn: plant out early sorts that are large enough before they get shanky.

Herbs, as Balm, Mint, Marjoram, Savory, Thyme, &c., may be increased by slips, offsets, or divisions of the roots: at this time they grow quickly after the operation: they must be well watered.

Other herbs, as Basil, potted Marjoram, Fennel, Dill, &c., may be sown on the open ground. They are not generally subject to the attacks of birds, as many other seeds are.

Flowers under Glass.—Conservatory.—A watchful eye must now be kept on, all house plants for insects, or the labours of months, perhaps even years, will be lost. Ply the syringe diligently upon all plants not in actual bloom, to keep away the red spider; wash off the scale with soft soap, and fumigate for aphis and thrips. Where fumigation is necessary for a few plants only, they can be removed to a close room and subjected to that process.

The occupants of the greenhouse are now being transferred, rendering the conservatory gay and lively. Roses will be coming forward from the forcing-houses. Camellias, their season of bloom being past, are now in their full growth, and will be benefited by being shaded from the bright sunshine. An application of weak manure-water will be of great use to them if the surface soil is getting dry.

Climbers must be trained, and the house kept thoroughly ventilated and moist.

Azaleas as they go out of bloom should be attended to, the old flowers and seed-vessels picked off. Should they require re-potting, it should be done when the new growth begins; the strong shoots of young plants stopped, except one to form a centre for a tall pyramidal-shaped plant, the best form for this, beautiful tribe of flowering-plants.

Pelargoniums trussing up for flowering require watching: tie out the shoots as far apart as possible, to admit air freely to the heart of the plant, keep them well watered and in a growing state.

Scarlet Geraniums should be encouraged to grow by liberal shifting; and when established, water them freely, giving liquid manure to those fully rooted. Stop those growing too rank, that they may become compact and bushy plants.

Fuchsias, shrubby *Calceolarias*, *Heliotropes*, and *Alonsoas*, like the geraniums, require liberal shifting in order to grow them properly.

Fruit under Glass.—The great enemies of fruit-forcing are insects. Strict watch must now be kept for the red spider; if allowed to establish itself on the vines now beginning to ripen their fruit, it will seriously compromise next year's crop.

Vinery.—The earlier crops now coming forward will be colouring: they must be kept perfectly dry, and have as much air as can be given safely, keeping the house at the temperature of 65°, or thereabouts. The most important part of the vine's growth is between the breaking and the setting of the fruit: for the formation of sound, healthy wood and perfect bunches, they should be assisted by artificial means during that stage of their growth.

Pinery.—Keep the atmosphere of the swelling fruit humid, and the earth about the roots moderately moist, using occasionally weak manure-water. Where extra heavy fruit is desired, all suckers should be removed as they appear. On warm afternoons syringe copiously, and close up with a temperature of 90°, giving air again towards evening. When there are indications of changing colour, withhold water, and see that the bottom-heat is kept steady at about 85°. The plants intended for autumn-fruiting should now be shifted into their fruiting-pots.

Orchard-house.—Ventilation must still be strictly attended to. Open all ventilators during the day, except in fierce north and east winds. Worsted netting of half-inch mesh may be placed over the ventilators with advantage in severe weather. If the caterpillar attacks the young shoots of the apricot, the ends must be pinched off and crushed. Summer-pruning of trees to be so treated to commence early this month.

Hotbed and Frame-cultivation.—Hotbeds may be made for starting cucumbers and melons with greater certainty of obtaining fruit, and also with far less labour and material than earlier, the weather being much warmer, and the sun aiding by his rays the efforts of the cultivator. Give plenty of air to growing plants, particularly in sunny weather. Neither cucumbers nor melons should be shaded—it is necessary that the stems be matured and ripened, in order to secure a good bearing condition: plants that are vigorous and healthy will bear the full light of the sun, if air is admitted proportioned to its influence.

JUNE.

Aspect of the Month;.—The direct power of the sun's rays, indeed, is now at its maximum, although the radiation of heat from the earth's surface, which decides the temperature of our atmosphere, does not attain its highest point till August. The variation of the temperature is still great, ranging, according to local circumstances, from a few degrees above freezing to 90°, the mean heat being 58°. The average mean temperature at Chiswick, for a period of ten years, at one foot below the surface, was 60°; at two feet, 58°; and on the surface, 60° 45′: the mean maximum and minimum being respectively 64° 13′, 63° 10′; 81° 13′. 45° 10′. The dryness of the atmosphere is also at its height in our moist climate, and vegetation now depends on the "Orient pearl;" the dew, with which the atmosphere is laden, is condensed, and every blade of grass and leaf saturated with it an hour or two after sunset and sunrise. Beautiful indeed are the mornings and evenings of June, when the dew hangs upon leaf and blossom, and beautiful the economy of Nature as displayed in this arrangement: for the formation of dew is an illustration of the law of attraction.

Flower-Garden and Shrubberies.—As soon as the beds, borders, &c., of the flower-garden are finished, the baskets and vases filled, and the general spring planting-out brought to a close, the remaining stock of bedding-plants should be looked over. A portion will be required for stock; and as a considerable number of plants will in all probability be required to make good failures, or to replace beds now occupied with short-blooming plants, and other demands through the season—these, with few exceptions, had better be kept in pots and therefore, if any unpotted cuttings yet remain, let them be potted off into clean pots. Re-pot others, also, getting too full of roots, plunging them afterwards in ashes, in a cool shady situation, and pinching off all early or premature blooms: they will soon be ready for turning out. A few kinds of annuals should also, for the same purpose, be sown on a light soil and shady border. By frequently transplanting and stopping, their tendency to bloom will be encouraged, and the formation of roots promoted, and they will soon bear removing to the permanent beds without injury.

The newly-planted beds require constant watching. All failures should be instantly made good, and the tying and staking of everything requiring support attended to. Where an early display of flowers is not wanted, the buds may be pinched off. Cuttings of *Iberis saxatilis* root readily under a hand-glass at this season: when placed in a shady situation, they form a beautiful edging, and may be cut like box, for a week or two, to encourage the plants to cover the ground. Pansies, anemones, double wallflowers, and other spring plants, should be removed as they go out of bloom, to make room for autumn-flowering ones, the beds being made up with fresh compost, in planting the later. Creepers against walls or trellises should be gone over and tied or nailed in.

Rose-Garden.—Standard and pillar roses should likewise be looked over to see that they are properly secured to their stakes. This being the month in which roses are in their glory, care should be taken that their effect is not destroyed by imperfect buds or deformed flowers. Weak-growing shoots should be tied up and regulated, and all fading flowers and seed-vessels removed, cutting back the perpetual or autumn-flower kinds, as soon as all the flowers of the branch are expanded, to the most prominent vertical eye, stirring the ground and saturating it with manure-water, or sprinkling the ground with guano and watering with soft rain-water.

Towards the end of the month many shoots will be firm enough for budding, and some sorts work best on the flowering shoots, provided the buds are taken before the flowering is over.

Watch the different annuals as they come into flower, and mark those varieties whose superior habit of growth, size of flower, and brilliancy of colour, make it desirable that seed should be saved from them.

Florists' Flowers.—Dahlias.—Dahlias already planted out should be watered in the evenings with soft water overhead, the soil being previously stirred, and others planted out for later bloom, taking care, in hot weather, to mulch, round the roots, where it can be done without being unsightly, with short well-decomposed dung. As the shoots advance, train and tie them up carefully, and search for earwigs and slugs in the mornings.

Ranunculuses will be making rapid growth. Always water in the evening and with water which has been exposed to the rays of the sun. When they begin to show colour, the awning, or other shade, should be placed over them.

Carnations, Picotees, and *Pinks,* as they advance, should be tied to their stakes, reducing the number of the shoots according to the strength of the plant. Care should be taken that the flower-pods of pinks do not burst; and those having ligatures round them will require easing and re-tying. Shade any forward flowers, giving plenty of water and liquid manure.

Auriculas and *Polyanthuses* should be removed into a northern aspect, and decayed petals taken away from the seed-pods, and as the capsules turn brown, they should be gathered. Water as they require it, and keep the pots free from weeds.

Pansies struck from cuttings in April and May will produce fine blooms if planted in shady situations, or potted into 6-inch pots, and shaded in very bright weather. Cuttings may still be taken from promising plants. Mark all seedlings having good or singular properties.

Reserve-Garden.—A shady piece of ground in the reserve-garden should now be prepared for cuttings of double wallflowers, rockets, sweetwilliams, pansies, and other plants required for next spring's bloom. Alreetias and many other spring-flowering plants may also be divided and planted out this month; and beds of annuals for autumn-flowering should be sown in the space left by the zinnias, China asters, and marigolds planted out.

Kitchen-Garden.—Many principal crops come in this month, and following suddenly upon a time when the supply from the kitchen-garden is somewhat scanty, show the real effect of cropping too abundantly in the early part of the year.

Asparagus.—Water newly-planted beds, and keep clear of weeds. Beds in bearing will be benefited by an application of liquid manure.

Seakale.—Thin out the crowns where they are at all thick. A few strong heads are better than many weak ones; young seedlings will be benefited by a sprinkling of wood-ashes. It being a marine plant, salt may be strewn between the rows. Keep the young plants well watered, and hoe frequently between.

Beans.—The last sowing of these should be made for the season; they seldom pay for sowing later. Top those in bloom before they become infested with aphis. Mulching will increase the quantity and quality of the crop.

Runner Beans do well sown any time before midsummer. On light ground they may be dibbled in—an expeditious method. Some recommend soaking them in water for a day before sowing, which may be advantageous in hot, dry weather; but it is as well to water the drills or holes at the time of sowing. Those sown last month should be earthed and staked before they begin to run.

Nasturtiums may still be sown, being very quick at this time. Those already up should have their supports about 4 or 5 feet high.

Peas.—After the second week this month it is not advisable to sow strong-growers—such sorts as Auvergne or Champion of Paris. The time from sowing to bearing is less, and there is more certainty of a crop.

Celery will probably be in condition for final planting towards the end of this month; the main crop had better be deferred till next month. Celery is generally considered a gross feeder, requiring a rich, highly-manured soil and abundance of water. It certainly cannot be grown to perfection without both.

Carrots.—Thin without delay, but not too closely, as some are apt to run, even under the best culture. From 9 inches to a foot is a good average. A succession may be sown any time before midsummer.

Onions should receive a final thinning, allowing 8 or 9 inches for the main crop. Use the small hoe as often as possible, and keep them clean. Onions for salading may still be sown. A shady border on the north side of a wall will suit them. Tree-onions, potato ditto, and those planted for seed, will require some support.

Leeks.—Plant in deep drills, to admit of earthing up; give an abundance of water in dry weather. Soot dredged over them will stimulate them, and prevent the attacks of insects in a great measure.

Potatoes.—Earth up before they get too tall, but leave the top of the ridges nearly flat, so that the tubers are not buried too deeply. It is a great error many fall into of drawing the earth as high as possible up the stems. They do not bear so well, from the greater exclusion of air from the roots. Potatoes that have been retarded may be planted this month; they will yield new potatoes in the autumn.

Turnips.—Sow a good breadth of these—they will come in well and be very useful in the autumn. Sow immediately after rain, or, if the ground is light, immediately after digging. They grow very quickly; but some slight protection from birds will be necessary the short time they are germinating. White worsted will generally be found efficient. Tread the seed in well, or use the wooden roller after sowing, but finish off with a rake.

Scorzoneras, &c.—Thin to about 10 inches or a foot, and stir the ground, well between them.

French Beans.—Sow a few rows of these for succession. There are many varieties; but it is immaterial what sort is sown, except on the question of flavour or productiveness. The larger-growing sorts are considered the best to sow now. The Royal Dun, the Negro and Cream-coloured, &c., unlike dwarfer sorts, continue in bearing a long time. Thin to 4 or 6 inches, and earth up, but give no manure.

Lettuce.—Sow on a north border, but plant in an open situation. It is necessary to sow often to insure a succession. Water the ground thoroughly, or not at all: surface watering is very injurious.

Endive may be sown this month, as it is less likely to run now than formerly. The seed grows very quickly, and birds do not seem to care about it; it may therefore be merely sown broadcast, trodden and raked. Plant out early to insure a good curl in the leaf.

Vegetable-Marrows and Pumpkins should be got out early this month. If good strong plants, they may be merely planted on a sunny border; but they are much better for having a little dung-heat; or dung without heat will suit them, for they delight in a loose bed of light but well-rotted dung that they can root into easily. Give plenty of water if the weather holds dry.

Capsicums and Tomatoes the beginning of this month.—Plant these against a south wall if possible—otherwise against a sloping bank: the full sun is necessary to induce them to bear well. Any vacancies under and between wall-trees may be well filled up by them.

Cress.—Sow American and Normandy for succession.

Brocoli.—Defer not later than the middle of this month the final sowing of late sorts. Walcheren sown now will very likely come in during the winter. Plant out those that are ready, and never allow them to draw up in the seed-bed; but prick them out temporarily: they will pay for it. If there is no room for them otherwise, transplant in drills made for the purpose.

Brussels Sprouts, Borecole, and Savoys.—Get these planted for good as early as possible; plant in drills two feet apart, and water freely. Puddling the roots in clay and soot mixed with water may be good for them and prevent clubbing in a great measure. Plant between rows of peas and beans that will soon be off the ground—no matter how firm the ground is. Judging from experience, this group do best if the ground has not been dug for several months before planting. Watering once a day, or oftener, will be necessary in dry weather.

Cabbage and Cauliflower should also be planted out when strong enough. The latter will prove very useful in August and September. A succession of these is an important matter.

Fruit-Garden.—If the fruit seems setting thickly, let it be partially thinned, reserving the main thinning, however, till after it has stoned. The trees will have been mulched last month to prevent evaporation, and should now be watered, and that so copiously that it does not require frequent repetition, pouring the water into the roots. *Apricots* will now require their final thinning, and stopping, and watering, also followed by mulching, which is important at this time for all fruit-trees where evaporation is active.

The beginning of the month is a busy period in this department, and much vigilance and perseverance will be requisite to keep pace with the advancing growth, in preventing and keeping down the different pests.

In disbudding pears, plums, and cherries, the fore-right shoots, and those not wanted for laying on, should remain for the present, as stopping them at this time would only cause a fresh breaking into wood, either of the eyes at the base of the stopped shoot, or some portion of the spurs ; as they, however, look unsightly on well-regulated trees, it will be better to tie them slightly to the main branches for the present: this will give a better appearance to the trees, and bending the shoots will in some measure stop the over free flow of the sap, and so help the object in view. The precise time at which shoots should be shortened must be regulated according to the vigour of the tree, and should be deferred till all danger of the remaining eyes again breaking into wood is over. Cherry-trees now progressing towards maturity should be gone carefully over, the shoots stopped and laid in, and the trees netted, to save the fruit and protect it from birds. If the black-fly appears, cut off the ends of the shoots, unless it is more convenient to wash them in tobacco-water.

Where a large number of strawberries are yearly forced, the plants, after the fruit is gathered, will be found valuable for planting out, producing a most abundant crop the following year; the later-forced ones will answer best, as they are not so liable to bloom again in the autumn.

Figs.—Stop all except the leading shoots, when they have made three or four joints, and lay on leaders and shoots required for filling up. Watering the roots with soap-suds is found greatly to benefit the fruit.

Vines will require going over. Thin out what wood is not wanted for bearing, and stop the bearing shoots at one joint above the shoot: nail in the leading shoots close to the wall. Where the long-rod system of pruning is adopted, a shoot must be selected and carried up from the bottom of each stem, to furnish bearing wood for next year. By careful attention to the vine-border and to pruning, the vine on open walls may be made much more productive, as well as ornamental, than it usually is.

Remove useless suckers from raspberry plantations, to admit more sun and air to the fruit. Begin to layer strawberries in 60-pots directly runners can be obtained for next season's forcing. Let the soil used be rich and rather light, to encourage the runners to root freely: when layered, do not let them suffer for want of water.

Flowers under Glass—The Conservatory.—The difficulty of furnishing the conservatory is now one of taste and selection. Every floral tribe will now be ready to furnish its quota, and discrimination only is required in selecting and arranging them. Avoid crowding; encourage variety and harmonious contrast in colour; remove all decayed or decaying blossom, and guard against insects of all kinds by cleanliness and timely fumigation. Regulate the luxurious growth of creepers and border-plants, watering copiously, occasionally using liquid manure. Ventilation is now of the utmost importance.

Large orange-trees grown for the flower-garden or grounds during the summer months may now be moved to the places they are to occupy. If they have been kept cool and airy, they will not have commenced their new growth, which should not take place till they are out of doors.

Fruit-culture under Glass.—Houses where the grapes are ripe should be kept dry, and succession crops encouraged by a little heat, according to their several stages. Although the nights are now getting warm, it will still be necessary to apply artificial heat, both in houses ripening and in later crops now in bloom.

Vines training in pots for next season's fruiting require daily attention and and stopping: when they have attained a proper length required for fruiting, stop the laterals and expose the principal leaves to the light. Water with liquid manure when the pots are full of roots.

Pinery.—The principal crop of summer pines, now swelling their fruit, must be encouraged by frequent waterings, using liquid manure alternately. Support each fruit in an upright position, and remove useless gills and suckers, reserving only sufficient of the latter for stock. Shade with some light material during the middle of bright sunny days, unless vines are grown over them; bearing in mind that the more light they get the better will be the colour and flavour of the fruit. Give air early, increasing it as the day advances, and close early in the afternoon, at which time the plants, beds, and interior walls should be damped over. When the nights become warmer, a little air may again be put on, which will assist the colouring of the fruit. To insure strong sturdy plants, maintain a uniform bottom-heat of 90° during the season of active growth. The frosty nights which occasionally occur, and cloudy or rainy days, require that this temperature should be kept up by fire-heat.

Peach-house.—The ripe fruit should be looked over each morning, to gather such as are likely to ripen in a day or two. The fruit will be higher in flavour than when allowed to ripen on the tree, and will be saved from getting bruised in falling, to which heavy fruit of the peach is very liable, with the best contrivances to catch them. As the crop is gathered, the young wood should be so exposed as to ripen well.

Melons.—As soon as the fruit is cut (if it is intended that there should be a second crop), prune back the shoots to where the fresh growth commences

Two or three inches of fresh loam should be spread over the surface of the bed, which should at the same time have a good soaking with manure water, to assist the plants to make a fresh growth; an additional stimulus at the same time should be given to the roots by slightly increasing the bottom-heat. Bring forward the succeeding crops.

Cucumbers at this season of the year do best with a considerable amount of shade: this should be attended to, and the necessary bottom-heat and moisture kept up. Keep the vines thin and regular by frequent stopping. In planting out at this season, use a rather poor, in preference to a rich soil, which in cold wet seasons produces canker.

Orchard-houses.—In hot and dry weather trees will require watering abundantly every evening; in all weathers syringe morning and evening, at 7 a.m. and 6 p.m. If the surface of the soil in the pots or border be dry, a new top-dressing may be added. Thin the fruit, pinching in all shoots to the third leaf.

Remove plum-trees and apricots into the open air to ripen their fruit. On the 10th, and again on the 25th, lift up the pots in order to break off the roots which have protruded through the drainage-holes, and attend to summer-pinching of pyramid and bush trees.

Ventilation and watering as in last month, syringing till the fruit begins to colour. Pinch the laterals, and at the end pinch off all leading shoots. The ripening of peaches, apricots and nectarines may be retarded by removal into the open air. Summer-pinch pyramidal peaches.

Hotbed and Frame-Cultivation.—Making hotbeds is seldom deferred till this time of the year; yet it may be done advantageously. Both cucumbers and melons, if started this month, will pay for cultivating; the directions for doing so being the same as in former months, it is unnecessary to repeat; but common brick pits will be very suitable for the purpose. Cucumbers in an advanced stage will want clearing of dead leaves, and the soil stirred about them, and probably fresh earth added. A toad kept in a frame will destroy a great many woodlice and other insects, and keep the plants cleaner than they otherwise would be.

Plants intended for open-air culture, if sown last month, will be ready for ridging out. A south border, or between rows of tall peas or scarlet runners, ranging north and south will suit them. Open a trench 4 or 5 feet wide, and fill with prepared stable-dung, to the thickness of 3 feet; cover this with a foot of soil; place the plants 5 or 6 feet apart, two or three together, and cover with hand-glasses.

Melons may be started for succession. As these are not generally continuous bearers, nothing is gained by endeavouring to induce old plants to bear again; it is more satisfactory to raise fresh plants and make new beds for them, unless, indeed, they are planted on old beds newly lined. With a tolerable bottom-heat, the growth of these plants is very rapid at this time of the year; and though they may be grown without it, still, for the production of fine fruit, heat is indispensable.

JULY.

The month of July is the hottest of all, the mean temperature being 61°, although the thermometer ranges from 82°, and sometimes falls to 42°. This high temperature is chiefly occasioned by the increased radiation of heat at the earth's surface; in consequence, the nights are much warmer than those of June. A period of rainy weather usually occurs about the middle of the month accompanied by thunder-storms.

Flower-garden and Shrubberies.—All strong-growing plants, such as asters, helianthuses, and solidagos, should be attended to. Hollyhocks planted on the lawn, whether singly or in groups, should be staked in time; in fact they should be staked when planted, and the leaves and plants kept in a healthy state by watering and syringing in hot and dry weather. Tie up œnotheras neatly. Speciosas, planted pretty thickly over the beds, will produce a fine mass of white flowers, if trained so that they have plenty of light and air, and watered abundantly in dry weather; metrocarpa and *Matricaria grandiflora* will well reward any labour bestowed on them. Beds of Verbenas, and similar plants, require occasional syringing with weak tobacco-water.

The first week or so in July will be chiefly occupied by the usual routine of pegging-down plants intended to be kept dwarf, tying others up, and keeping the surface of the beds free from weeds until it is covered by the growing plants. If pinks are attacked by wireworm, place pieces of potato just below the surface of the soil. Examine them every morning, and a great number can be thus caught and destroyed. Pinks should now be propagated; cuttings may likewise be put in of tea and China roses, selecting wood of the present year when it becomes a little firm at the base. Roots, bulbs, anemones, tulips, crocuses, scillas, fritillarias, &c., which have been out of the ground for some time to dry, may now be re-planted.

Fuchsias, Geraniums, and other plants in flower, will require regular supplies of water.

Florists' Flowers.—Take up tulips whenever the weather will permit. When lifted, do not separate the offsets from the parent bulb, nor remove the roots or skin; these had better remain till a later period. When lifted, ridge up the soil of the beds for exposure to the air. In taking up seedlings, great care must be used, as their bulbs will often strike down from 4 to 6 inches. If possible, keep the stock of each separate; this will save an immense deal of trouble hereafter. Tie carefully the splindling shoots of carnations and picotees—not too tightly; keep the pots free from weeds, and in dry weather do not let them suffer from drought. Attend to the fertilization of pinks; a very little attention to this interesting operation will insure a good crop of seeds, and by selecting only excellent varieties instead of trusting to chance and gathering promiscuously, abundant success will be the result.

By the end of the month seedling ranunculuses should be taken from the pans or boxes in which they may have been grown; but as many are so minute, and so like the colour of the soil, that without great precaution, they may be overlooked, the best way is to put soil and roots together in a fine wire sieve, and by holding it under a tap, or pumping into it, the soil will be washed away and the roots left; they must then be placed in the sun for an hour, and afterwards removed to an airy shady place to dry gradually. The large roots of named varieties must be taken up at once, for should they start again, which they are very apt to do previous to removal, death is inevitable. Continue to put in pink pipings; disbud carnations and picotees, giving occasional doses of liquid manure. Attend sedulously to dahlias; tie as they require it, and give a good supply of water.

Rose-Garden.—Autumn-flowering roses now require a liberal supply of liquid manure; guano sown on the ground, and thoroughly soaked with rain-water, will serve the purpose. Remove faded flowers and seed-capsules every morning; plants which have flowered in pots keep growing freely, as the future bloom depends on their vigorous growth at this season. Climbing roses should now be pushing out strong shoots from the roots and main stem; if not required for future training, these should be taken off entirely, or have their tops pinched off a foot or so from the stem. Budding should be in full operation, watering the roots and plants freely in dry weather, both before and after budding. Cut back perpetual-blooming roses, and water them with the richest manure-water to encourage a second growth and bloom.

Baskets, vases, &c., will require an occasional regulating; those having plants in them which require to be tied up, should be examined. Convolvuluses, maurandias, lophospermums, &c., after being pegged over the surface of the soil, should be left to grow over the sides of the vases, or allowed to ramble among the more formal plants which fill up the centre. Baskets, cases, or other contrivances containing plants in bloom, will require frequent attention to keep them fresh. Remove everything in the shape of decayed bloom or leaves, and take advantage, when fresh plants are wanted, to effect a change in the arrangement, which will be found more pleasing than adhering to one plan.

Kitchen-Garden.—Probably this is the busiest month of the year in the kitchen-garden, both on account of everything growing so fast, and because many crops have ceased to be useful, and must be removed and give place to others. We have to look forward to a long winter and spring, when vegitation is stationary or very slow; yet at that time it is necessary to have suitable crops, and to prepare the ground and get them in their places. It may be observed, that where rows of vegetables have previously grown, the ground is usually dry and hard. However moist the season has been, it will always be found different to that eighteen inches or so on either side; it is not, therefore, advisable to crop immediately over the same spot; the difference will soon be observable between the rows planted exactly where peas have grown and those planted at the distance indicated. It is best not to plant winter crops on ground that has been newly-dug or trenched.

Asparagus.—Cease cutting early this month, unless some part of the bed can be spared for late use, in which case it must have a rest the following season. Late cutting has the effect of weakening the roots, but they will recover after a season's rest, if they have not been cut too closely. Hoe frequently between the rows.

Artichokes will now be in bearing. Cut when the heads are about three parts open. These root deeply and scarcely require water.

Seakale should have an abundance of water, particularly young plants. Soot or wood-ashes strewn about them will, in a great measure, prevent the attacks of insects.

Beans.—Pull up early crops as soon as they have done bearing; those advancing will produce better for being well watered. Make a groove each side of the rows, and give enough to soak the ground to a considerable depth.

Runner-beans.—Apply strong sticks, if not already done. These beans may be kept dwarf by picking off the runners as fast as they appear; but it is much better to let them have full play by providing supports: the produce is tenfold greater.

Peas.—If any are sown this month, let it be sorts that bear equally, or the shortening days will prevent their bearing at all. Dwarf early sorts are best. Clear away any that have ceased to be productive, and stake any that are just above ground; copious waterings will greatly benefit these.

Celery.—The main crop should be got out directly. If this is planted where peas had previously grown, make the trenches *between*, not *on* the rows.

Cardoons, like celery, should be got out in the trenches, remembering that they require a soil highly enriched with manure; they should also have plenty of room and abundance of water: be not hasty in earthing-up.

Beet.—See that this crop is properly thinned, and keep the ground well hoed between.

Carrots may be sown any time this month; they will be useful in winter and spring. Sow on an open spot, and do not dig the ground deep; look over the main crop, and pull up any runners: they will be of no use if left. Take care that no weeds are allowed to grow.

Onions may be sown now for salad in the autumn. Towards the end of this month some of the main crop will be showing signs of maturity, when they may be pulled up and laid on their sides, and thick-necked ones may be pinched.

Leeks.—Plant out the main crop on well-manured ground in deep drills or shallow trenches, for the convenience of earthing. Leeks are strong feeders. and should be well-watered.

Potatoes.—Pick off the flowers, if possible: if allowed to seed, they diminish the produce, the tubers growing less in proportion.

Turnips.—At the beginning of this month, a principal sowing should be made for autumn and early winter use, and again towards the end, another sowing should be made. These will be useful in winter and following spring.

Conservatory.—Remove from the conservatory all plants which show by their faded blooms, that they are past their best; their prolonged presence would detract from the freshness essential to beauty and good order.

Achimenes, gloxinias, &c., out of bloom, should be removed to a pit to ripen their bulbs. Clerodendrons, &c., in the same way may be transferred to vineries, or any place where there is a dry, cool atmosphere.

At no period of the year do heaths and hard-wooded plants require more care than the present, particularly such as have been recently potted. To keep the old ball sufficiently moist to preserve the plant in health in the high temperature without getting the new soil in a sour state, requires great nicety in watering, supposing the plants to be under glass. At this season all the air possible should be given to the greenhouse and stove plants, keeping it on all night.

Greenhouse plants, after they have done blooming, should have a comparatively cool temperature, and no structure presents so many advantages for this purpose, as well as for growing delicate-leaved plants through the summer, as houses having a north aspect.

Camellias, whenever the young wood appears getting ripe, may be removed to the open air; they thrive best in the shade; they must be placed on a dry bottom to prevent worms from getting into the pots.

Chinese azaleas should also be turned out. Unlike camellias, they require full exposure to sun and air, and should be placed in an open situation, that their wood may become thoroughly ripened. It will, however, perhaps be necessary to place them for a week or two in a partially-shaded situation, to harden their foliage sufficiently to bear the full sun; otherwise the sudden change from a house to full sunshine might cause their leaves to turn brown or burn.

Orange and lemon-trees will now be in bloom, and should be supplied with water at least three times a week in dry weather, and occasionally supplied with liquid manure after stirring the surface of the soil and top-dressing.

Fruit-culture under Glass.—Vinery.—Ripe grapes, if required to be kept, must be shaded during hot sun, to prevent shrivelling. The Cannon Hall, Muscat, Sweetwater, and Frontignans, having tender leaves, are most liable to burn, either from bad glass or imperfect ventilation. They must be well watched, as the injury done to the foliage not only affects the present crop but the succeeding one. Any heat given now should be given during the day, to forward them before the season gets too far on.

As the houses are cleared of their fruit and the wood is ripened, the vines will be much benefited by having the lights off, and by being freely exposed to the atmosphere.

Peaches and Nectarines.—Any tendency to the leaves decaying, when the fruit has been gathered, should be prevented by syringing and watering the roots. Fruit coming to maturity will be the more delicious for a comparatively cool temperature while ripening. Examine daily and gather before it is over-ripe. The great object now is to get the wood properly ripened; and that will be best promoted by a full exposure to the sun, the air, the rain, and the dews, by removing the sashes and top-lights.

Pinery.—Still continue to supply swelling fruit with water, and syringe frequently, but not during bright sunshine, unless the shading is immediately put on. Young plants growing fast will require liberal watering, in addition to air in large quantities by day; the temperature will allow them to have a good portion by night. During hot weather forced fruits of all descriptions will be benefited by this practice.

About the second week the plants selected in the spring for autumn and winter fruiting will be showing fruit; and if they are in a pit by themselves will require, if a steady bottom-heat is kept up, but little attention for some time, except slight shading, plenty of air, and a liberal allowance of water. On no account let the plants be wetted while in bloom.

Figs swelling off their second crop should be assisted with liquid manure, more especially if growing in pots or tubs. As the fruit ripens, care must be taken to preserve it from damp.

Proceed to pot strawberries for forcing: as soon as the pots in which the runners were layered become filled with roots, pot them in 6 or 7 inch pots, using rich loam of medium texture, and well-rotted dung, with plenty of drainers.

Melons.—Keep a steady bottom-heat and free ventilation, more especially in wet weather. Watch for red spider and mildew; for both of which, sulphur, properly applied, is the best remedy: also keep the roots in action by a well-adjusted bottom-heat. Sow Lord Kenyon's, or any other good house cucumber, for autumn supply, following the same directions as for melons.

Orchard-house.—Ventilation is now the greatest care; fasten back and front shutters down, so that they cannot be closed; syringing night and morning, and watering copiously when dry. If any trees are growing too rapidly, tilt up the pots, and cut off all the roots on that side which are making their way into the soil. A week later, serve the other side in the same way.

Hotbed and Frame-Cultivation.—The purposes of hotbeds are limited at this time of the year, at least in most places. Cucumbers, melons, &c., are usually grown in houses and pits that are otherwise unoccupied at present; and as their culture is more cleanly in this way, hotbeds may be dispensed with; those, however, already in operation will require attention. The weather is usually hot this month and next, but is often changeable, and the manager of frames must be ruled by it.

Cucumbers in bearing should be watered. If the soil is not sufficiently moist, take care to water plentifully, as at this time of the year there is less danger of overwatering. If the weather is hot, and not too dry, the lights may be pulled quite off for an hour or two before 8 a.m. and after 5 p.m.; but be careful that the plants are not chilled before closing. Pickling cucumbers may be planted in the open ground at the beginning of this month. The soil should be well dug, and made pretty firm again, and well mulched after the plants are put in. Choose a warm sheltered spot for them, and place hand-glasses over them if they can be spared.

Melons planted at the beginning of this month may be put out in the ordinary manner in a common melon-pit, with a good body of dung; but, if planted later, it must be so that heat can be applied to ripen the fruit, which will be required, as the days are short. Plants that are ripening their fruit must have very little water.

Surface Crops.—French Beans.—A late sowing of these may be made now any time: dwarf kinds, as the Newington Wonder, are best.

Lettuce sown now do well on a shady border, provided the spot is not too much overhung by trees. An open well-manured spot is best for them, if kept well watered.

Endive.—Two sowings should be made this month; one at the beginning, another towards the end. Sow in the same way as lettuce, and plant out as soon as large enough to handle.

Tomatoes should be carefully trained, and stopped as they grow. Stop just over a bunch of flowers, and leave no more shoots than can be conveniently trained. Unless the ground is very dry, they do not require watering.

Vegetable-Marrows will be in active growth; and, planted on a manure-heap, they will grow freely enough without watering; but if on the common soil, they should be freely watered in the morning.

Spinach—It is not advisable to sow this month; but the ground should be prepared for sowing next month.

Brassicas (*Brocoli, Brussels Sprouts, and Savoy*).—The principal crops of these should be got out. Plant them in drills two feet apart, and 18 inches in the rows. If liable to club, dip the roots in a puddle of clay and soot before planting, or fill up the holes with wood-ashes, which will prevent it in a great measure.

Cabbage.—Sow for coleworts early in the month, and for early cabbaging about the end of it: strew lime in soot over the young plants, to drive away the fly. This should be done in the morning, while the dew is on them. Plant out for autumn use.

Cauliflower sown now may be useful late in the autumn.

Mint, and such-like herbs, should be cut for drying just as they begin to flower; *Savory*, *Sage*, and others, may be now propagated by cuttings or division; *Parsley* and *Chervil* may be sown for winter use.

Fruit-Garden.—Peaches and Nectarines should receive their final thinning. Some prefer allowing them to get large enough to use for pies, &c., before doing so; but the sooner the surplus fruit is taken off the better for the crop Some little judgment must be exercised in thinning both wood and fruit; the object being to regulate both, so that a fair balance be maintained: if too much fruit is left on, the present year's crop will not be so good, nor will the strength of the tree be maintained for future bearing; if too much wood is left, the fruit is unduly shaded, and the wood itself becomes weak.

Apricots, Plums, Cherries, and Figs, on walls, should be carefully looked over, and all shoots that are not really useful, or any that are ill-placed and cannot be properly nailed in, should be removed.

Vines out of doors should be closely stopped and trained in. All the heat of the sun is necessary to the well-doing of this fruit, which cannot be expected to ripen in our short seasons, unless every care is taken to secure it all the light and warmth of the sun.

Gooseberries, Currants, Raspberries, and other bush-fruit will require some protection from birds.

Strawberries.—This is by far the best month of the year for making new plantations. There are various methods of doing it; perhaps the best is as follows:—The earliest runners are laid in 3-inch pots, fixed in their place by means of small pegs; in three weeks they have rooted into the soil with which the pots are filled. During this time they require an occasional watering, but may be planted out permanently as soon as rooted, placing them eighteen inches apart in rows three feet apart.

Flower-cultivation under Glass. — *Conservatory.* — The "sere and yellow leaf" is now apparent here: the work of decay has commenced: exotic bulbs have nearly finished flowering, and require a state of rest; those whose stems are still green should have water, in order to mature the bulbs. When done flowering, keep them in dry earth or sand, and in a warm situation, to ripen. Cinerarias and calceolarias require as cool an atmosphere as the house admits. Cuttings of geraniums and most greenhouse shrubs may now be struck, and forwarded by plunging in a gentle hotbed.

Fuchsias, geraniums, achimenes, and salvias, requiring larger pots, should be shifted, removing the entire ball, and placing them in the centre of the new pot, properly drained and half-filled with fresh compost.

Hard-wooded plants, including most of the genera from New Holland, which bloom early in the spring, will about the middle of the month be so far advanced in their new growth, that any requiring re-potting should at once have a shift. After turning them out, loosen the outside roots before placing them in their new pots, to enable them to take up the fresh soil more readily. Keep them close for a few days, especially if the roots have been much disturbed, and damp them once or twice daily overhead.

Attention at this season should be directed to the stock of plants intended to furnish the supply of bloom through the winter, as it is requisite plants should complete their growth early for this purpose. Of heaths, those which flower through the winter should also be encouraged to complete their growth. Keep epacrises under glass till their growth is complete; but more air and light must be allowed them as the wood gets firmer. Towards the end of the month they may be placed out of doors in an open situation, where they can be protected from heavy rains.

Such stove-plants as are intended to flower in the winter, as justicias, *Eranthemum pulchellum*, euphorbias, jasmines, &c., should be looked to. Many of these things require to be kept in small pots, and should bo watered with liquid manure.

Balsums, thunbergias, and other annuals intended to decorate the conservatory and show-house for the next two months, should be finally potted, using soil of a light and rich description. Keep down spider with the syringe Ipomœas, thunbergias, and other creepers, should be neatly trained to their respective trellises as they advance.

Brugmansias, and similar plants of vigorous habit, should be frequently assisted with manure-water: as they are often troubled with red spider, the engine and syringe must be kept constantly at work, taking care, however not to injure the fine foliage. Succulent plants, as cactuses, euphorbias, cereuses, sedums, and others of similar habit, require to be abundantly supplied with water, and also a full exposure to the sun, in order to obtain a fine b loom.

AUGUST.

Aspect of the Month.—The eighth month of the Julian year received its name from Augustus Cæsar, as July commemorated that of the greater Julius. Less rain falls this month than in July, and the mean temperature is a little higher; the nights are certainly hotter. Mean temperature of August—61·28° at the surface, 61·808 at a foot, and 61·268 two feet below.

Flower-garden and Shrubberies.—The flower-garden will now be in its greatest beauty, and every means must be taken to keep turf, gravel, and edgings of all kinds in neat order; dead flowers should be picked off daily, and stray growth reduced within proper limits.

Autumnal Bulbs, such as colchicums, narcissuses, Guernsey lilies, and amaryllis, may still be planted in borders, beds, or pots, in light sandy loam.

Florists' Flowers.—*Carnations* and *Picotees* should now be layered, but without shortening the grass. Where seed is required, pick off all decaying petals to prevent damp injuring the pods. If not wanted for seed, cut the old plant down.

Dahlias require constant watering and attention in tying out lateral shoots, removing superfluous ones, and relaxing the ties.

Hollyhocks require the same attention as to staking and selecting.

Pinks.—First-struck pipings may now be planted out, potting a quantity in order to fill up vacancies which may be caused from the ravages of the wire-worm, &c. Make pansy-beds.

Rose-Garden.—Perpetual-flowering roses in dry weather require copious supplies of water. If mildew appear, syringe the plants with soft water in the evening, and dust the affected parts with flour of sulphur. Towards the end of the month any roses budded last month may have their bandages removed. Cuttings of Tea-scented, Noisette, China, Bourbon, and Hybrid perpetuals, may be struck in light sandy soil, over a gentle hotbed.

Kitchen Garden.—The kitchen gardener who would have everything thrive and prosper, must exercise the greatest vigilance during this month. Apart from the necessity of cropping and removing such crops as have ceased to become profitable, his attention will be drawn towards the multitudes of garden pests, which exhibit their effects at this time of the year more than any other.

Artichokes.—Cut these down as the heads are gathered, and fork the ground between,—they will come up again before winter.

Asparagus.—Keep the beds clear from weeds, especially where there are young plants. Unless seed is wanted, it is advisable to cut off most of the bearing heads.

Beans.—Pull up the haum of any that have done bearing: lay the stalks together, and they will soon rot, or dry them, and they will burn. Some may be cut in lengths, and dried for earwig-traps, to place among flowering plants.

Beans, Running, should be stopped after reaching the top of the sticks: they will set quicker than if left to grow as they please. Give plenty of water at the roots if necessary, but none overhead.

Beans, French.—A row or two should be left for seed. It is not advisable to leave any to ripen on bearing plants, as they cease to yield for the table while ripening seed.

Cabbages.—Sow early this month for a full crop of summer cabbage. Sow thinly on an open spot, that the plants may come up strong, and scatter lime on the ground to protect from birds and insects; dust also the young plants when up; get out a supply of early coleworts; they will most likely make small head in November.

Carrots.—Early sowings may be taken up and stowed away for use; but if the ground is not particularly wanted for other crops, it is quite as well to let them remain till required for use. A little seed may be sown early this month to stand the winter; plants will be useful in the spring, when the winter store is exhausted.

Cauliflowers should be sown two or three times this month: seed sown at the beginning, about the middle, and at the end of the month, will give a succession. Sow in the same way as cabbages. It will be necessary to give them the protection of frames or hand-lights during the winter; but the sowing may be in the open ground.

Celery.—This may be got out in any quantity.

Endive.—Sow early this month for the last time this season; prick out the plants as soon as large enough.

Leeks may still be transplanted; but the sooner the better, or they will not get any size before winter. Plant in deep drills two feet or so apart, and water freely; draw earth up to those in full growth. Liquid manure given occasionally will benefit them.

Lettuce.—The first week in this month sow cabbage-lettuce for winter use and from that time onwards, both cos and cabbage-lettuce may be sown to stand the winter for spring use.

Onions will most likely be arriving at maturity, and had better be pulled up as soon as this is the case, and laid on their sides.

Parsnips.—Stir the ground well between, so that the rain may penetrate quickly. Destroy weeds, and keep the crops clean.

Radish may be sown any time this month. The Black Spanish should be sown early for winter use. This takes rather longer than other sorts.

Spinach.—About the second week in this month, sow the main crop of winter spinach.

Tomatoes.—Attend to these as directed last month. To have them bear well in our short seasons, it is necessary to aid them as much as possible, by pinching out all superfluous growth, exposing the flowers well, and training close.

Turnips may be sown any time this month: they will not, probably, grow large, but will be useful in February and March for the early greens which they yield; they may be left thicker than the early sowings.

The gathering and drying of herbs should not be left later than this month.

Fruit-Garden.—Look well over wall-trees; for snails, wasps, and flies are as fond of choice fruit as man himself. Snails will attack peaches, nectarines, &c., before they are ripe, and spoil the appearance of every fruit they approach. Find out their haunts, and pick them out with the hand.

Tack in all useful wood. This should not be omitted this month, and the trees will scarcely require it after; remove every shoot that is not really wanted; every scrap of wood that is not useful may as well be removed now as at any other time. As it is not proper to drench the trees when the fruit is ripe, or ripening, any shoots infested with aphis should be cleaned with a brush.

Standard trees, where a regular thinning is not adopted, should be shaken occasionally, to bring down any fruit that may be blighted. These can be no good on the trees, and the sooner got rid of the better.

Strawberry beds may be planted.

Plant-Culture under Glass.—Conservatory.—Camellias and acacias now require copious watering, care being taken that they are not started into second growth. Sprinkle borders daily, and keep up a moist atmosphere. Train and prune all climbing plants. Strong-growing plants, such as diosmas, the epacridæ, rochozemas, which have been in shade to prolong their flowering, should now be placed in a bright sunny place. Late-flowering *Azaleas* require shifting and training, so that the foliage draws out properly before winter.

The principal plants that decorate the conservatory at this season will be some of the more common—fuchsias, scarlet geraniums, with achimenes; but where there is room, a considerable number of stove plants and orchids may be safely introduced.

Achimenes, as they go out of bloom, may be placed in a frame to ripen their tubers, exposing them fully to the sun, but keeping them rather dry. If the different varieties of epiphyllum have made their growth under glass, they may be removed to a sunny spot out of doors. Pot off seedling cinerarias, Chinese primroses, and calceolarias from the seed-pans when the plants are large enough for the purpose.

Fruit under Glass.—Vinery.—Whenever the leaves in the early-house show indications of ripening, the sashes should be removed and the vines fully exposed: beyond stopping any late laterals, vines should not be touched until the leaves fall.

Fires, especially to houses containing Muscat grapes, should be made each evening, and during wet, dull days, that abundant ventilation may be kept on.

Vines in pots, intended to fruit next season, must be closely watched to get the wood perfectly ripened.

Pinery.—As soon as the house for next season's fruiting is ready, the plants should be transferred there at once. As a rule, early fruiting pines should have their pots well filled with roots by the middle of September; and, while growing, allow them all the light you can command, and a proportionate quantity of air. The best pines for very early forcing are the black Antigua, common Queen, and the Providence.

Peaches.—As the houses are cleared of fruit, the trees should be gone over, and the wood not required for fruit next season should be cut away.

Melons.—The late crop will be advancing; and as light is decreasing, keep the vines further apart, that the leaves, as they are formed, may not crowd each other. Attend carefully to bottom-heat, which must not be allowed to decline. Red spider must be kept in check.

Cucumbers, as the nights get colder, may have a slight covering, and the bottom-heat, if declining, should be renewed.

Hotbed and Frame-Cultivation.—This is a good time to strike the winter stock of bedding-plants, for raising cinerarias, &c. Frames without hotbeds are very useful. Mignonette, nemophila, and other annuals sown now in pots, and kept in cold frames, will flower in the winter.

SEPTEMBER.

The reduction of temperature begins to be felt this month, less, however, by night than by day, the mean temperature of the air being 66·14°; one foot below the surface, 57·54°; and two feet, 57·89°; being from this month till April, warmer at two feet than at one foot. The average fall of rain is also increased considerably.

Flower-Garden and Shrubberies.—Now that the beds are thoroughly covered, nothing contributes more to a high style of keeping than the removal of every dead flower and leaf.

Dahlias will require careful tying, disbudding, and thinning off the shoots, where first-rate flowers are required.

Roses.—Perpetuals may still be cut back with the hope of a third bloom; and late-budded plants will require looking after, watering, and training to stakes.

Pot off layers of carnations as fast as rooted, water sparingly, and place in a cold frame for a few days until they make a fresh start.

Plant out in beds early-rooted pansy cuttings, insert a succession of cuttings, and prick out seedlings in the reserve-garden. Here also let seedling polyanthuses, offsets of these and auriculas, be planted on rich, shady beds.

Kitchen-garden.—To secure a supply of vegetables in the winter and early spring, all arrangements not already completed should be made without delay.

Celery.—The earthing-up of this useful vegetable now demands special attention. The sowings made in July and August will be ready for transplanting.

Cauliflowers may still be sown in some situations, and those sown last month pricked out under hand-glasses, or in frames, as they advance: if the season is mild, they may be planted out under a south wall.

Cabbages.—Prepare a piece of ground by deep trenching and heavy manuring, for spring cabbages, savoys, and winter greens, and keep it over forked regularly until the plants are sufficiently advanced for planting out. When ready, plant in rows two feet apart, watering well to settle the earth at their roots. Savoys and spring cabbages, in particular, require a rich soil thoroughly manured.

Brocoli also require a good soil richly manured. Plant them out from the beds in rows where they are to grow, two feet apart each way: water till the plants have rooted. Mr. Errington finds an advantage in dibbling a large hole to receive the plants, and filling it up with calcined wood or vegetable ashes. This crop may follow peas with advantage, or be set between the rows of late sorts.

Brocoli seeds may be sown to stand the winter, and come up for a late spring crop.

Brussels Sprouts and Winter Greens.—Plant out for autumn use.

Endive seeds sown now will come in to supply plants for winter use; the green curled being the best for main crops. Water the beds in dry weather, and tie up to blanch plants advancing to maturity.

Small Salading.—Sow cresses, mustard, radishes, and other small salads, every seven days, choosing a shady border and sowing in very shallow drills: water daily.

Lettuces.—Sow cos and cabbage lettuces in a bed of rich mellow ground; in the first, second, and fourth weeks, prick out on nursery-beds the plants last sown.

Spinach, for winter use, sown late in July or early in August, should now be planted out. The prickly-seeded or triangular-leaved is the hardiest for winter use.

Turnip-radishes (black and white) should now be sown for winter use; also some small Italian (white and red) for autumn use.

Onions may be sown, of the Strasburg and Welsh sorts, early this month; the former to transplant in the spring, the latter for use in salads. The general crop will be ready for harvesting.

Carrots should be sown in an open situation, and on light soil, sowing them as soon as the bed will work after digging.

Turnips may still be sown for autumn and winter use, the Early Stone being a good sort: sow immediately after digging. Hoe the crops sown in May and June in dry weather, and thin out till the plants are seven or eight inches apart.

Fruit-garden.—The chief work to be done in the fruit-garden and orchard is harvesting the fruit and preparation for planting. Early apples and pears should be gathered a day or two before they are ripe: it is a very good plan to make two or three gatherings from the same tree.

Peaches and Nectarines require to have the future bearing-shoots nailed in closely, and all laterals not required removed, so that the fruit may have the full benefit of the sun, from which it derives colour and flavour. A few of the leaves may also be removed where they shade the fruit too much. As the fruit approaches its ripened state, nets should be extended beneath to catch any that fall.

Gooseberries.—Mat over where necessary, to retard ripening.

Strawberries.—Alpines and other late sorts are now in full bearing. This is also the season for saving seed, if seedlings are desired for planting.

Flower-culture under Glass.—Conservatory and Greenhouse.—Both houses must at once be got ready for winter occupants. Many of these, such as ericas, epacrises, azaleas, camellias, have probably been in cold pits, or sheltered situations out of doors, for the last four months. In ordinary seasons they will be safe enough there until the end of September. Meantime, however, if the house requires painting or cleaning, the sooner it is done the better.

Cinerarias.—Pot-off suckers from old shoots; prick-off, pot, and shift seedling plants, and push forward the first batch for flowering from November to February. Calceolarias require the same general treatment. Shift chrysanthemums, liberally water at top for late blooms, and stake.

Provide plenty of linums, *Salvia splendens*, oxalis, &c., for winter or spring; likewise hyacinths, narcissus, tulips, &c., &c.

Stephanotis, passion-flowers, jasmines, &c., on the roof, must be carefully trained, cleaned, and regulated. Allamandas often make a splendid display when trained as semi-climbers on the roof of a stove. The new violacea would look well near to the cathartica or other varieties. Archimenes and other plants, suspended from the roof in elegant wire baskets, have a charming effect among climbers, and make the roof at least as showy as shelves and beds. Gloxinias and gesnerias will also make a splendid display here during the month.

Orchid-house.—More light and air, and less water, must be the rule here. Those plants, however, that are in full growth must not be stinted, by any means, as the natural growing season of most orchids is the rainy season,—the season when it rains every day and night, for perhaps six weeks, without ceasing. Rapid growth, long seasons of perfect repose, and sudden excitement, seem to be the essentials of successful orchid-culture.

Fruit under Glass.—Orchard-houses cannot have too much air. Where no fire is used, sometimes late varieties of peaches, &c., are grown here, to come in after the fruit out of doors.

Vines.—Care must be taken, in preserving the foliage of grape-vines, not to allow too many leaves on the lateral shoots. It is the large leaves at the base of the fruiting-branches, near to the main stem, that are of most consequence. The buds at their base will yield next year's crop, and the fuller, rounder, and more plump they become, the larger that crop will be. The great point is to maintain these leaves in health without *inducing new growth* or causing the buds to break. A comparatively dry atmosphere and cool temperature are needed.

Ripe grapes must be frequently looked over, and every specked berry be at once removed. If mildew makes its appearance in the late houses, paint the pipes with a mixture of equal parts lime and sulphur, and sprinkle the infested parts with dry sulphur.

Peach-house.—The lights may now be removed for six weeks from the early peach-house; or, if this is not practicable, as much air as possible should be given night and day.

Figs require plenty of water when in full growth: in fact, in this state they should be treated almost like aquatics. The second crop of fruit will now be ripening, and those who wish for a third crop in November and December should have stopped the shoots in the middle of August; but where a very early crop is required, the shoots must not be stopped after this period.

Pines.—Keep a genial atmosphere of from 70° to 83° among fruiting plants;

water them with clear manure-water, and refrain from syringing plants in flower and ripe fruit. Providences, and the black varieties for winter fruiting, would be best in a house by themselves from this time. Maintain a steady bottom-heat of 85° to fruiting plants, and 75° to succession plants.

OCTOBER.

Flower Garden and Shrubberies.—The mean temperature of the month is nearly 7° lower than that of September, and frost is by no means uncommon towards the end of it. The moisture in the atmosphere increases, and evaporation diminishes considerably; the mean average temperature being, at one foot below the surface, 51·52°; at two feet, 52·78°; and at the surface 49·35°.

Maintain scrupulous cleanliness in the flower-garden, and continue the beauty as long as possible.

The great business of propagating for next year should now be completed. Nevertheless, such things as verbenas, calceolarias, &c., &c., may still be put in with the certainty of success.

Florists' Flowers.—Place auriculas, polyanthuses, pinks, carnations, &c., if not already done, in their winter quarters.

Kitchen-Garden.—Towards the end of the month the asparagus-beds may be cleared of their haum, but not till it is yellow and the seed ripe, and a portion of the soil forked into the alleys; then mix some good manure with a little salt, and lay a good coating of it over the plants, covering the whole with the soil thrown into the alleys.

Celery.—Earth-up as often as it becomes necessary.

Seakale will be ready to force towards the end of the month.

Cardoons should now receive a general earthing-up.

Carrots, Potatoes, and Parsnips are now at maturity. Dig them up and store.

Small crops of *Mazagan Beans* may be planted with a chance of their standing the winter, and coming in in May or June. A crop of *Early Peas* may be sown.

All the *Cabbage* tribes require the greatest attention this month in weeding and warring with caterpillars, which infest them.

Cauliflowers sown in August will require pricking out, not less than four or five inches apart, where some kind of protection can be given.

Lettuces for a spring supply may be pricked out under a frame; the hardier kinds will frequently stand the winter on a warm border.

Fruit-Garden.—The planting of fruit-trees should be proceeded with if the necessary preparations have been made.

Apples and Pears are now ripening fast. Gather on fine days, taking care that pears especially are tenderly handled.

Peaches and Nectarine-trees should have all superfluous shoots removed, and the young wood left exposed to as much sun as possible, to ripen the shoots.

Plums.—In wet seasons gather the late sorts, with their stalks attached; suspend them in the fruit-room, or wrap them in thin paper, and they will keep for several weeks. Quinces, medlars, and all sorts of nuts, are now fit to gather.

Raspberries of the autumn-bearing kind should be bearing fruit.

Strawberries.—Remove all runners from the plants, and manure and dig between the rows; using the three-tined fork, so as to avoid injuring the roots.

Flowers under Glass.—The chief duty of the month is to see that all tender and all hardy plants intended to be bloomed in winter are placed under requisite shelter.

It is good practice, in fine weather, to place out, in sheltered situations, many of our heaths, azaleas, camellias, &c., &c.; but better far to keep them entirely under glass than leave them out too late in the season.

The Conservatory.—The conservatory is always dependent for three-fourths of its charms upon the taste and skill displayed in its arrangement. Beautiful objects beautifully placed, lovely climbers neatly festooned or gracefully trained, and the preservation of all this beauty as long as possible, are the grand desiderata here. In managing the house, two things must be equally guarded against,—a moist stagnant atmosphere and a sharp current of frosty air. All shading may now be dispensed with, and the foliage of the climbers gradually reduced, however beautiful they may be. They should be gone over two or three times until they are finally cut in to the smallest compass by the middle of November. Every ray of light at this period is alike necessary for the health of the plants and the colour of the flowers. Little or no syringing or sprinkling of paths will be necessary, except over chrysanthemum and camellia leaves for the first week or fortnight after their introduction from out of doors. Generally a sufficiency of vapour will rise from the surface of pots and borders, without having recourse to either sprinkling leaves or paths.

Plants in full growth coming into bloom always require more water than plants past the meridian. Therefore, all those already named, and early cinerarias, &c., will require much more copious supplies than late-flowering fuchsias, geraniums, begonias, &c., &c. Semi-stove plants, such as gesnerias, gloxinias, globe amaranths, achimenes, &c., which it is desirable to keep in bloom throughout the month, will now require very little water.

Greenhouse and Heathery require similar general treatment, but should be kept five or ten degrees cooler than the conservatory. They will also bear sharper currents of air with impunity. Leschenaultias, chorizemas, &c., in this house, must be carefully examined for green-fly. Hand-picking is the most effective remedy.

Stove Plants.—Vincas, clerodendrons, &c., that have finished flowering, should now be cut back, and, after they have slightly started, be shaken out of the pots, and inserted in pots as small as possible, for they seldom winter well in large pots. Water liberally poinsettias, justicias, begonias, gesnerias, &c., coming into bloom,—other plants going out of flower water scantily.

Forcing-pit.—Introduce the first batch of rhododendrons, kalmias, Ghent and Indian azaleas, &c.; also some tea and hybrid perpetual roses, and early-flowering and sweet-scented geraniums, white and Anne Boleyn pinks, perpetual carnations, and lily of the valley; also *Salvia gesneraflora*, late gesnerias, and *Euphorbia polygonifolia.* Towards the end of the month, some hyacinths, tulips, &c., potted late in September, should now be pushed forward.

Cold Pit and Frames.—Give all the air possible, unless it actually freezes: guard against damp and over-crowding.

Culture of Fruit under Glass.—In our climate, fruit must be cultivated under glass, if at all, at this season of the year; and those who wish for peaches or grapes on their table in May must begin this month.

Peach-house.—Supposing that the fruit was gathered in May or June, the lights removed in July, the trees pruned in August or September, they may now be thoroughly painted over with a composition consisting of equal parts of sulphur, clay, cow-dung, and soot. The borders should also be forked up, six inches or one foot of the old soil removed, and the same quantity of turfy maiden loam substituted in its place.

Vinery.—All preliminary matters may proceed here exactly on the same principle as for the peach-house. In all forcing, either of flowers or fruit, let cleanliness be your first care. Before the painting, &c., let every bit of loose bark that will rub off with your hand be removed; severe barking, scraping it off with knives, &c., is not desirable.

Pineries.—It is a good plan to cover the pots of succession and other pines two inches thick with partially decayed tan or leaves for the winter. Pines planted out in beds might be mulched over in a similar way. The less water given in any stage during winter, consistent with health, the better.

Melons.—A second or third crop may occasionally be well ripened during this month. If in pits, renew linings, &c., to maintain a brisk heat; if in houses, keep the fires moving to secure a bottom-heat of 80°, and a surface one of 70°. Beware of watering to excess.

Figs.—These may possibly be ripening their third crop; if so, a brisk temperature of 65° or 70° must be kept up. If the second crop is gathered, and a third is not wanted, reduce the supply of water and the temperature to a minimum, to induce rest or hasten maturity.

Orchard-house.—This house, unless used to ripen fruit that has been retarded behind a north wall, must now stand open night and day.

Strawberries.—These ought to have completed their growth for the season; the sooner they go to rest the better.

NOVEMBER.

The atmosphere during this month is saturated with moisture; dense mists and fogs abound, and gloomy, boisterous weather, as a rule, prevails. The leaves of the beeches are changing their green hue for the purplish chestnut autumnal shade. The mean temperature is about 42° Fahr., but the thermometer ranges between 23° and 62°. The mean temperature of the earth one foot deep is 46·01°; at two feet, 47·28°; and that of the air being 42·98°, on an average of the years 1844 to 1853.

Flower-Garden and Shrubberies.—The glory of the flower-garden is waning, and it will soon be desolate, in spite of the gardener's care. Meanwhile, keep the beds neat by the timely removal of decaying foliage, and the grass and gravel walks clean and smooth by frequent rolling. Plants to be taken up and potted should be attended to immediately, or at least protected during the nights, for fear of sudden frosts.

The stock of cuttings should be looked over, and additional heat applied when the roots are not fully formed. Late geranium-cuttings may be removed to the kerbs of the pine-pits, to make roots. In storing the stock away for the winter, endeavour to keep all those plants together which require similar treatment. Some kinds will stand more damp than others, and may be wintered in common frames; but the better kinds of bedding-out geraniums, and some other tender things, will require a moderately dry house or pit. To preserve verbenas, petunias, &c., &c., properly through the winter, they must be kept dry to prevent mildew.

The herbaceous ground will now require a thorough clearing.

Shrubberies should now be thinned out, alterations completed, and the formation of new shrubberies brought to a close during the month.

Rose-Garden.—Planting and transplanting are now the chief employment; if very dry during the month, give a good watering to each plant before the soil is fully filled in. Stocks should also be collected and planted for budding on next season. The true dog-rose makes the best stock, and may be distinguished from sweet-brier by the large white thorns which thickly cover the stem of the latter towards the base; and from those of climbing habit, by the dark green colour of the bark and weakness in the stem.

Florists' Flowers.—At this season of the year the amateur cannot do better than get together those soils, &c., which are indispensable for the proper growth of his favourite flowers. Turf, pared two inches thick from a loamy pasture or a green-lane side, stacked together to decompose, will be the foundation of his composts. A large heap of melon-bed manure should also be secured, not forgetting as large a quantity of fallen leaves as possible. A cartload of sharp river-sand is an indispensable adjunct, and the florist should look out for willow-dust and decayed and rotten sticks. A quantity of excellent food for plants may be scraped out from hedge-bottoms. Allow auriculas to have abundance of air, but little or no moisture; the plants being in a state of rest, require but little. Tulips out of the ground now suffer every day.

Auriculas are in their winter quarters; they require abundant air, and occasional inspection to see that no worms are in the pots.

Carnations and Picotees, layered in previous months, should be potted off. *Pinks* planted last month must be protected from the wind.

Dahlias are still fresh and gay, if the weather has been tolerably mild: should frost appear, no time should be lost in taking them up.

Hollyhocks.—Cut down and propagate from the old stools. Take eyes from the flowering stems, but without forcing.

Pansies should be potted off as reserves for filling up vacancies.

Polyanthuses in beds will be benefited if the surface of the soil is stirred, and a top dressing of equal parts of maiden loam, leaf-mould, and cow-dung.

Tulips, not yet planted, should be got in without delay, taking care, however, that the soil is not wet; hoop them over, and let matting be prepared against rainy weather.

Kitchen-Garden.—Approaching winter bids vegetation prepare for a rest. In the kitchen-garden the crops will make little progress for the next four months. At this time it is very necessary to give attention to the state of the drainage.

Asparagus, if not already done, should be cut down, and the beds receive a dressing of very rotten dung.

Artichokes.—A good mulching of leaves will be of considerable benefit to these in protecting the crowns from the frost.

Seakale.—As the leaves decay and detach themselves, they may be removed.

Beans.—On light ground and sunny borders, these may be put in without fear of failure.

Runner-Beans.—Pull up these, as they will produce nothing more.

Peas, like beans, may be sown, and with the same proviso as to the nature of the ground.

Celery.—It is advisable to give the final earthing-up during this month; it grows much slower now, and must be allowed time to blanch.

Cardoons.—Treat in a like manner.

Beet.—Get this crop housed or pitted at once; it will not stand frost. Cut off the leaves not too close to the roots, which should lay a couple of days to heal or callow; then stow them where they will not mould or damp till pitted.

Carrots.—Treat in a similar manner to beet. It is advisable to get them housed before there is any danger of very severe frost. Young crops to stand the winter should be carefully thinned and hoed between.

Onions.—The autumn-sown should be treated in a similar manner.

Leeks ought to be earthed up, if not done before, when they can be taken up as wanted; they will continue to grow in mild weather.

Parsnips are as well left in the ground till wanted.

Potatoes, if any are left in the ground, should be taken up without delay, and stored: they keep well in the ground by taking up every other row, and placing an additional layer of earth over each ridge left.

Turnips should be hoed and kept clean.

Scorzoneras, &c., are best left in the ground till wanted.

Lettuce, if tied up for blanching, should be kept dry, if possible, or they will soon rot. The advantage of good cabbaging sorts will be apparent at this time. Some may yet be planted out to stand the winter.

Endive.—Continue to blanch in succession. If this is done with flower-pots, these, as they are removed, can be placed on others.

Spinach.—If this has been properly thinned and kept clean, it will continue to grow, the leaves alone being picked for use. If the plants stand nine inches or a foot apart it will be all the better.

Brocoli.—Such as are coming in now should be watched. Remove dead leaves, and use the hoe between them.

Brussels Sprouts, *Borecole*, and *Savoys* are best kept free from dead leaves, which in damp weather become unpleasant.

Cabbage may still be planted out for the next summer's crop; but the earlier it is done now the better. Use the hoe freely amongst those planted last month: they will be much better for it.

Cauliflowers.—Stir the soil about those in hand-glasses, and keep the lights off unless there is fear of frost.

Fruit-Garden.—Let the bulk of kitchen and dessert apples in the fruit-room be often looked over to remove decaying fruit. In doing this be careful not to bruise the others. Clear off the remaining leaves from wall-trees; and now that the greater part of the fruit-tree leaves have fallen, the whole should be cleared off the ground preparatory to pruning and turning up the borders for winter. Figs against walls should have any odd remaining fruit taken off. Thin out superfluous shoots, and pinch out the points of the wood selected for bearing, when the branches should be tied together and matted, or protected by haybands, fern, &c., for the winter.

Towards the end of the month is the best time to commence prunning dwarf apples and pears.

Young plantations of strawberries should have some short dung spread between the rows, to preserve their shallow roots from frosts, which otherwise might lift them out of the ground.

Plums.—All late sorts should be gathered before the frost sets in, and either wrapped in paper or hung by the stalk in the fruit-room. Pruning should follow.

Currants and Gooseberries.—Plant and prune both while the weather is favourable.

Flowers under Glass.—Embrace every opportunity of admitting the external air to conservatories and greenhouses when it is of a temperature of 45°: also change the air of stoves, &c., during the few hours of sunshine that often come to chase away a November fog. In fine, the more fresh air the better, provided it be warm and genial;—the less the better when it is otherwise.

As the quantity of external air admitted may now be safely reduced to its minimum, so may also be the quantity of water.

Greenhouses occupied by heaths, azaleas, &c., not in flower, must be kept cool, dry, and clean. They may also have more air than the conservatory, and, a temperature of 40° will suffice.

Camellias.—Where these have a house devoted to them, they require careful management. The buds will just be swelling, and a sudden change of temperature, a scarcity or excess of water, or a cutting draught of cold air, will often cause the buds to drop. Be extremely careful not to give an excess, when fire-heat becomes necessary, on account of frost, and maintain a genial growing atmosphere of 45°.

Orchids.—See that they are clean: keep them in a dry temperature of 65°, and let them sleep.

Plant-stove.—Here the *Poinsettia pulcherrima*, *Euphorbia jacquiniflora*, *Begonia nitida*, *Gesneria cinnabarina*, will be lighting up by their dazzling beauty, masses of ferns, palms, and variegated plants. Late caladiums must now be watered with great care, as the bulbs are impatient of damp during winter. Those beginning to go off must have scarcely any water, and as soon as the leaves are matured, water should be entirely withheld, and the pots

16

turned on their sides for the winter. Remove dead leaves and flowers as soon as they appear; water in the morning for the next three months; keep a temperature of 65°; frequently change the arrangement of the plants, and only admit air in fine weather.

Fruit under Glass.—One of the chief duties here is fruit-preservation. This is just the very worst month in the whole year for keeping ripe fruit of any kind, and especially trying for ripe grapes: one speck of decay or mildew will soon increase under the influence of a November fog. Houses of ripe fruit must therefore be examined daily, and every specked berry or decayed leaf removed. Brisk fires must also be lighted in the morning, to enable you to give air both at front and back, to agitate the atmosphere and expel damp.

Vines in pots may be started in a bottom-heat of 53° in hotbeds, unless means are found for giving them bottom-heat over flues, &c., in the houses in which they are to be fruited. After they have fairly started, they can be carefully moved to their fruiting quarters; in many places the first vinery will now be started. Proceed as recommended last month.

Peaches.—If these are wanted next May, trees in pots must now be started. They should already have been untied, pruned, washed, &c. Examine the borders thoroughly; water, and top-dress with good maiden loam, if necessary See that the house, as well as the trees, is scrupulously clean, so that you do not have to battle with vermin as well as dark skies and inclement weather for the next six months. The Royal George, Noblesse, Galande, and Vanguard peach, and the Red Roman and Violette hâtive nectarine, are best for early forcing. Proceed slowly; give no fire unless compelled, and do not exceed 45° by fire-heat during the month.

Orchard-houses.—If these are either open or unroofed, see that the hungry birds do not destroy your next year's crop. They seem fond of model standard trees, and in a single day will often mar the hopes of a twelvemonth. The lights should be placed on these houses; it is a dangerous practice to allow standard trees to be much frozen. The cold is also much more intense here than on the surface of a south or west wall.

Fig-trees grown under glass should never be frozen. The embryo fruit will most likely be destroyed, and a whole month's or six weeks' forcing lost in consequence. This is a good time to examine the wood thoroughly for scale, &c., and to paint the trees over with the composition recommended for vines.

Pines.—Those swelling off must be assisted by a warm genial atmosphere of 75°, and be watered when necessary. Bottom-heat will require to be examined and fermenting material renewed, possibly. Plants intended to fruit next spring and summer must be guarded against any sudden check, to be kept rather dry, and *rest* for the next three months in a temperature of 60° to 65°.

Similar treatment, except the *resting*, will suit the general stock of succession plants. They must be kept slowly moving in a dryish atmosphere.

Much attention will be necessary to renewing linings, &c., to those in pits, to maintain the requisite temperature. Coverings of mats or reed-frames, must also be applied in severe weather, and all sudden changes guarded against. Occasionally, too, some of the strongest succession plants will require water at the roots, although the air may still be very damp.

Cold Frames.—Plants in these are often treated as if they were more tender than they really are. The object is not so much to stimulate them into growth, as to protect them from such injury from frost and storms as they would be exposed to in the open air. Corn salad, endive, lettuce, cauliflower, parsley, carrots, radishes, onions, and many more light crops, are not so tender but that they will stand out of doors; but then they keep so much better and fresher under the protection of frames, that it is well worth while to have a few lights devoted to them. They also begin to grow rather earlier in the spring, and continue growing later in the autumn, than they would do if quite exposed; but it should be strictly observed not to keep them in any way close, so as to breed mould. If any mouldiness accrue, it is a sure sign that they are kept too close.

Seeds of radishes, lettuce, and small salading may be sown any time during the month, or any time in the winter. They will germinate slowly, but may come in very useful in the spring; the latter will be ready in about a month. In frosty weather the protection of a mat will, in addition to the lights, be sufficient for most of these things.

Garden-frames are very useful for protecting other plants than those above named. Many plants, as pinks and pansies, stocks and chrysanthemums, and indeed many plants generally accounted hardy when planted in the borders, will, when in pots, require the protection of a frame, or, if planted in a bed of soil placed within the frame, they will flower earlier and stronger.

DECEMBER.

December, the last month of our year, was, as its name indicates, the tenth of the Roman calendar. Although the thermometer often sinks below freezing during this month, the frosts are seldom of long continuance. Rain and wind abound. The variations of temperature are smaller than in November. The mean temperature of the earth, one foot beneath the surface, is now 41·13°; at two feet, 42·83°; that of the atmosphere being 38·14°.

Though November is the best month for transplanting and for laying down turf for lawns, whatever in this respect has been left undone may still be accomplished if the weather remains open.

Flower-Garden and Shrubbery.—The pruning of deciduous trees and shrubs should also be proceeded with, unless during severe frost. Most evergreens are best pruned in April. Nevertheless, as that is a busy, and this a comparatively leisure season, the hardiest evergreens, such as laurels, &c., may be pruned now; any, however, that require cutting down, had better be left till that period.

The beds in the flower-garden, now disrobed of their summer beauty, will either be furnished with shrubs, herbaceous plants, annuals, or bulbs, or simply roughed up for the winter. Previous to either being done, it is hoped that they received a liberal top-dressing of manure.

All newly-planted roses should be mulched over with 3 or 4 inches of light dungy litter on the surface.

Reserve Garden.—Annuals, to stand the winter here, will probably require some slight protection. Carefully watch against the inroads of mice, rats, snails, &c. Not only bulbs, but even young plants, are often devoured by the two former; and in mild winters a little black slug will clear off whole beds of annuals. Let every vacant space be roughly dug up, manured, &c.; choice beds of tulips, anemones, &c., sheltered in bad weather, and the whole examined daily to guard against accident and ward off disease.

Florists' Flowers.—Dahlia roots stowed away in cellars, &c., must be carefully and frequently examined to see how they are keeping, and any scarce sorts placed in heat towards the end of the month, to insure a large stock before May.

Pinks and carnations in beds will require pressing firmly into the earth after severe frost. Examine the beds for slugs in mild weather, and see that the plants are not destroyed by rats and mice.

The same precautions are necessary with pansies in beds in the open ground.

Tulips.—If planted early last month, some of them may be peeping through the soil; if so, they must be protected by having a light pyramid of sandy peat-earth or leaf-mould placed over them. During very frosty weather the beds or rows must be covered with mats, woollen nets, &c., as nothing injures these bulbs more than severe frosts on their crowns just as they are coming through the ground.

Kitchen-Garden.—The experienced gardener knows the importance of winter operations. Respecting crops, individually, little can be done; but collectively some attention should be given to the various stores of seeds and vegetables; the latter should be looked over occasionally, turned, sorted, and cleaned; kept moist without being damp, cool without frost, and where there is a free circulation of air. As to seeds, it is well to have them ready for sowing; that is, thoroughly dried and rubbed out, every particle of husk and light seed blown out, and carefully papered and labelled. Those that have to be purchased should be procured early.

With regard to particular crops:—

Seakale.—Some may be covered for forcing. Place the kale-pot over a bunch of crowns; see that enough is covered; then, having previously prepared and shaken out the dung, and got it into a condition to maintain a moderate heat, cover the pots to a thickness of about three feet from the ground; too great a body of dung is apt to heat too violently and spoil the crowns. Give just enough to maintain a moderate heat, and no more; it will be ready to cut in about three weeks, according to the amount of heat. Some gardeners cover with leaves, which answer the purpose; but in collecting leaves, a great many slugs and other vermin are collected also. These do mischief to the kale; in other respects the effect is the same.

Rhubarb may be treated in a precisely similar manner, but requires larger pots; and none but the earliest sorts should be forced.

Peas and Beans of the early kinds may be sown on light ground; but it is not advisable to sow many. Those sown in February will be as early with in a few days, and much more certain.

Celery.—Cover with litter in frosty weather. It will be so much better to take up, besides keeping it fresh and uninjured.

Parsnips and other crops that remain in the ground ought also to be covered with litter or leaves. The slightest covering will make a vast difference in case of sharp frost, which should always be bargained for at this time.

Endive.—Blanch with pots, and cover with litter; and a good supply may be kept up the whole winter without having recourse to frames, the litter helping to blanch it before the pots are put on; but a dusting of lime should be given occasionally to destroy slugs, which are very fond of endive.

Brocoli, &c.—It will now be seen what advantage there is in giving the various sorts of brassicas plenty of space. Those planted among other crops are shanky, and more exposed to the frost, while those planted open are short, firm, and stocky, and far more likely to stand severe frost. Let this be considered in cropping next year.

This is the best time to make any general alterations. Where old bushes are to be grubbed up, and the ground prepared for cropping, or where young bushes are to be planted, also where drainage is necessary, now is a good time to do it before the winter rains make a swamp of the garden. Set the edgings and paths in order, and carefully remove any accumulation of rubbish which is likely to harbour vermin. *Herb beds* should be cleared, and made as neat as possible, both for the appearance and well-doing of the herbs.

Fruit-Garden.—December is the month of rest here as in other departments of the garden; but there is much to be done which is too often left undone. Planting may now be presumed to be over; at least, unless the weather is unusually mild, when it will be well to prepare the ground, and leave the planting until the early spring.

Peaches, nectarines, and other wall-trees now require pruning, and the shoots selected nailed in; but both operations should be avoided in frosty weather.

Standard apples and pears should receive their final autumn pruning and thinning out, chiefly of the interior branches, so as to admit of a free current of air; badly-placed shoots should be removed. Espalier trees, and trees planted against a wall for horizontal training, do best when the shoots are tied down; in the absence of trellis on the wall, studs should be driven into it at convenient distances, in order to avoid the distortion the branches undergo in the old process of nailing with shreds.

Plants under Glass.—The re-labelling, cleaning, and arranging of all plants should be diligently forwarded, so that every plant should have its proper name and its best dress before Christmas. Climbers on roofs and pillars may also have their final pruning, cleaning, and tying; every dead leaf and every particle of dirt removed. This, so essential to health everywhere, is especially necessary in the conservatory, where nothing offensive to good taste should ever be seen. The interest of this house is often much increased at this season by introducing some pots of Christmas roses, hyacinths, narcissuses, &c., from the forcing-pit.

The same principles will apply to giving air, &c., as in November; only, as we have generally more sun in December, more may be admitted. Care must, however, be taken to prevent a cold draught cutting off the beauty of any plant that may have recently come from a warmer house or forcing-pit. Frequently re-arrange the plants in this house in winter.

Where a temperature of 47° is maintained, the various sorts of *Epiphyllum truncatum*, so often met with in stoves in winter, will bloom well in this house. There are now several varieties of this charming winter-flowering plant grown as dwarfs in suspended baskets, or as tall plants, umbrella-fashion, or as pyramids: worked on the Pereskia, it is exquisitely beautiful. Perhaps it flowers best in a cool stove, but it will flower for six weeks or two months in a warm conservatory. Potted in a rough mixture of peat, leaf-mould, loam, brickbats, old plaster, and charcoal, and kept in a temperature of 60°, its progress is rapid. During the summer and autumn months, the plants should be fairly exposed to the sun in an airy house. Place them in a temperature of 55° towards the middle of October, and now every leaf will terminate in one, two, or three beautiful flowers.

In the stove the *Poinsettia* should now be in flower. No description can convey an exaggerated or sufficiently strong idea of the regal beauty of this plant. Half a dozen of them, from four to five feet high, with eight shoots each, terminated by bunches of enormous scarlet braces, set off by the peculiar shape and colour of the true leaves, is a sight worth going twenty miles on any December morning to see. Any one with a plant-stove may have this treat at home, as few plants are more easily propagated or grown. They can be had in flower from October to March.

Euphorbia jacquiniflora should also be flowering. The brilliant effect of this plant among begonias, ferns, and other things, must be seen to be appreciated. Some late caladiums, grown for this purpose, will now be intermixing their beautiful leaves with the bright flower-stems of *Gesneria cinnabrina*, and others. The bright red berries of the *Ardisia crenulata* will also be exhibiting themselves in striking contrast with the shining green leaves.

Orchids.—Rest here also should be the order of the day; nevertheless that rest will be partially broken by the flowering of some or all of the following plants:—*Pholænopsis amabilis*, several Oncidiums, Cymbidiums, Epidendrums, Cattleyas, Zygopetalums, &c. Maintain a temperature of 60° or 70°, and avoid all stimulating treatment.

Greenhouses.—Preserve a minimum temperature of 40°; give as much air as possible; see that the stock is kept perfectly clean by occasional smoking, washings, dippings, &c.; put on a fire on dull mornings to expel damp; remove heaths, epacrises, &c., to the conservatory as they come in flower; shift young plants of kalosanthus into their blooming-pots as they require it, and keep everything in a quiet semi-dormant state until the new year awakens them.

Pelargoniums.—Early varieties for cut flowers may be forced into bloom in a vinery or peach-house at work. The general stock will require careful treatment this month. The latest-flowering specimens may receive their final shift, and all will require careful training, a genial temperature of 45°, and great skill in watering and ventilating. Fancy varieties often show a disposition to bloom prematurely. These early flowers must be perseveringly removed, to throw the strength into the shoots, for a perfect inflorescence at the proper season. Keep the plants within a yard of the glass, if possible, to prevent their drawing, and fumigate as soon as one green-fly is visible.

Cinerarias.—The earliest of these will now be in flower, and few plants are more effective for decorative purposes.

Cold Pits and Frames.—Water and cover with care; give all the air possible in mild weather. During severe weather these may remain hermetically sealed for a week with impunity at a temperature of 35° to 40°.

Forcing-pit.—Keep up a growing temperature of 55° to 60°. Introduce fresh batches of shrubs, roses, bulbs, and everything that will flower early, to supply the place of those draughted off for other service.

Fruit-culture under Glass.—The Vines are now showing branches. Bring artificial heat to your aid, and keep a night temperature of 60° and a day one of 70°; admit every possible ray of light; keep the leaves within nine inches of the glass, and create a midsummer climate in December. Examine frequently the state of the borders, and keep the roots as warm as the tops. Stop the shoots a joint beyond the branches; damp the floors, paths, and pipes, if these are used, during bright days; admit air whenever it is practicable, and try to secure strength rather than length. Where a succession of grapes is wanted start another vinery; proceed as directed in November.

Continue to look over and preserve late grapes, and maintain for these a cool and equable temperature of 40° to 50°. If a gentle current of air can be kept up through the house by night and by day, the fruit will keep all the better. Vine-borders may be formed, or old borders renewed during this month.

Pines.—Keep fruiting pines almost entirely dry if you wish them to start in January. Maintain a day temperature of 70°; night, 60° to 63° This dryness is necessary to throw the plants into bloom. Nothing does this more effectually than a check, although the check must neither be too severe nor too long-continued. Beware of moisture settling upon any pines that may now be in flower, as it often prevents the proper fructification of the blossoms; and deformed fruit is the consequence.

Peach-house.—In many places the fruit-house is started this month. For instructions in the preliminary stages, see November. Beware of frost. The wood of trees under glass cannot bear cold so well as that nurtured out of doors.

Figs.—Lee's Perpetual fig is the best sort grown; the white Ischia and Marsella are perhaps a little better flavoured, but they are small. The first batch of strawberries should now be introduced. Cuthill's Black Prince the best for the early season, and if they can have a little bottom-heat, so much the better. Plants in 48-sized pots do best for early work, the cramping principle inducing fruitfulness.

Hotbed and Frame.—Little can be added to what has been already said Let the weather be the principal guide as to giving air, &c.; be careful that the frames are ventilated without causing draught, which might injure the plants considerably. See that the heat is maintained, and cover with mats a night; but do not shorten the days more than they are. The mats should be taken off as soon as it is light in the morning, and not put on till it is getting dark at night, so that the plants may have all the daylight they can get.

London: W. W. HEAD, Printer, 2 and 3, Plough Court, Fetter Lane, E.C.

INDEX.

[*The figures indicate the Number affixed to each paragraph, and not the page of the book, except where the letter p. stands before the figures.*]

WARD, LOCK, & TYLER, Paternoster Row, E.C.

WARD, LOCK, & TYLER, Paternoster Row E.C.

BIOGRAPHY.—WORDSWORTH.

HOME PETS.—MALTESE DOG.

GEOGRAPHY.—ALGIERS.

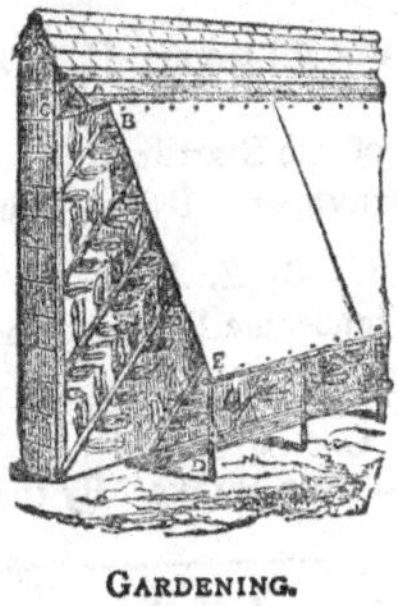

GARDENING.

HOME PETS.—GUINEA PIG.

WARD, LOCK, & TYLER, Paternoster Row, E.C.

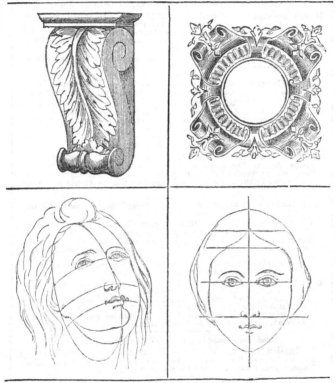

NEW VOLUMES—SEE NEXT PAGE.

Printed in the United States
By Bookmasters